EMERG ENERGY ALTERNATIVES FOR SUSTAINABLE ENVIRONMENT

EMERGING ENERGY ALTERNATIVES FOR SUSTAINABLE ENVIRONMENT

EDITORS
D P Singh • Richa Kothari • V V Tyagi

CRC Press
Taylor & Francis Group
Boca Raton London New York

CRC Press is an imprint of the
Taylor & Francis Group, an **informa** business

The Energy and Resources Institute

CRC Press
Taylor & Francis Group
6000 Broken Sound Parkway NW, Suite 300
Boca Raton, FL 33487-2742

First issued in paperback 2023

© 2019 by D P Singh, Richa Kothari, V V Tyagi and The Energy and Resources Institute
CRC Press is an imprint of the Taylor & Francis Group, an informa business

No claim to original U.S. Government works

ISBN 13: 978-1-03-265387-7 (pbk)
ISBN 13: 978-0-367-17889-5 (hbk)
ISBN 13: 978-0-429-05827-1 (ebk)

DOI: 10.1201/9780429058271

Print edition not for sale in South Asia (India, Sri Lanka, Nepal, Bangladesh, Pakistan or Bhutan)

Publisher's Note
The publisher has gone to great lengths to ensure the quality of this reprint but points out that some imperfections in the original copies may be apparent.

Library of Congress Cataloging in Publication Data
A catalog record has been requested

Visit the Taylor & Francis Web site at
http://www.taylorandfrancis.com

and the CRC Press Web site at
http://www.crcpress.com

Preface

Sustainability of environment is an emerging global issue at present. Unsustainable or deteriorating environment is a matter of concern as it has threatened the survival of living creatures. Recently, climate change has been matter of great concern at a global platform owing to imbalances in natural environment. Increasing population has increased the demand for energy, which has ultimately put pressure on natural resources and caused a paradigm shift from resource generation to exploitation. Fossil fuels are limited reserves of energy on the earth which are now declining at an accelerated rate, and soon it will be a resource of the past. The lion's share of energy generation depends on fossil fuels, and hence we are moving fast towards an arena of energy crisis. Rapid industrialization and huge consumption of fossil fuels have adversely affected our environment and become a major contributor to global warming. The paradigm shift towards renewable energy technology is the major innovation of the decades. The thrust on renewable energy-based power generation will not only be environmentally friendly, but it will also ensure the energy security. In the past years, various dimensions of renewable energy have been developed, such as biomass, solar, wind, hydro, and geothermal energy. Although there is immense potential for energy generation from renewable energy, various issues related to these sources of energy still need to be resolved. Emerging green energy technologies are the healing technology for environment that covers a wide spectrum of materials and methods, leading energy generation to non-toxic, clean, and green products. Government of India is also making every effort to achieve the clean development mission through energy generation by solar and other green energy sources. Green chemistry and engineering are the most advanced fields of technology, which have the potential to boost the nation's economy in coming years.

Emerging Energy Alternatives for Sustainable Environment aims to address the role of sustainable technologies in energy generation options for clean environment. It covers a wide spectrum of energy generation approaches, with an emphasis on five key topics: (i) renewable energy sources and recent advances, (ii) emerging green technologies for sustainable development, (iii) assessment of biomass for sustainable bioenergy production, (iv) solid waste management and its potential for energy generation, and (v) solar energy applications, storage system, and heat transfer. This book provides essential and comprehensive knowledge of green energy technologies with different aspects for engineers, technocrats and researchers working in the industry, universities, and research institutions. The book is also very useful for senior undergraduate and graduate students of science and engineering who are keen to know about the development of renewable energy products and their corresponding processes.

We acknowledge the help and cooperation by all the contributing authors. We are especially grateful to all the team members, Mr Vinayak Vandan Pathak, Mr Shamshad Ahmed, and Mr Alok Rai, who provided their valuable suggestions and help from time to time. We would especially like to thank Dr Adarsh Kumar Pandey, UMPEDAC, University of Malaya, Malaysia and Mr Ravi Sharma, Department of Mechanical Engineering, Jaypee University of Engineering and Technology, Guna, India, for their helpful suggestions to improve the quality of the book.

We are also very thankful to Ms Sushmita Ghosh and all team members of TERI Press for their hard work and timely publication of this book.

D. P. Singh
Richa Kothari
V. V. Tyagi

Contents

1

Biogas Potential in India: Production, Policies, Problems, and Future Prospects

Sohini Singh, Barbiee Choudhary, Stebin Xavier, Pranita Roy, Neeta Bhagat, and Tanu Allen[*]

Amity Institute of Biotechnology, Amity University, Noida
[*]*E-mail: tallen@amity.edu*

1.1 INTRODUCTION

India has a high requirement for energy because of its large population. Although energy generation in India is less than energy demand, till now these demands were met by the use of forests resources. Moreover, with the global crunch in fossil fuel reserves, this demand is growing at a yearly rate of 4.6% (Ramachandra, Vijay, Parchuri, *et al.* 2006). According to Mourad, Ambrogi, and Guerra (2004), around 10%–14% global energy is contributed through biomass. Although the government is exploring the field of energy source and production, security of energy supply, and reduction in carbon dioxide (CO_2) emission, biomass appears to be the most promising energy source. First, biomass is a renewable source of energy. Second, transforming biomass to bioenergy such as biogas through anaerobic digestion method is precisely known, and in developing countries it has been used as a source for cooking and lighting (Omer and Fadalla 2003). Anaerobic digestion not only provides biogas energy but also manages and stabilizes organic wastes, converting them into nutrient-rich matter which can be used as a natural fertilizer and soil conditioner.

Biogas is a cheap, clean, renewable, naturally produced, and underutilized energy source (Verstraete, Morgan-Sagastume, Aiyuk, *et al.* 2005). It weighs 20% less than air and ignites at the temperature range from 650°C to 750°C (Kohler, Hellweg, Recan, *et al.* 2007). It burns giving a blue flame and is a colourless and odourless gas (Mandal, Kiran, and

Mandal 1999). Its caloric value is 20 MJ/m^3 (FAO 1997) and usually burns in a conventional biogas stove with 60% efficiency.

Therefore, the need of the hour is a renewable, eco-friendly, green energy source that can also cater to the energy demands of ever-growing population. This chapter reviews the scenario of Indian biogas generation, policies, and problems faced. Applications and future prospects of biogas technology have also been discussed.

1.2   PRODUCTION OF BIOGAS

Biogas is generated by anaerobic digestion. It is important to convert biodegradable waste into useful fuel, thus reducing the volume of waste products. Anaerobic digestion also assists in killing disease-causing pathogens. In anaerobic digestion, microorganisms digest organic materials in the absence of oxygen, airtight condition, and at a certain level of moisture, temperature, and pH (Angelidaki, Ellegaard, and Ahring 2003; Buren 1983). It is a multi-step biological process of acidogenesis, acetogenesis, hydrolysis, and methanogenesis in which the organic carbon gets converted into CO_2 and methane (CH_4) (Figure 1.1).

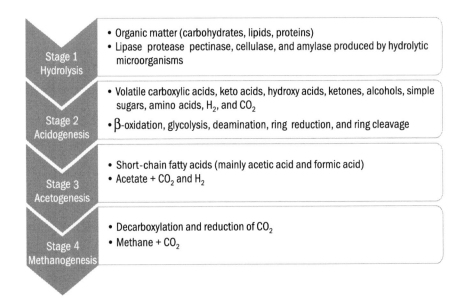

Fig. 1.1 *Scheme of anaerobic digestion*

Source *Angelidaki, Ellegaard, and Ahring (2003)*

1.2.1 Hydrolysis

It is the primary or initial stage of biogas production through anaerobic assimilation. Facultative anaerobes (hydrolytic bacteria) break down complex organic compounds into soluble organic molecules, for example, proteins into amino acids, lipids to fatty acids, and carbohydrates to sugars, by extracellular amylase, cellulase, lipase, or protease enzymes (Parawira, Murto, Read, *et al.* 2005).

During anaerobic digestion, as reported by Vavilin, Rytov, and Lokshina (1996), the presence of large particulate molecules with reduced surface to volume relation/ratio in the substrate may constrain the digestion reaction and make the hydrolysis itself a rate-limiting step.

1.2.2 Acidogenesis

The second step of anaerobic digestion is acidogenesis (Vavilin, Rytov, and Lokshina 1996) in which facultative and obligate anaerobes or anaerobic oxidizers utilize the soluble organic molecules obtained from hydrolysis (Garcia-Heras 2003) and break them further into acetate, hydrogen, and CO_2, which yields higher energy for microorganisms. According to Schink (1997) and Angelidaki, Ellegaard, Sorensen, *et al.* (2002), the products of acidogenesis consists of approximately 19% H_2/CO_2, 51% acetate, 30% intermediate reduced products, such as higher alcohols, lactate, or volatile fatty acids (VFAs) which can be used directly by methanogens.

1.2.3 Acetogenesis

In acetogenesis, breakdown of short-chained, alcohols, aromatic fatty acids, and higher VFAs to acetate and H_2 occurs. The products undergo the last step of the biogas production known as methanogenesis.

1.2.4 Methanogenesis

Methanogenesis is the terminal step of anaerobic digestion in which methanogenic archaea converts H_2/CO_2 and acetate to CO_2 and CH_4 (Kotsyurbenko 2005). Kotsyurbenko also reported the use of homoacetogenic bacteria in the CH_4 formation pathway, which is dependent upon the concentration of hydrogen in the system that oxidizes or synthesizes acetate.

1.3 SOURCES/MATERIALS FOR BIOGAS PRODUCTION

Resource that is generated from biological organisms and can be tapped for energy uses such as biofuels and biogas is called biomass. A number of substrates such as peel of corn, soy, wheat and rice, discarded parts of straw or manure from animals or industrial wastes, and even energy crops can be utilized for biogas generation (Ofoefule and Uzodinma 2008; Uzodinma, Ofoefule, Onwuka, *et al.* 2007). As anaerobic digestion process uses facultative anaerobes for the decomposition of organic matter, environmental conditions and nutrient availability play important roles in the biogas production. The components of biogas (Figure 1.2) may differ depending on the material being decomposed (Anunputtikul and Rodtong 2004). Some of the sources are discussed next.

1.3.1   Food Processing Industrial Wastes

In food processing industries such as fruits, dairy, vegetables, sugar, meat or oil processing, a large amount of waste is generated in both solid and liquid forms. These wastes are organic in nature and have a high content of carbohydrates, lipids, and proteins. With a suitable chemical composition, waste can be used as a substrate for microbiological fermentations. Anaerobic digestion is a popular way of treating these types of wastes (Kaushik, Satya, and Naik 2009).

1.3.2   Sugar Processing Waste

Sugar is produced in 121 countries. About 70% of the produced sugar is from sugar cane and 30% from sugar beets. Processed and unprocessed by-products of beet sugar industry include cut-offs of beet leaves and beet top, molasses, and beet pulp. The sugar extraction process generates a syrup residue known as molasses, and after the desugaring process,

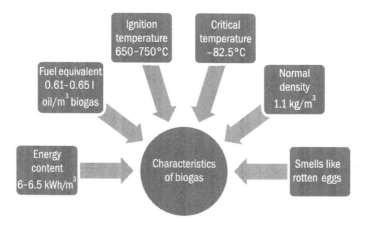

Fig. 1.2 *Characteristics of biogas*
Source *Angelidaki, Ellegaard, and Ahring (2003)*

this residue is known as desugared molasses (DM) (Satyawali and Balakrishnan 2007; Olbrich 1963). Thus, every single lot (tonne) of beet sugar yields 0.33 tonne grass cut-offs, 0.24 tonne DM, and 0.33 tonne beet pulp (Danisco Sugar 2001).

1.3.3   Potato Starch Processing Waste

There has been an increase of about 46 million tonnes in the potato production over the past 40 years in Africa, Asia, and Latin America (FAOSTAT 2010). China and India account for one-thirds of the total potato production in the world. According to Karup Kartoffelmelsfabrik (2002), from per tonne potato flour, which is 20% water and 80% potato starch, 0.73 tonne of potato pulp and 6.6 m^3 of potato juice are produced as by-products, rich in both protein and starch. Since these are a biodegradable constituents, they can be utilized for generation of biogas.

1.3.4   Paper Waste

In India, around 18 lakh tonnes of paper is produced every year. Office, schools, factories, printing presses, and so on produce huge amounts of waste paper. Paper waste is considered as a good feedstock for biogas production; therefore, it can be utilized as a source of cheaper energy generation and waste management. Paper waste is treated with animal waste for constant gas flame and this is called co-digestion process (Mshandete, Kivaisi, Rubindamayugi, *et al.* 2004; Nielsen and Angelidaki 2008).

1.3.5   Poultry Waste

Brazil is the world's original aviary maker. The poultry litter is a mixture of aviary excreta, uneaten feed and fowl quills. The quality of poultry litter and the amount delivered depend upon the base material used, the season of the year, the creation time, and the populace thickness of the flying creature (Costa, Barbosa, Alves, *et al.* 2009). The biogas generated from poultry litter can be used as a vitality source. The same gas can be used for inner ignition motors as fuel to produce power. The effluents and strong squanders from anaerobic ageing procedure can be used as compost for obtaining high quantities of vitamins, minerals, and proteins (Rondon 2008).

1.4   ROLES OF MICROBES IN PRODUCTION OF BIOGAS

Nagamani and Ramasamy (1991) compared the microbic diversity in biogas digesters to that of cow rumen and reported about 17 species of fermentative microorganisms that play an important role in the biogas production. Moreover, it is the behaviour of a substrate that determines the kind and extent of the fermentative bacterium that will be within the biodigesters. Much of that bacterium adheres to the substrate before intensive chemical reaction. Some bacteria species involved in fermentation, such as *Ruminococcus albus, Bacteroides succinogenes, Clostridium cellobioparum*, and *Butyrivibrio fibrisolvens* are predominant. The cellulase enzymes in the biogas digesters were characterized by Sivakumaran, Nagamani and Ramasamy (1992) and it was reported that *Acetivibrio* sp. showed higher cellulase activity than that shown by *Clostridium* and *Bacteroides* spp. Nagamani, Chitra, and Ramasamy (1994) reported that though obligate hydrogen-producing acetogenic bacterium is less in extent, it is among the important strains in biogas reactors. These organisms release energy as CH_4 by oxidizing the longer chain fatty acids.

Anaerobic degradation of aromatic compounds depends upon hydrogen-consuming bacterium which acts as the minor species in the biogas digesters during fermentative reactions. Mackie and Bryant (1981) reported a considerable rise in biogas formation by the activity of these organisms, especially *Methanosarcina barkeri*, a predominant archaebacterium, as acetate is the most important substrate in biodigesters. Among the methanogenic genera, *Methanosaeta* sp. and *Methanosarcina* sp., besides producing biogas by acetoclastic reaction, are also useful in stabilizing of the digester.

1.5   ISSUES AND CONCERNS FOR THE BIOGAS PRODUCTION PROCESS

Concerns related to the generation of biogas largely depend not only on the nature of the raw material but also on the operational aspect of how the process is carried out. According to Angelidaki and Ellegaard (2003), at times the raw material may be loaded with cations that may act as inhibitors at larger concentrations, or toxic complexes, such as VFAs, which are initially absent in the feed may be generated during the anaerobic course of the digestion process (Angelidaki, Ellegaard, Ahring 2003; Angelidaki, Ellegaard, Sorensen, *et al.* 2002). Feedstock factors (that is, pH, nutrients, inhibitory compounds, and buffering capacity) and operational settings [that is, both OLR (organic loading rate) and temperature] affect the functioning of microorganisms.

1.5.1   Temperature

Temperature has a major influence on anaerobic digestion as the procedure can be used for temperatures ranging widely between psychrophilic (<20°C) and acute thermophilic (>60°C) as reported by Kashyap, Dadhich, and Sharma (2003) and Lepistö and Rintala (1999). According to Boe (2006), the advantages of increasing temperature include increased organic compound solubility; increased rate of biological and chemical reaction; improved diffusivity of soluble substrates; increased death rate of pathogenic bacteria, especially when conditions are thermophilic; and increased long-chain fatty acids and VFAs degradation. The disadvantages of high temperature are effects on kinetics and thermodynamics of the biological processes. Two temperature ranges that can provide good digestion conditions for biogas generation are as follows:

(i) **Mesophilic range:** Microbes are called mesophilic when they can function around the temperature range 30–38°C or 20–45°C.

(ii) **Thermophilic range:** Thermophilic bacteria act around temperature range of 49–57°C or up to 70°C (Kangle, Kore, and Kulkarni 2012).

1.5.2 pH and Buffering Capacity (Alkanility)

pH is an important factor and has an effect on the microbial growth throughout the anaerobic fermentation. Microorganisms of different types may act at a similar range of pH or each type of microorganisms has a specific range of pH to grow optimally. According to Hwang, Jang, Hyum, *et al.* (2004), fermentative bacteria act in the pH range of 4–8.5 (Horiuchi, Shimizu, Tada, *et al.* 2003). In contrast, methanogenic species, which grow in a specific acidic condition, can work in the relatively limited pH range of 5.5–8.5 (Boe 2006), otherwise their growth can be inhibited in acidic conditions within the digester (Kangle, Kore, and Kulkarni 2012). Only *Methanosarcina* is able to grow in lower pH values (pH 6.5 and below) compared to others, as the metabolism gets considerably suppressed at pH <6.7 (Deublein and Steinhauser 2008). In an anaerobic digester, for diverse species, pH range from 6.6 to 7.8 is considered ideal (Lay, Li, and Noike 1997), which can be achieved by OLR (Poliafico 2007). As acetogenesis occurs for organic matter, the pH initially decreases, but due to the presence of methanogens in the digester, those acids will be consumed rapidly, thus increasing the pH and stabilizing the reaction in the bioreactor. Kangle, Kore, and Kulkarni (2012) suggested that as the anaerobic digestion continues, the pH values of the contents in the digester are influenced by the quantity of CO_2 produced and VFA present; therefore, the concentration of VFAs should be maintained below 2000 mg/L, particularly acetic acid, for the effective anaerobic digestion.

Another important factor which stabilizes reactions in context to pH change is the buffering capacity (also called alkalinity). Bicarbonate (HCO_3^-) ion having pK_a value 6.3 is the main buffer, whereas VFAs with pK_a 4.8 are the acids mainly generated in the anaerobic digester

(Boe 2006). Besides bicarbonates and VFAs, high concentrations of other complexes, such as ammonium ion (pK_a 9.3), hydrogen sulphide (pK_a 7.1), and dihydrogen phosphate (pK_a 7.2) present in the digester also affect the pH balance (Björnsson 2000).

1.5.3 Nutrients

Sufficient nutrients important for microbial cell growth are required for the efficient biodegradation of organic matter. Macro- and micro-nutrients such as nitrogen, phosphorus, carbon, sulphur, potassium, Ni, Fe, Co, and Zn are needed in a lesser amount for the optimal growth of anaerobic microbes (Kayhanian and Rich 1995; Cresson, Carrere, Delgenes, et al. 2006).

1.5.4 Volatile Fatty Acids

In the anaerobic biogas production process, VFAs are probably the key intermediates and the conversion of these intermediates into CO_2 and CH_4 is most essential as reported by Pind, Angelidaki, and Ahring (2003). The increased concentration of VFA is assumed as an after-effect progression difference during biogas generation. Therefore, as suggested by Björnsson (2000) and Boe (2007), VFA is recommended as a marker during the anaerobic digestion process.

1.5.5 Organic Loading Rate

The biological transformation extent of the anaerobic digestion system is the OLR. Biological matter flows into the digester per unit time, stated as the weight of organic matter by the volume of the digester over time. OLR values generally range between 0.5 and 3 kg VS/(m^3/day) (Poliafico 2007). Almost all industrial wastes comprising a high segment of easily degradable organic substances that bring about high CH_4 yield, contrarily, result in the production of high VFAs. VFA accumulation occurs generally owing to overloading of the reactor. Subsequently, VFA at a higher concentration is more noxious to methanogens and also reduces the pH, which can end the anaerobic digestion process. Therefore, both underload and overload of VFA will introduce disparity in the biodigesters (Stamatelatou, Lyberatos, Tsiligiannis, et al. 1997).

1.6 AUGMENTATION/UPGRADATION OF BIOGAS PROCESS

The process of biogas production can be enhanced not only by tactful and expert control but also through efficient monitoring as suggested by Boe (2006). The substrate can be pretreated to discharge more biodegradable substances or co-digested with animal dung or various other wastes to increase the biogas production.

1.6.1 Pretreatment of Substrate

Baharuddin, Hock, Yusuf, *et al.* (2010) reported that enormous amounts of wastes, generally solids, such as shells (5%), mesocarp fibres (12%), and empty fruit bunch (EFB) (23%), are yielded by oil palm refinery for each fruit bunch lot processed within the refinery. According to Chew and Bhatia (2008), in 2005, around 17 million tonnes of EFB produced by the Malaysia oil palm industry was used to enhance the biogas production (Misson, Haron, Kamaroddin, *et al.* 2009; Tan, Lee, and Mohamed 2010).

1.6.2 Treatment by Chemicals

This procedure consists of treatment using bases, acids, and oxidants. Treatment by acid considerably enhances the reaction rate for hydrolysis of cellulose step, whereas on treatment with base, the interior surface of lignin swells and increases. This results in annihilation of linkages and bonds between different polymers and lignin and declination in the polymerization gradation and crystallinity and, thus, breakdown of lignin (Sanchez and Cardona 2008).

1.6.3 Hydrothermal Treatment

Treatment by steam at high pressure and temperature is known as hydrothermal treatment. Throughout the pretreatment, regular mixing of biomass with water is carried out, followed by heating for around 5–15 min at about 180–200°C, which results in annihilation of the protective lignin cover and the accessibility of enzymes to cellulose, thus leading to increased production of biogas (Bruni 2010).

1.6.4 Combined Chemical and Hydrothermal Treatment

Biomass undergoes steam treatment in the presence of catalyst NaOH, and this process leads to 26% greater CH_4 production than unprocessed biomass (Bruni 2010). Besides, hydrothermal treatment of classified municipal solid waste presoaked in NaOH has been reported to increase the biogas production by 50% (Wang, Wang, Lu, *et al.* 2009).

1.6.5 Co-digestion

The co-digestion model is a combination of digesting organic wastes and feed (Raven and Gregersen 2005), and it is effectively used for biogas production from industry and household waste (Tafdrup 1994). It is environmentally and financially profitable because it handles many categories of waste in a cost-effective manner (Margarita, Spyros, Katerina, *et al.* 2009). There are various benefits of co-digestion with animal dung. First, the dung is a source of vitamins, nutrients, trace metals, and different compounds required by microbes to grow and survive. Second, because of co-digestion, pH neutralizes, which enhances the alkalinity in the digester (Angelidaki and Ellegaard 2005). Third, dilution of more concentrated biological matter happens owing to higher moisture level in feed, otherwise it would inhibit the biogas production.

1.6.6   Pretreatment Combined with Co-digestion

If industrial waste before co-digestion is pretreated with activated sludge by alkaline hydrolysis, the CH_4 production is increased up to 67%, and if the industrial waste is digested with kitchen waste, the production of CH_4 is found to increase by 61% (Neves, Ribeiro, Oliveira, *et al.* 2006).

The biogas plant types used in developing countries are as follows:
 (i) Bag digester plant
 (ii) Fixed dome digester plant
 (iii) Floating drum digester plant
 (iv) Vacvina biogas plant

Bag digester plant

Bag digester plant is also known as a balloon plant. This digester was first built in Taiwan in 1960. It usually consists of a plastic or rubber digester bag which is UV resistant and made up of materials such as neoprene, rubber, and RMP (red mud plastic). The biodigester consists of inlet and outlet ducts through which organic waste along with slurry is supplied for degradation. The biogas formed gather at the top portion of the bag, even though, in a few cases, it is possible to collect the gas in another bag. The residue is collected in the lower part. The bag digester primarily acts as plug flow reactor (PFR), in which materials added each day, theoretically, will move as unit mass through the digester until hydraulic retention time (HRT) is attained. Then, the digested mass will flow from the digester bag as a unit. A portion of the effluent is inserted again into the inlet so that it will act as a seed to re-inoculate the bag digester (GTZ 1999). The advantages of using the bag (balloon) digester are follows:
 (i) It is cheaper compared to all other digester types.
 (ii) The fitting and fixing is simple and can be done quickly within hours.
 (iii) It offers uncomplicated cleaning mechanisms.

(iv) The digester heats along with its contents by sun or any outside supply/resource as partitions in the reactor are not very thick.

The disadvantages of the bag (balloon) digester are as follows:

(i) Its lifetime is short (about 5 years).

(ii) The digester certainly can be damaged and would be difficult to restore.

(iii) Sludge removal and transportation to the field require a lot of labour.

(iv) It requires a constant temperature.

(v) Insulation is difficult.

(vi) It needs the best quality of plastic (PVC).

Fixed dome digester

It is a Chinese model biogas plant. It was constructed in China around 1936. In this model, the digestion chamber is combined with gas holder as one single unit. Some examples suggested by ISAT; GTZ (1999) are Deenbandhu, Chinese fixed dome, CAMARTEC, and Janata and I Janata II models.

The biomass is converted to a liquid known as slurry (digested waste) and biogas by bacteria in the digester. The gas consists of CH_4 and CO_2, with traces of other gases. After the slurry is collected in the gas holder, which is of dome shape, it is transferred to the offset tank. The amount of slurry produced depends upon the loading rate of the feed, its utilization, and gas generated. At the time of gas generation, pressing of slurry backwards and sideways happens and then the slurry is moved to the offset tank. As the gas is used, slurry is transferred back into the digester from the offset tank. Thus, owing to these interchanges, altered phases of slurry by gradation mixing are achieved, and, therefore, according to Stalin (2007), this model is considered as a mixed digester reactor (CSTR, continuously stirred tank reactor).

The fixed dome digester is relatively cheap and, thus, economical with a lifespan approximately up to 20 years. The fixed dome digester is safe to withstand cold temperatures because the majority part remains below the surface of earth. Also, variation occurs between day and night temperatures inside the biodigester, which is better for the biogas production by methanogens. The advantages of the fixed dome digester are as follows:

(i) It is low cost compared to the advantage it presents when matched to other kinds.

(ii) It is reliable and can be used up to 20 years.

(iii) It is totally insulated and underground construction. It is the best type of digester for generating biogas at colder climate in countries like Bolivia.

The disadvantages of the fixed dome digester are as follows:

(i) It is very difficult to build in bedrock areas.

(ii) To prevent gas seepage due to design failure, technical skills are essential for fixed dome construction and sealants for the gas holder.

(iii) The fixed dome digester's lifespan is longer than Khadi and Village Industries Commission (KVIC) plant.

Floating drum digester

This model is also called KVIC model as it was supported and consented by KVIC (Jashu Bhai J. Patel developed the design of floating drum biogas plant in 1962) (GTZ 1999) and the use of this model is widespread in India and all over the world. A concrete wall surrounds the main or reactor tank. It has two parts: (i) the inlet, which supplies slurry to the tank, and (ii) a stainless steel cylindrical dome, which hovers on the slurry and consists of an outlet pipe for collection of the gas produced. Expansion of the decomposed matter makes the slurry overflow into the succeeding compartment, which can be then used as a natural fertilizer. The advantage of the floating drum digester is that the gas pressure remains constant in this digester because of the drum weight. The disadvantage of the floating drum digester is that stainless steel is used in the construction of the floating chamber, which is expensive. It needs constant maintenance and observation to avoid rusting.

VACVINA biogas plant

The VACVINA model is an upgradation of the previous biogas designs, namely, fixed dome and plastic container types. The digester is rectangle in shape. It is constructed with a volume capacity of over 5 m^3 apt to a little animal farm. The VACVINA model is supplied animal waste as feed from a trench and set behind or below the animal shed. The output, that is, biogas, from the digester is collected and stored in two or three plastic bags and used as a fuel for the kitchen range (GTZ 2007). The advantages of the VACVINA model are as follows:

(i) It is basic in design with very few defects.

(ii) It is somewhat less inexpensive to maintain and can be constructed in less space.

(iii) It is suitable for colder climates since the biogas digester is underground and the gas reservoir is outside and made of plastic.

(iv) It has a longer lifespan probability.

Biofilters/Fixed film bioreactors

This method uses filters of a considerable size usually for the substrates having low solid content. Young and Song (1984) suggested that several bases should be inspected for selecting the relevant materials and for longer lifespan of the matrix of fixed film. Non-biodegradable items should be used. The structure should be stable mechanically. The materials should be affordable and readily available in nearby markets. According to Wilkie, Faherty and Colleran (1984), for the matrix of fixed film reactors, several materials such as clay pipes, PVC [poly (vinyl chloride)], sponges, and nylon are used.

1.7 TYPES OF BIOGAS REACTORS

According to Angelidaki and Sanders (2004) and Parawira, Murto, Zvauya, *et al.* (2004), biodegradability of the organic matter and generation of surplus biogas in batch can be achieved by the conventional anaerobic method of digestion. This process is effective owing to the regulated and stable supply and the steady state of the bioreactor, thus maximizing the production. The mixing of fluid (sludge and substrate) or the particulate solid contents in the reactor determines the selection and classification of the reactors as reported by Stalin (2007). Organic soluble matter is treated in high-rate biofilms such as EGSB and UASB, while slurry and solid wastes are treated in CSTRs (Kato, Field, and Lettinga 1997; Angelidaki, Ellegaard, Sorensen, *et al.* 2002).

1.7.1 Continuously Stirred Tank Reactor

Demirel and Scherer (2008) reported the successful use of CSTR in anaerobic digestion of energy crops and food remains (Fu, Achu, Kreuger, *et al.* 2010). In the main liquid, biomass is suspended in the CSTR process and removed along with the effluent. By doing so and maintaining this step up to 10–20 days, the hydraulic and sludge retention time becomes same or equal, which in turn prevents the slow growing methanogens from being washed out (Boe 2006). Boe and Angelidaki (2009) reported that a single CSTR yields less CH_4 compared to serial CSTRs, with the same industrial slurries or the same volume of manure. The CSTR process is practical and simple to work and, thus, has many advantages (Kaparaju, Serranoa, and Angelidaki 2009).

1.7.2   Upflow Anaerobic Sludge Blanket

The UASB (upflow anaerobic sludge blanket) reactor was developed in the 1970s (Lettinga, van Nelsen, Hobma, *et al.* 1980) and is now widely used for the production of biogas by treatment of several wastewater types (Shastry, Nandy, Wate, *et al.* 2010; Sevilla-Espinosa, Solorzano-Campo, and Bello-Mendoza 2010). In the UASB reactor, an immobilized cell is used, where the biomass is retained while the substrate is pumped through, giving rise to a high organic loading rate (Kaparaju, Serranoa, and Angelidaki 2009). The success of the UASB concept depends on the development of a dense sludge bed at the bottom of the reactor in which biological processes take place. This sludge bed is formed by the accumulation of incoming suspended solids and the growth of bacteria (Seghezzo, Zeeman, and van Lier 1998). In UASB systems, natural turbulence is caused by the influent flow and the biogas formed inside the reactor provides a good wastewater biomass contact.

Developed in the 1970s (Lettinga, van Nelsen, Hobma, *et al.* 1980), the UASB reactor is now extensively used for biogas production by treating different wastewater types (Sevilla-Espinosa, Solorzano-Campo, and Bello-Mendoza 2010; Shastry, Nandy, Wate, *et al.* 2010). As wastewater is pumped through the UASB reactor, the loading rate increases and retention of the biomass through immobilized cells takes place (Kaparaju, Serranoa, and Angelidaki 2009).

The success of the UASB application depends on the sludge layer formed by the solids suspended in wastewater because it is here that the degradation and digestion of organic matter by microbes take place (Seghezzo, Zeeman, van Lier, *et al.* 1998). The biogas, thus, generated offers a significant interaction between wastewater and biomass, and influent flow causes natural turbulence in the UASB systems.

1.7.3   Expanded Granular Sludge Bed

De Man, Vander Last, and Lettinga (1988) introduced the concept of expanded granular sludge bed (EGSB), a modification to the traditional UASB reactor. The granular sludge is inoculated in both EGSB and UASB, but changes in hydrodynamic (superficial velocity, K_s) settings enhance the mixing and contact among sludge and wastewater in the EGSB reactor (Puyol, Mohedano, Sanz, *et al.* 2009; Kalogo and Verstraete 1999).

1.8   POLICIES RELATED TO BIOGAS PRODUCTION IN INDIA

Since the 1990s, a lot of research, experiments, and analyses have been carried out in the country for the production of renewable energy.

A course of political initiatives and plan of action have been designed to assist research, generation, and employment of this kind of energy. Biogas technology (anaerobic digestion) has utmost importance in this regard (Chanakya, Ramachandra, and Vijayachamundeeswari 2007). The potential and role of bioenergy in the Indian context have been realized by MNRE and a few other agencies. The Bio-energy Technology Development Group of MNRE has the distribution and propagation programmes for biogas such as National Biogas and Manure Management Programme (NBMMP) under its umbrella over the past 10 years which cater to rural cooking needs by constructing family-size biodigesters (MNRE 2010c). This biomass power acquiring a yearly venture capital of more than USD 125 million (₹600 crore) has become an industry, producing annually about 5000 million units of electricity (MNRE 2010b). Its goals are as follows:

(i) Make available the gaseous fuel that is harmless and used mainly for cooking and organic manure in rural and semi-urban houses using family-type biogas plants.

(ii) Decrease the hardship and toil of rural women, lessen the dependence on forests, and emphasize on social values.

(iii) Develop better sanitation system in villages by linking biogas plants with toilets.

(iv) Make available bio-digested slurry (liquid/semi-solid and dried) as an improved origin of enrichment for manure to lessen and complement the utilization of chemical fertilizers by combining enrichment units such as worm composting plants with biogas digested slurry and other organic enrichment facilities of slurry.

(v) Help minimize climate change by avoiding discharge of CO_2 and CH_4 into the atmosphere.

The programme is implemented by State Nodal Departments/State Nodal Agencies and the KVIC located in Mumbai. Moreover, issues related to cook stoves at the individual household or the community level and the family-size biogas plants are also included in this programme.

Cook stoves are a huge menace as biomass is not burnt sufficiently in it, apart from creating fumes and being hazardous to health. Sometimes, price of the stove is high and unaffordable for rural people. A National Biomass Cook-stove Programme was launched by the ministry (i) to create awareness for the right kind of stoves to be utilized by households and (ii) to provide expertise for the marketability of improvised community cooking stoves. To achieve the second objective of the programme, an experimental model is initiated that focuses on places such as Anganwadis, government schools providing mid-day meals, hostels, governmental institutions, and private dhabas using community cooking stoves. It is expected that till 2022 about half a million society stoves and over 10 million household stoves need to be manufactured and distributed in Indian market with support from Department of Women and Child Welfare, Ministry of Tribal Affairs, and Department of Elementary Education.

Besides cooking stoves, biogas-based power units under the Biogas Based Distributed/Grid Power Generation Programme of MNRE can emerge as a reliable power production system in the country. This process can be speeded up by undertaking projects of variable capacities between 2.5 and 255 kW depending on the availability of raw materials or feedstock from different sources such as agro- or food processing industries, kitchen or forestry, or animal wastes. This would lead to infrastructure development, skilled manpower, upgradation of technical issues, maintenance, and dissemination at large scale. If the biogas generated electricity from these projects is supplied or sold to any rural area, community, or individual grid, or even to Remote Village Electrification (RVE) Programme of MNRE, it should be done on mutual terms.

During 2008–09, a new initiative under the R&D policy of MNRE was promoted to display the entrepreneurial mode of Integrated Technology-package for construction of medium-sized biogas mixed feed fertilizer plants (BGFP) for production, purification or enrichment, and bottling and piped biogas distribution. The construction of these sustainable business model-type plants aims at production of compressed natural gas (CNG) quality of compressed biogas (CBG) to be used as a vehicular fuel in the near future. There is a huge potential of installation of medium-sized biogas fertilizer plants in numerous villages and other regions and agro- and food processing industries of the country. Till now around 22 BGFP-related projects together with an aggregate capability of 37,016 m^3/day have been sanctioned in 10 states, namely, Chhattisgarh, Andhra Pradesh, Punjab, Gujarat, Haryana, Karnataka, Maharashtra, Madhya Pradesh, Uttar Pradesh, and Rajasthan, for their implementation.

1.9   BENEFITS OF BIOGAS ENERGY PRODUCTION— POTENTIAL IN INDIA

At present, in India, people rely on old methods for cooking and heating. These methods consume huge amounts of energy, which leads to overutilization of resources. The procedure of making dung cakes and burning them are both uneconomical and unhygienic. This brings us to understand the current consumption rate of energy and analyse other available resources so as to secure resource sustainability.

One kg of animal excreta generates 40 L of biogas and a biogas plant that can fulfil the requirement of a small family (2–4 m^3) requires 50 kg of animal excreta and the same amount of water to yield 2000 L of biogas per day. The approximate calorific value of the biogas is obtained by multiplying the calorific value of CH_4 with the volume fraction of CH_4 in the biogas. The estimated calorific value of CH_4 is 8548 kcal/m^3 (Ravindranath and Hall 1995). A family of 3 members of a farmer owns 4 animals and collects 40 kg of animal waste per day. When agricultural and kitchen wastes are added to it, the mixture yields 0.8 m^3 of pure biogas. This amount of biogas is sufficient to light 1 burner for almost 2 h. The liberated biogas meets the necessities of the farmer. 1 m^3 biogas plant costs approximately ₹20,000. This biogas plant is more than sufficient to process the excreta of four animals and some kitchen waste (Sharma 2011). Also, the quantity of excreta yield per cattle varies from one place to another.

An approximate calculation indicates that India has the capability to yield 6.38 × 1010 m^3 of biogas by utilizing 980 million tonnes of animal excreta per annum. Heat value of the biogas measures 1.3 × 1012 MJ. Also, 350 million tonnes of compost is also yielded (Hagen and Polman 2001). When other feasible organic wastes such as sewage waste, municipal solid wastes, industrial wastes, and so on are used as the feedstock for biogas production, the sum total biogas capacity would further increase (Kishore and Srinivas 2003).

In India, biogas is having the capacity to overtake the use of both fossil fuels and chemically synthesized fertilizers. For example, during 2001–02, 22.7 billion m^3 of natural gas (excluding natural gas from liquefied petroleum gas, LPG, shrinkage) and 7.3 million tonnes of LPG were utilized (Mittal 1996). With respect to heat energy, the amount is 1.08 × 1012 MJ per year, when compared to the potential heat value of 1.3 × 1012 MJ per year accessible from biogas yielded per annum from animal excreta. Similarly, when confronted with chemically synthesized fertilizers, the utilization of 19.4 million tonnes during 2001–02, 350 million tonnes of compost could be generated by biomethanation of animal excreta per annum (Hagen and Polman 2001).

Biogas is better than other gases as it has a higher heating value compared to natural gas and coal, indicating higher consumptions. It is very cost effective and extremely convenient to use as a cooking fuel. On the basis of its effective heat generated, a 2 m^3 biogas plant could overtake fuel equivalent to 26 kg LPG (almost two standard cylinders), 37 L kerosene, 88 kg charcoal, 210 kg fuelwood, 740 kg cattle excreta in a month. Also, it does not have any bad odour, does not cause health hazards, and burns producing bluish coloured sootless flame; therefore, it is non-messy. Biogas is cost-effective and economical compared to the conventional biomass fuel (excreta or dung cakes, fuelwood, crop wastes, etc.) and LPG. Though it is costlier than kerosene, LPG and kerosene have issues related to their supply in villages (Soma 1997). The technology of biogas increases energy supply through decentralization, thus making its access feasible in villages to meet energy demands. A comparison between using excreta directly and using it in the form of biogas showed that 25 kg of fresh excreta will generate about 5 kg dry excreta, which is almost equal to 1 m^3 of biogas.

The technology used in biogas production is more useful in agro-ecosystem and can be applied to different uses. The biogas is also useful as a fuel as it is a substitute for firewood, agricultural residue, electricity, and so on, relying upon the type of work, supply circumstances, and constraints. Thus, it provides energy for both cooking and lighting. After anaerobic digestion, biogas plants also generate the residue organic waste which has greater nutrient qualities over the normal organic fertilizers and cattle excreta, as it is in the form of ammonia. The biogas can also be used to supply power to engines and help in pumping water in irrigation systems.

There are other advantages of biogas as well which are associated with renewable sources. This includes biogas having the ability to overtake the use of biomass-based fuels, especially wood. Biogas could also decrease the demand of wood from forests and establish a vacuum in markets, which can be at the minimal for firewood. There is more than 85% reliance on bioresources to fulfil the everyday necessities of fuel and fodder in villages in many parts of India (Ramachandra, Kamakshi, and Shruthi 2004). The effectiveness of transforming cattle residue and their excreta could be increased to 60% by allowing their digestion anaerobically (to generate biogas). The biogas production would also prevent the everlasting dispute between energy recovery and nutrient usage as the effluents produced by the digester could be reused in fields.

A survey was conducted from January 2005 to July 2006 in 1500 households. It showed that excreta collected per cattle was about 3–7.5 kg per adult cattle, excreta per buffalo was about 12–15 kg, excreta of stall-fed buffalo was about 15–18 kg, and excreta of hybrid variety was about 15–18 kg. By taking the minimum values (such as 3 kg dung/cattle/day and 12 kg dung/buffalo/day), the total excreta obtained per day per village was calculated. With the assumption of 0.036 m^3 of biogas produced per kg of dung of cattle/buffalo, the total amount of biogas produced was calculated. It gave the quantity of biogas produced per village (Ramachandra, Krishna, and Shruthi 2005). Similarly, on taking the maximum values, that is,

7.5 kg dung/cattle/day and 15 kg dung/buffalo/day, the total biogas was calculated per village per day, by assuming 0.042 m^3 of biogas produced per kg of cattle/buffalo excreta. It was presumed that the per capita demand of biogas for the domestic purposes was about 0.34–0.43 m^3/day. The biogas need was calculated by multiplying the adult equivalent of the village population by the per capita biogas demands. Need of 0.34–0.43 m^3 per day could be considered for estimating the minimum and maximum values of biogas required in the village (Ramachandra, Kamakshi, and Shruthi 2004).

1.10 PROBLEMS AND ISSUES ENCOUNTERED

In village economy, biogas was found to be an essential part, though its uses are quite challenging. Coordinated disseminations have ushered in immense rates of non-functional plants and also threatened its uptake further. In spite of well-planned pursuit to deal with the free energy demands of villages in India, especially the poverty ridden, the biogas programmes have failed to fulfil the demands of people (Mohan and Kumar 2005).

China also initiated biogas implementation in villages in 1973, and in a short period of 7 years, it overtook India's biogas production by a ratio of 1:10 (80,000 in India and 720,000 in China). By 1998, China set up a total of 6,900,000 plants compared to India's installation of 2,750,000 plants (NABARD 2007). Although China was having a less number of animals compared to India (where in 2010 there were approximately 529.7 million cattle compared to 300 million animals in 1950), it overtook India's biogas production by nearly a ratio of 3:1. Some of the problems were related to agencies responsible for implementation, while the rest were associated with the diffusion of technology and also ground realities. The Swedish researchers Porras and Gebresenbet (2003) stated that these were "in discrepancy with the decentralized behaviour of the biogas". The above-mentioned categories of issues and problems were assumed to lead to failure.

1.10.1 Technical

With many biogas plants being non-functional, problems related to techniques are the reason why biogas plants are inoperative. Technical issues arise in many ways—plants lying non-operational, carelessness, or rusting over time because of issues related to both the apparatus and the unit. In a study of India's biogas plants in 1995, using a sample of 24,501 biogas plants in 432 villages, it was observed that approximately 53% of the total biogas plants were operational (Tomar 1995). As there are several technical problems, the policies and planning discussed next can be considered.

Input problems

In India, excreta is the major feedstock in biogas digesters, and it is commonly collected from cattle; pig excreta and human excreta are not employed as they are not acceptable (UNAPCAEM 2007). For the proper functioning of a biogas plant relying purely on cattle dung, there will be the minimum requirement of 4 to 5 cattle per household, but in India only 20% of the households in villages own 4 to 5 animals. Therefore, biogas plants are left underfed and this problem of shortage becomes more difficult as people living in villages are reluctant to use wastes other than the cattle excreta (Gustavsson 2000). For poverty-ridden farmers with not more than 3 cattle, it is a major issue, as underfeeding the digester of a biogas plant disrupts its normal operation. A study of biogas plants in India conducted in 1995 found the shortage of excreta fed to digesters as the second major reason for inlet damage (Tomar 1995). When studying the digester of the biogas plant it was found that in order to start a plant, an initial feeding of 2–3 bullock carts loaded excreta was necessary (Porras and Gebresenbet 2003). For well-off farmers, this is not an issue, but poverty-ridden farmers might need to buy excreta or collect it with the help of their family members and friends. In areas having limited excreta, buying dung is costly. Excreta are, at present, a "free" source of fuel, popularly used in villages in India. However, Porras and Gebresenbet (2003) observed that the increasing demand for excreta as a feedstock in biogas plants in Indian villages may end up in "commercialization of the excreta economy".

Another crucial input component that is often not given importance is water. Water is also to be mixed with cow's excreta in the ratio of 1:1 (dung:water). This can be a problematic activity and can require up to several hours per day to obtain the amount of water needed in dry areas in India. Women living in villages spend several hours every day making long trips to obtain water, needed for both drinking and cooking purposes, except that required to feed digesters. Porras and Gebresenbet (2003) observed that ironically fuelwood is scarcely available in dry areas, and so alternative fuels such as biogas may be welcomed in such places.

Assessing techno-economic feasibility

Deficiency in the availability as observed by NABARD (2007) is a reason responsible for technical failures because beneficiaries are not able to maintain the unit properly. Detailed reports including the plant size, related to the quantity (m^3) of biogas produced, are necessary to ensure that the biogas does not remain unutilized. Also, the capacity applications that change rely upon the biogas plant size (MNRE 2007). Safeguarding the beneficiaries could also reconcile the customized needs of the biogas plant, and it is also an important feature of fruitful dissemination. This is important when the total cost of the biogas plant and also the expense of working are given, and they are huge in spite of being sponsored. NABARD (2007) observed that a 4 m^3 biogas plant worth ₹4000 (CAD 87) (after loan) would value between ₹40 (CAD 0.87) and ₹50 (CAD 1.09) per month (presuming a 10% interest on the net amount). NABARD observed that "there are minimum chances that villagers would spend that much amount of money (in cash) for the purpose of household cooking," thus increasing the stress of repaying on the poverty-ridden beneficiaries without concerning if they are having proper means for regularly paying, and this also contributes to its failure (NABARD 2007).

Technical parameters for target setting

The Indian Government lays down a few targets and also determines the "capability" of the biogas application purely on technical limits, for example, bovine inhabitants, excreta availability, and animal possession among the households (PEO 2002). Since the real values of excreta availability, inhabitants of bovine, and animal possession differ from place to place, variations are observed among the things that the PEO (2002) terms as "technically derived capacity" or "realizable capacity." The PEO, evaluating by using the MNRE's benchmark of animal possession and also by taking information from 1991 to 1992 animal census, approximates 24 million biogas plants as the technically derived capacity. However, in the survey which considered more accurate values of the head of animals necessary for the functioning of biogas plants and divided in the total earnings, education level, and worth of alternative resources, the approximation of FTBP (family-type biogas plants) was found to be less than 11.7 million, that is, realizable capacity (PEO 2002).

1.10.2 Political/Bureaucratic

In India, the FTBP obedience is a government-funded programme which includes many states as well as district bureaus, funding institution, centres for training and NGOs. This might be adversely disposed that the contribution of the government related to biogas, which at the base, might as well be referred to the failure in identifying the real demands of the poverty-ridden people living in the villages. There are subsequent political or bureaucratic inefficiencies:

Agency multiplicity and procedural delays

There are many entities, for example, state and district terminal bureaus, organizations related to financing, NGO, and KVIC which are entrusted to enforce various parts of NBMMPs. The distribution of fund and other financial loans requires too much time. NABARD (2007) observed that the discharge of loans by the state government and also by the KVIC, and the amount of time taken by the banks to pass the loan have constantly been a reason for controversy. It suggests that, first, the loan by district rural development bureau should be streamlined and, second, the local procedural planning should include funding bureaus from the beginning so as to reduce the time taken to authenticate and sanction the subsidy application. Bureaus responsible for implementation are also worried about the procedural time lag that they face in enforcing and dispatching of funds (Rao and Ravindranath 2002). In addition, manufacturers also experience time lags in getting technical approval for raw materials, for example, steel and cement, as the approval process is too tedious and bureaucratic. Steel and cement, which are insufficient in India, are controlled by government and transported by the Indian Government as per the quota prices. Apart from the prices being too high, which discourage manufacturers, there are institutional restraints also, for example, sanctions drawn out among the government bureaus that add bureaucratic tediousness to the construction of these plants (Porras and Gebresenbet 2003).

1.10.3   Training/Monitoring

To ensure that the troops included in NBMMP are having prior information and are satisfied in discharging the duties, training plans are important. RBDTC centres deal with different types of training plans, which are for users, TKW, mason, and staff members, who are included in the NBMMP (PEO 2002).

Lack of systematic monitoring

In the guidelines for exercising NBMMP, MNRE states that supervision includes not only substantial authentication but also involves making supervision reports, inspecting the area by both MNRE and RBDTC officers, evaluating the studies done by institutions, and submitting authorizing certificates along with the investigated statement of expense (NABARD 2004). Moreover, each RBDTC is required to carry out random case verification of 500 biogas plants set up in a particular area at a specific time; however, this goal is rarely fulfilled as administrative staff members are very few. The problems with the unsuccessful orderly reporting system is that the report is adversely affected, and even the advancement reports of these biogas plants are made without the actual supervision and substantial authorization. Ironically, advancement reports of the biogas plants determine the future plan for the NBMMP.

1.10.4 Reporting

To deal with the problems related to reporting, NBMMP has highlighted the reporting procedure to safeguard regular reporting of the advance that consists of the quarterly advance reports, the monthly advance reports, and also the material regarding the annual advance report of the MNRE (NABARD 2004). There were issues related to the reporting function. These were mostly related to the accuracy of the reports, the absence of proper supervision and the non-precise reporting of the data, such as inauthentic reporting of achievement and reports not properly maintained.

Inauthentic reporting of achievement

The variations in the reporting of data and also in the achievement of goals in the block, district and state levels were observed by PEO (2002). The PEO put forward that at upper level, the maintenance of the record is not authentic. In addition, the advance reports for the purpose of submitting it to the higher levels are made randomly and without any substantial authentication of biogas plants (PEO 2002). In a study of these plants in the 1990s, Tomar (1995) observed that the analysis by researchers independently indicated that the mean percentage of working biogas plants was merely 52%. The MNRE proposed that according to the records, the mean percentage of working biogas plants was 77.1% (Tomar 1995). Such vast variation shows, according to PEO's inference, that these records at the upper level in the NBMMP are not authentic.

Reports not properly maintained

The PEO (2002), at the state level, observed that the records were not regularly maintained. Monthly advance reports that were needed to be submitted to the MNRE have to include the performance, relative, and not comparative to the annual goals, and all the details such as the number of biogas plants inspected, their working, loans utilized, and advertisements. When these quarterly, monthly, and annual advance reports were needed by all the states, not only the reports were regularly recorded but also the worthiness of these reports was ensured.

1.10.5   Human Resources

For the success and sustainability of NBMMP, it is important that well-trained labourers build the biogas plants. However, biogas dissemination programmes in India find themselves lacking proper staff to train, overlook, report, and head the programmes.

Lack of staff

It was mentioned that despite the state level being sufficiently staffed, there was absence of technical manpower at the lower level in many of the states (Kaniyamparambil 2011). Odisha Renewable Energy Development Agency (ODEDA) (Odisha's exercising bureau) observed that because of the absence of technical manpower, junior engineers who used to be the representatives of the other fields supervised the biogas plants. In Tamil Nadu, which is one of the most densely populated states in India, among 28 districts, there were only 5 technical persons recruited in 5 districts, aside from the district administrator, who used to look after the biogas exertion. In Uttar Pradesh, it was observed that no technical manpower was accessible for the biogas plant introspection that was prior to the loan approval, and in the state of Andhra Pradesh, due to manpower shortages, in most of the cases, loans were sanctioned without substantial authentication of biogas plants. As a consequence, in India, biogas was not a preference among those who help in implementing it—marking a disconnection between high-level preparation and low-level exertion.

Lack of staff for training and developing local capacity

To tackle the problem of the lack of staff for training, there was a need of a large number of industrialists and technicians for building up biogas plants and maintaining the small biomass energy machine. The present and future supply providers of Bio-energy Technologies (BET) need to know all important techniques and skills to merge this machinery into functioning. Alongside BET, it was reported that when the machinery was ready and displayed, the lack of skills became a hurdle for its fruitful exertion. The setting up of training programmes could show a path to lessen the lack of skillful manpower. It was necessary to ensure that the manpower involved in the training and advancement was properly trained (Kaniyamparambil 2011).

1.11   FUTURE PROSPECTS FOR BIOGAS IN INDIA

Biogas consists of 60%–65% CH_4, 35%–40% CO_2, 0.5%–1.0% H_2S, some amount of water vapour, and so on. It was found to be approximately 20% lighter compared to air. Biogas, like LPG, could not be converted to liquid form under usual pressure and temperature. Nevertheless, after removing CO_2, H_2S, and moisture and then condensing it into cylinders, it was made easy for transport purpose and for immobile uses. On the basis of the availability of animal excreta from approximately 304 million animals, it was found to have an approximate capacity of 18,240 million m^3 of biogas production per annum. The growing number of pullet farms can be the alternative source as they could produce 2173 million m^3 of biogas per annum with 649 million birds. In addition, waste from kitchen, universities, institutes, restaurants, industries, and so on in both urban areas and semi-urban areas and also non-edible de-oiled cakes from *jatropha* provide immense capacity.

The biogas bottling project was planned to substitute manure and fuel worth ₹40 lakh per annum, in a span of 4–5 years. The extraction and bottling of CO_2 and separation of humic acid from the sludge would enhance the viability of Bio Gas Bottling Plants (BGBP). The BGBP gives three-in-one resolution of gaseous fuel production, organic fertilizer generation, and wet biomass removal. The left sludge can be used as the organic fertilizer so as to improve the fertility of the soil.

Biogas is eco-friendly as it is devoid of foul odour, weed seed, and pathogen. It contains many nutrients such as nitrogen, potassium, and sodium (NPK) and micronutrients such as iron and zinc. The biogas plants prevent the emission of black carbon normally observed in households using *chulhas*. Biogas is a hygienic cooking fuel as it prevents CH_4 emission from unprocessed animal excreta and biomass ruins can be prevented. The biogas containing nutrients can be sealed in CNG cylinders and used as and when required. These biogas plants are eco-friendly, and so these plants play the role as the most effective tool in checking the changes in climate.

1.12 CONCLUSION

Biogas has been used as the most appropriate machinery for several decades, allowing suitable usage of the available resource. It has been found to be a clean, hygienic, and easy-to-use fuel at minimal cost, in addition to being completely environment friendly. Women living in villages no longer need to spend hours and travel long distances to gather firewood for cooking and burning purposes. They can now use this time for other activities. A smokeless and also a soot-free kitchen would mean that the women are not at risk of lung and throat infections and can live a longer, healthier life. Biogas has the capacity to meet all the fuel demands of houses, agricultural lands, and industries. For example, biogas can be used for cooking, heating, lighting, and so on. In farms, it can be utilized to dry crops, pump water during irrigation, and so on. A noticeable advantage of this is that it saves firewood. With the setting up of biogas plants, employment chances are also created in villages. The utilization of biogas increases as the biogas plant generates fuel and fertilizers. The most important advantage of the biogas plants is that it could digest any wet mixture of waste, fertilizer, and plant leftovers because of complicated bacterial methods. It does not cause reduction in the ammonia nitrogen content at the time of anaerobic digestion and helps in killing pathogens and weed seeds. In total, India has an immense capacity to produce electricity and heat from the waste as biogas. Till date, only a small part of the total capacity is used, and more investment in the field could increase the exploitation and help in realizing its true capacity in the near future.

REFERENCES

Angelidaki, I. and L. Ellegaard. 2005. Anaerobic digestion in Denmark: Past, present and future. In *Proceedings of FAO, Anaerobic Digestion 2002, IWA Workshop*, Moscow, 18–22 May 2005

Angelidaki, I. and W.T.M. Sanders. 2004. Assessment of the anaerobic biodegradability of macro-pollutants. *Reviews in Environmental Science and Biotechnology* 3: 141–158

Angelidaki, I. and L. Ellegaard. 2003. Co-digestion of manure and organic wastes in centralized biogas plants: Status and future trends. *Applied Biochemistry and Biotechnology* 109(1–3): 95–106

Angelidaki, I., L. Ellegaard, and B.K. Ahring. 2003. Applications of the anaerobic digestion process. In *Bio-methanation II*, edited by B. K. Ahring, pp. 1–33. Berlin: Springer

Angelidaki, I., L. Ellegaard, A.H. Sorensen, and J.E. Schmidt. 2002. Anaerobic processes. In *Environmental Biotechnology*, edited by I. Angelidaki, pp. 1–114. Institute of Environment and Resources, Technical University of Denmark (DTU)

Anunputtikul, W. and S. Rodtong. 2004. Laboratory Scale Experiments for Biogas Production from Cassava Tubers. *The Joint International Conference on Sustainable Energy and Environment (SEE)*, Hua Hin, Thailand

Ashden. 2004. Biogas cooking stoves for villages on the fringes of the tiger reserve in Ranthambhore Park. The Ashden Awards for Sustainable Energy. Details available at http://www.ashdenawards.org/files/reports/PrakratikIndia2004Technicalreport.pdf

Baharuddin, A.S., L.S. Hock, M.Z.M. Yusuf, N.A.A. Rahman, U.K.M. Shah, M.A. Hassan, M. Wakisaka, K. Sakai, and Y. Shirai. 2010. Effects of palm oil mill effluent (POME) anaerobic sludge from 500 m^3 of closed anaerobic methane digested tank on pressed-shredded empty fruit bunch (EFB) composting process. *African Journal of Biotechnology* 9(16): 2427–2436

Björnsson, L. 2000. Intensification of biogas process by improved process monitoring and biomass retention. *Ph.D. Thesis*, Department of Biotechnology, Lund University, Sweden

Björnsson, L., M. Murto, and B. Mattiasson. 2000. Evaluation of parameters for monitoring an anaerobic co-digestion process. *Applied Microbiology and Biotechnology* 54: 844–849

Boe, K. and I. Angelidaki. 2009. Serial CSTR digester configuration for improving biogas production from manure. *Water Research* 43: 166–172

Boe, K. 2006. Online monitoring and control of the biogas process. Ph.D. Thesis, Institute of Environment and Resources, Technical University of Denmark (DTU)

Boe, K., D.J. Batstone, and I. Angelidaki. 2007. An innovative online VFA monitoring system for the anaerobic process, based on headspace gas chromatography. *Biotechnology and Bioengineering*. 96(4):712–721

Bruni, E. 2010. Online improved anaerobic digestion of energy crops and agricultural residues. Department of Environmental Engineering, Technical University of Denmark (DTU)

Buren, V. 1983. *A Chinese Biogas Manual*. Popularizing Technology. Country side. West Yorkshire: Intermediate Technology Publications Ltd

Chew, T.L. and S. Bhatia. 2008. Catalytic processes towards the production of biofuels in a palm oil and oil palm biomass-based biorefinery. *Bioresource Technology* 99: 7911–7922

Costa, J.C., S.G. Barbosa, M.M. Alves, and D.Z. Sousa. 2012. Thermochemical pre- and biological co-treatments to improve hydrolysis and methane production from poultry wastes. *Bioresource Technology* 111: 141–147

Cresson, R., H. Carrere, J.P. Delgenes, and N. Bernet. 2006. Biofilm formation during the start-up period of an anaerobic biofilm reactor—Impact of nutrient complementation. *Biochemical Engineering Journal* 30: 55–62

Danisco Sugar. 2001. Sugar production, Green accounts. Details available at http://www.lcafood.dk/processes/industry/sugarproduction.htm

De Man, A.W.A., A.R.M. Vander Last, and G. Lettinga 1988. The use of EGSB and UASB anaerobic systems for low strength soluble and complex wastewaters at temperatures ranging from 8 to 30 °C. *In Proceedings of the Fifth International Conference on Anaerobic Digestion*, edited by E.R. Hall and P.N. Hobson, pp. 197–209

Demirel, B. and P. Scherer. 2008. Production of methane from sugar beet silage without manure addition by a single-stage anaerobic digestion process. *Biomass and Bioenergy* 32: 203–209

Deublein, D. and A. Steinhauser. 2008. *Biogas from Waste and Renewable Resources*. Weinheim: Willey

Dutta, S. 1997. Role of women in rural energy programs: Issues, problems and opportunities. In *Rural and Renewable Energy: Perspectives from Developing Countries*, edited by P. Venkataraman. New Delhi: Tata Energy Research Institute

FAO. 1997. A system approach to biogas technology. Details available at http://www.fao.org/sd/Egdirect/Egre0022.htm

FAOSTAT. Potato starch processing waste. Details available at http://faostat.fao.org/site/567/DesktopDefault.aspx?PageID=567#ancor, last accessed on 2 October 2010

Fu, X., N.I. Achu, E. Kreuger, and L. Björnsson. 2010. Comparison of reactor configurations for biogas production from energy crops. In Power and Energy Engineering Conference (APPEEC), Asia-Pacific, 28-31 March 2010, pp. 1–4

Garcia-Heras, J.L. 2003. Reactor sizing, process kinetics and modelling of anaerobic digestion of complex wastes. In *Bio-methanization of the Organic Fraction of Municipal Solid Wastes*, edited by J. Mata-Alvarez, pp. 21–58. London: IWA Publisher

GTZ. 1999. Biogas - application and product development. Information and Advisory Service on Appropiate Technology. *In Biogas Digest Volume II*. Details available at http://www.gtz.de/de/dokumente/en-biogas-volume2.pdf

GTZ. 2007. MDG monitoring for urban water supply and sanitation. Catching up with reality in Sub-Saharan Africa.Edited by German Technical Cooperation (GTZ).

Gustavsson, M. 2000 Biogas technology—solution in search of its problem: A study of introduction and integration of small-scale rural technologies. Department of Interdisciplinary Study, Goteberg University. Details available at www.ted-biogas.org/assets/download/Gustavsson2000.pdf

Hagen M. and E. Polman. 2001. Adding gas from biomass to the gas grid. Final report submitted to Danish Gas Agency

Horiuchi, J., T. Shimizu, K. Tada, T. Kanno, and M. Kobayashi. 2003. Selective production of organic acids in anaerobic acid reactor by pH control. *Bioresource Technology*. 82(3): 209–213

Hwang, M. H., N.J. Jang, S.H. Hyum, and I.S. Kim. 2004. Anaerobic bio-hydrogen production from ethanol fermentation: the role of pH. *Journal of Biotechnology*. 111(3):297– 309

ISAT and GTZ (Eds). 1999. *Biogas Basics*. Information and Advisory Services on Appropriate Technology (ISAT) and German Agency for Technical Cooperation GmbH (GTZ).

Kalogo, Y. and W. Verstraete. 1999. Development of anaerobic sludge bed (ASB) reactor technologies for domestic wastewater treatment: Motives and perspectives. *World Journal of Microbiology and Biotechnology* 15: 523–534

Kangle, K., S.V. Kore. and G.S. Kulkarni. 2012. Recent Trends in Anaerobic Codigestion: A Review. *Universal Journal of Environmental Research and Technology.* 2 (4): 210–219

Kaniyamparambil, J.S. 2011. Master of Engineering and Public Policy—Inquiry. Details available at http://wbooth.mcmaster.ca/epp/publications/student/ Joshua%20Samuel%20Kaniyamparambil.pdf

Kaparaju, P., M. Serranoa, and I. Angelidaki. 2009. Effect of reactor configuration on biogas production from wheat straw hydrolysate. *Bioresource Technology* 100: 6317–6323

Karup Kartoffelmelsfabrik. 2002. Green accounts. Details available at http:// www.lcafood.dk/processes/industry/potatoflourproduction.htm, last accessed on 09 November 2010.

Kashyap, D.R., K.S. Dadhich, and S.K. Sharma. 2003. Biomethanation under psychrophilic conditions: A review. *Bioresource Technology* 87: 147–153

Kato, T.M., J.A. Field, and G. Lettinga. 1997. The anaerobic treatment of low strength wastewaters in UASB and EGSB reactors. *Water Science and Technology* 36(6–7): 375–382

Kaushik, G., S. Satya, and S.N. Naik. 2009. Food processing a tool to pesticide residue dissipation—A review. *Food Research International* 42(1): 26–40

Kayhanian, M. and D. Rich. 1995. Pilot-scale high solids thermophilic anaerobic digestion of municipal solid waste with an emphasis on nutrient requirement. *Biomass and Bioenergy* 8(6): 433–444

Kishore, V.V.N. and S.N. Srinivas. 2003. Biofuels of India. *Journal of Scientific and Industrial Research* 62(1–2): 106–123

Klass, D.L. 1984. Methane from anaerobic fermentation. *Science* 223(4640): 1021–1028

Kohler, A., S. Hellweg, E. Recan, and K. Hungerbuhler. 2007. Input-dependent life-cycle inventory model of industrial wastewater-treatment processes in the chemical sector. *Enviro. Sci. Technol* 41: 5515–5522

Kotsyurbenko, O.R. 2005. Trophic interactions in the methanogenic microbial community of low-temperature terrestrial ecosystems. *FEMS Microbial Ecology* 53(1): 3–13

Lay, J.J., Y.Y. Li, and T. Noike. 1997. Influences of pH and moisture content on the methane production in high-solids sludge digestion. *Water Research* 31(6): 1518–1524

Lepistö, R. and J. Rintala. 1999. Kinetics and characteristics of 70°C, VFA-grown, UASB granular sludge. *Applied Microbiology and Biotechnology* 52(5): 730–736

Lettinga, G., A.F.M. van Nelsen, S.W. Hobma, W. de Zeeuw, and A. Klapwijk. 1980. Use of the upflow sludge blanket (USB) reactor concept for biological waste water treatment, especially for anaerobic treatment. *Biotechnology and Bioengineering* 22: 699–734

Mackie, R.I. and M.P. Bryant. 1981. Metabolic activity of fatty acid-oxidizing bacteria and the contribution of acetate, propionate, butyrate, and CO_2 to methanogenesis in cattle waste at 40°C and 60°C. *Appl Environ Microbiol.* 41: 1363–1373

Mandal, T.B., A. Kiran, and N.K. Mandal. 1999. Determination of the quality of biogas by flame temperature measurement. *Energy, Conversion & Management* 40: 1225–1228

Margarita, A.D., N.D. Spyros, S. Katerina, Z. Constantina, and K. Michael. 2009. Biogas production from anaerobic co-digestion of agro-industrial wastewaters under mesophilic conditions in a two-stage process. *Desalination* 248: 891–906

Ministry of New and Renewable Energy (MNRE). 2007. Family Type Biogas Plants Programme—NBMMP, Ministry of New and Renewable Energy. Details available at http://www.mnre.gov.in/prog-ftbp.htm

Ministry of New and Renewable Energy (MNRE). 2010a. National Biomass Programme (NBMMP) Details available at http://www.mnre.gov.in/prog-biomasspower.htm

Ministry of New and Renewable Energy (MNRE). 2010b. National Biogas and Manure Management Programme (NBMMP). Details available at http://mnre.gov.in/prog-ftbp.htm

Misson, M., R. Haron, M.F.A. Kamaroddin, and N.A.S. Amin. 2009. Pre-treatment of empty palm fruit bunch for production of chemicals via catalytic pyrolysis. *Bioresource Technology* 100: 2867–2873

Mittal, K.M. 1996. *Biogas Systems: principles and applications.* New Delhi: New Age International (P) Limited

Mohan, M.P.R. and L. Kumar. 2005. Biofuels: A key to India's sustainable energy needs. In *Proceedings of the RISO International Energy Conference: technologies for sustainable energy development in the long run*, Roskilde, Denmark.

Mourad, L., V.S. Ambrogi and S.M. Guerra. 2004. Potencial de utilizaçãoenergética de biomassa residual de grãos. In *Proceedings of the 5th Encontro de Energia no Meio Rural, Campinas, São Paulo, Brazil.*

Mshandete, A., A. Kivaisi, M. Rubindamayugi, and B. Mattiasson. 2004. Anaerobic batch co-digestion of sisal pulp and fish wastes. *Bioresource Technology.* 95(1):19–24

Nagamani, B. and K. Ramasamy. 1991. Biogas production technology: An Indian perspective. *31st Annual Conference of Association of Microbiologists of India* held at Tamil Nadu Agricultural University, Coimbatore

Nagamani, B., V. Chitra, and K. Ramasamy. 1994. Biogas production technology: An Indian perspective. *Proceedings of 35th Annual Conference of Association of Microbiologists of India* 13: 33–35

National Bank for Agriculture and Rural Development (NABARD). 2007. Model Bankable Projects: Agricultural Engineering—Biogas. Details available at http://www.nabard.org/modelbankprojects/biogas.asp

National Bank for Agriculture and Rural Development (NABARD). 2004. Implementation of the National Biogas and Manure Management Programme. NABARD Technical Digest, Vol. 8. Details available at http://www.nabard. org/fileupload/ContentDisplay_TechnicalDigest.aspx?file_id=3&file_ title=Volume%208&file _year=2004

Neves, L., R. Ribeiro, R. Oliveira, and M.M. Alves. 2006. Enhancement of methane production from barley waste. *Biomass and Bioenergy* 30: 599–603

Nielsen, H.B. and I. Angelidaki. 2008. Codigestion of manure and industrial organic waste at centralized biogas plants: process imbalances and limitations. *Water Sci. Technol.* 58(7): 1521–1528

Ofoefule, A.U. and E.O. Uzodinma. 2008. Effect of Chemical and Biological treatment on Pre-decayed field grass (panicum maximum) for biogas production. *Niger. J. Solar Energy.* 19: 57–62

Olbrich, H. 1963. *The Molasses*. Berlin, Germany: Biotechnologie-Kempe GmbH

Omer, A.M. and Y. Fadalla. 2003. Biogas technology in Sudan, Technical note. *Renewable Energy* 28: 499–507

Parawira, W., M. Murto, J.S. Read, and B. Mattiasson. 2005. Profile of hydrolases and biogas production during two-stage mesophilic anaerobic digestion of solid potato waste. *Process Biochemistry* 40(9): 2945–2952

Parawira, W., M. Murto, R. Zvauya, and B. Mattiasson. 2004. Anaerobic batch digestion of solid potato waste alone and in combination with sugar beet leaves. *Renewable Energy* 29(11): 1811–1823

Pind, P. F., I. Angelidaki, and B.K. Ahring. 2003. A new VFA sensor technique for anaerobic reactor systems. *Biotechnology and Bioengineering* 82(1): 54–61

Poliafico, M. 2007. Anaerobic digestion: decision support software. *Mastersthesis.* Cork, Ireland: Department of Civil, Structural and Environmental Engineering, Cork Institute of Technology

Porras, J.P. and G. Gebresenbet. 2003. Review of biogas development in developing countries with special emphasis in India. Technical Report Faculty of Natural Resources and Agricultural Sciences, Department of Energy and Technology, Sveriges Lantbruksuniversitet Uppsala. Details available at http://pub.epsilon.slu.se/3941/

Programme Evaluation Organisation (PEO). 2002. Evaluation Study on National Project on Biogas Development. Details available at planningcommission.nic. in/reports/peoreport/peoevalu/peo_npbd.pdf

Puyol, D., A.F. Mohedano, J.L. Sanz, and J.J. Rodriguez. 2009. Comparison of UASB and EGSB performance on the anaerobic biodegradation of 2, 4-dichlorophenol. *Chemosphere* 76: 1192–1198

Ramachandra, T. V., G. Kamakshi, and B.V. Shruthi. 2004. Bioresource status in Karnataka. *Renewable and Sustainable Energy Reviews* 8: 1–47

Ramachandra, T.V., S.V. Krishna, and B.V. Shruthi. 2005. Decision support system for regional domestic energy planning. *Journal of Scientific and Industrial Research* 64: 163–174

Ramachandra, T.V., K. Vijay, M. Parchuri, and V. Subbrao. 2006. A study on biogas generation from non-edible oil seed cakes: Potential and Prospects in India. *2nd Joint International Conference on Sustainable Energy and Environment (SEE), Bangkok, Thailand*

Rao, K.U. and N.H. Ravindranath. 2002. Policies to overcome barriers to the spread of bioenergy technologies in India. *Energy for Sustainable Development* 6(3): 59–73

Raven, R.P.J.M. and K.H. Gregersen. 2005. Biogas plants in Denmark: Successes and setbacks. *Renewable and Sustainable Energy Reviews* 11(1): 116–132

Ravindranath, N.H. and D.O. Hall. 1995. *Biomass Energy and Environment: a developing countries perspective from India.* Oxford: Oxford University Press

Rondón, E.O. 2008. Tecnologias para mitigar o impactoambiental da produção de frangos de corte. *Revista Brasileira de Zootecnia V.* 37: 239–252

Russell, J.B. and F. Diez-Gonzalez. 1997. The effects of fermentation acids on bacterial growth. *Advances in Microbial Physiology* 39: 205–234

Sanchez, O.J. and C.A. Cardona. 2008. Trends in biotechnological production of fuel ethanol from different feedstocks. *Bioresource Technology* 99: 5270–5295

Satyawali, Y. and M. Balakrishnan. 2007. Removal of color from biomethanated distillery spent wash by treatment with activated carbons. *Bioresource Technology.* 98: 2629–2635

Schink, B. 1997. Energetics of syntrophic cooperation in methanogenic degradation. *Microbiology and Molecular Biology Reviews* 61(2): 262–280

Seghezzo, L., G. Zeeman, J.B. van Lier, H.V.M. Hamelers, and G. Lettinga. 1998. A review: the anaerobic treatment of sewage in UASB and EGSB reactors. *Bioresource Technology* 65: 175–190

Sevilla-Espinosa, S., M. Solorzano-Campo, and R. Bello-Mendoza. 2010. Performance of staged and non-staged up-flow anaerobic sludge bed (USSB and UASB) reactors treating low strength complex wastewater. *Biodegradation* 21(5): 737–751

Sharma, K. 2011. *Nursing Research and Statistics.* New Delhi: Elsevier

Sivakumaran, S., B. Nagamani, and K. Ramasamy. 1992. *Biological Nitrogen Fixation and Biogas Technology,* S. Kannaiyan, K. Ramasamy, K. Ilamurugu, and K. Kumar (eds), pp. 101–110. Coimbatore: Tamil Nadu Agricultural University

Shastry, S., T. Nandy, S.R. Wate, and S.N. Kaul. 2010. Hydrogenated vegetable oil industry wastewater treatment using UASB reactor system with recourse to energy recovery. *Water, Air, Soil Pollution* 208(1–4): 323–333

Stalin, N. Prabhu. 2007. Performance evaluation of Partial Mixing Anaerobic Digester. *ARPN Journal of Applied Sciences* 2(3):1–6

Stamatelatou, K., G. Lyberatos, C. Tsiligiannis, S. Pavlou, P. Pullammanappallil, and S.A. Svoronos. 1997. Optimal and suboptimal control of anaerobic digesters. Environmental Modeling and Assessment 2: 355–363

Tafdrup, S. 1994. Centralized biogas plants combine agricultural and environmental benefits with energy production. *Water Science and Technology* 30(12): 133–141

Tan, H.T., K.T. Lee, and A.R. Mohamed. 2010. Second-generation bio-ethanol (SGB) from Malaysian palm empty fruit bunch: Energy and exergy analyses. *Bioresource Technology.* 101: 5719–5727

Tomar, S.S. 1995. Status of biogas plants in India—an overview. *Energy for Sustainable Development* 1(5): 53–56

United Nations Asian and Pacific Centre for Agricultural Engineering and Machinery (UNAPCAEM). 2007. Recent developments in biogas technology for poverty reduction and sustainable development. *Proceedings of International Seminar on Biogas Technology for Poverty Reduction and Sustainable Development.* Details available at http://www.unapcaem.org/publication/pub_biogas.htm

Uzodinma, E.O., A.U. Ofoefule, N.D.Onwuka, and J.I. Eze. 2007. Biogas production from blends of Agro-industrial wastes. *Trends Appl. Sci. Res.* 2 (6): 554–558

Vavilin, V.A., S.V. Rytov, and L.Y. Lokshina. 1996. A description of hydrolysis kinetics in anaerobic degradation of particulate organic matter. *Bioresource Technology* 56(2–3): 229–237

Verstraete, W., F. Morgan-Sagastume, S. Aiyuk, M. Waweru, K. Rabaey, and G. Lissens. 2005. Anaerobic digestion as a core technology in sustainable management of organic matter. *Water Science and Technology* 52(1–2): 59–66

Wang, H., H.Wang, W. Lu, and Y. Zhao. 2009. Digestibility improvement of sorted waste with alkaline hydrothermal pre-treatment. *Tsinghua Science and Technology* 14: 378–382

Wilkie, A., G. Faherty and E. Colleran. 1984. The effect of varying the support matrix on the anaerobic digestion of pig slurry in the upflow anaerobic filter design. *In Energy from Biomass, 2nd E.C. Conference.* pp. 531–535

Young, J.C. and K.H. Song. 1984. Factors affecting selection of media for anaerobic filters. *Proceedings of III International Conference on Fixed Film Biological Processes,* Arlington, Virginia, pp. 229–245

2

Membrane-Less Microbial Fuel Cell: A Low-cost Sustainable Approach for Clean Energy and Environment

Atin Kumar Pathak[a], V. V. Tyagi[a,*], Har Mohan Singh[b], Vinayak V. Pathak[b,c], and Richa Kothari[b,c,*]

[a]*School of Energy Management, Faculty of Engineering, Shri Mata Vaishno Devi University, Katra, Jammu and Kashmir 182320*

[b]*Bioenergy and Wastewater Treatment Laboratory, Department of Environmental Science, Babasaheb Bhimrao Ambedkar University, Lucknow, Uttar Pradesh 226025*

[c]*DST-Centre for Policy Research, Babasaheb Bhimrao Ambedkar University, Lucknow, Uttar Pradesh 226025*

[*]*E-mail: vtyagi16@gmail.com; kothariricha21@gmail.com*

2.1 INTRODUCTION

Rapid industrialization and urbanization have accelerated environmental pollution and energy crisis, a major concern nowadays, at global level (Li, Zhang, Zeng, *et al.* 2009). In the last two decades, the demand for energy has increased about three times (Rahimnejad, Ghoreyshi, Najafpour, *et al.* 2011). The gap between demand and supply of oil, coal, petroleum, and electrical power is ever increasing; even water is not adequately available to be used for energy production in both developing and developed countries. Exhaustive consumption of petroleum products has led to depletion of fossil fuel reserves and hike in fuel price. Greenhouse gas emissions have also increased as a result. Countries are making efforts to obtain a sustainable solution against energy crisis and pollution. Renewable energy such as solar energy, waste-to-energy, wind energy, bio-energy, small hydro energy are being adopted widely as a sustainable tool.

Waste-to-energy provides an attractive opportunity for concurrent energy generation and pollution reduction as it minimizes the quantity of waste and can be used as an effective waste management practice. Fuel cell has many advantages over the other energy generators such as less emission of polluting gases (SO_x, NO_x, CO_2, and CO) and higher conversion efficiency and so forth (Peighambardoust, Rowshanzamir, and Amjadi 2010). In this context, microbial fuel cell (MFC) became the focus of many researchers as a new approach in the late-1980s and mid-1990s. Scientists discussed component materials, fuel cell design, operational parameters and its reliability and the effect of using immobilized bacteria on electrode surfaces. In the late 1990s, a chain of investigations concluded that certain bacteria can be applied directly to MFC for production of electricity without involving a hydrogen-mediated process (Choi, Adams, Stahlhut, et al. 1999). To produce bioelectricity, certain bacterial strains are used in anodic chamber of MFC which contains wastewater. Bacterial species present in the chamber decomposes the organic waste and produces electrons, and due to movement of electrons from anode to cathode electricity is produced and wastewater gets treated (Rahimnejad, Ghoreyshi, Najafpour, et al. 2011). The process involves oxidation of organic matter by microorganisms, while resulting electrons from their metabolism are transferred to an electrode (anode). By using wastewater as a substrate, it reduces the load on the environment. The efficiency and further application of MFC is dependent on the process of microbial reactor construction (Li, Zhang, Zeng, et al. 2009).

Electron is transferred in two ways: (i) by direct contact with cytochrome c or (ii) through mediators. Produced electrons are passed through circuit from anode to cathode to a potential electron acceptor such as oxygen or metals (Leong, Daud, Ghasemi, et al. 2013). Oxygen is widely used as an acceptor. This electron acceptor is reduced and combined with a proton of the anodic chamber to form water. Both simple substrates, such as acetate, lactate, or glucose, and complex substrates, including industrial or domestic wastewater, can produce electricity (Logan 2004).

An MFC basically comprises anode and cathode (Figure 2.1), separated by a proton exchange membrane (PEM), and is called a dual chambered MFC. It has some fundamental disadvantages (Fornero, Rosenbaum, Cotta, et al. 2008), such as non-affordable cost of exchange membranes (proton, cation and anion), potential for bio-fouling, and high internal resistance (Liu, Cheng, Huang, et al. 2008). Different types of MFCs available are as follows:

(a) **Mediator-less MFC:** As the name suggests, this type of MFC does not require metal to transfer electrons from the anode chamber to the cathode chamber due to which potential difference develops

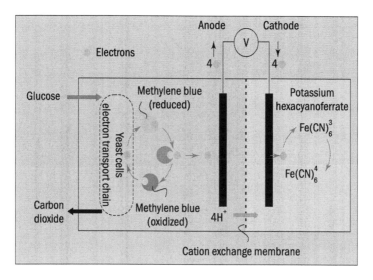

Fig. 2.1 *Basic design of dual chambered MFC*

across the chambers and electricity is produced. In place of metals, bacterial species directly transfer electrons from one chamber to the other. In the absence of a mediator for electron transfer to electrodes, it is known mediator-less MFCs. It has also been found that certain metal-reducing bacterial species mainly belonging to the *Geobacteraceae* family can transfer electrons directly to the electrodes because of the presence of electrochemically active enzymes, such as cytochromes on their outer membranes (Kaufmann and Lovely 2001). Schematic representation of mediator-less MFC is given in Figure 2.2. Two electrodes (anode and cathode) are placed in water in two compartments separated by a PEM.

(b) **Membrane-less MFC:** Jang, Pham, Chang, *et al.* (2004) revolutionized the designing of MFC by developing a less expensive membrane-less MFC. In this, enriched electrochemically active microbial species, which utilize organic wastes, were used and electricity is produced in the form of a by-product. The design, however, has a basic drawback of poor cathode reaction. A large quantity of oxygen diffuses towards the anode chamber making the chamber less anaerobic in nature. To minimize the flow of oxygen from the cathode chamber to the anode chamber, electricity flow was decreased. As non-membrane means more oxygen diffusion to the anode chamber. Viridis, Rabaey, Yuan, *et al.* (2008) made a loop model MFC to remove carbon and nitrogen compounds from wastewater. Further studies are required to improve the design

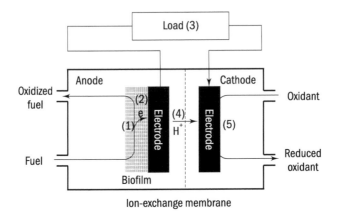

Fig. 2.2 *A mediator-less MFC: (i) fuel oxidation, (ii) transfer of electron from microbes process to electrode, (iii) load in form of resistance, (iv) transfer of proton from anode to cathode compartment, and (v) reduction taking place with the help of oxygen supply*

Source *Gil, Chang, Kim, et al. (2003)*

of membrane-less microbial fuel cell (ML-MFC) for better current yield. Figure 2.3 gives a basic modelling design of ML-MFC used by Tardast, Rahimnejad, Najafpour, *et al.* (2012).

In this chapter, the authors focus on the basic designing, types, and applications of ML-MFC. The integration of ML-MFC with wastewater treatment provides an opportunity to produce energy as well as treat wastewater without membranes.

2.2 BASIC REQUIREMENTS FOR MEMBRANE-LESS MICROBIAL FUEL CELL

The MFC device treats organic pollutants present in wastewater with the help of microbial communities and reduces up to 90% of the chemical oxygen

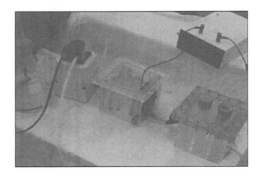

Fig. 2.3 *Membrane-less microbial fuel cell*
Source *Tardast, Rahimnejad, Najafpour, et al. (2012)*

demand (COD) and biochemical oxygen demand (BOD) of wastewater. Despite the advantage of treating wastewater and generating electricity, MFC has some limitations such as high membrane cost, which can be resolved by using ML-MFC. It utilizes bacteria as a biocatalyst for treating wastewater rich in organic matter and concurrently generates bioelectricity (Guerrini, Cristiani, and Trasatti 2013). In the absence of a membrane, ML-MFC technology becomes cheap. However, there are some drawbacks too. Nowadays some computational models are in progress that provide the model for power generation by wastewater from different industries/sewage which is organically rich in nature. In a simple MFC, the membrane works as an electrolyte and as an electronic insulator which only allows protons to move through. This electrolyte insulator is an expensive material and makes the application of MFC a limited technology for wastewater treatment and bioelectricity generation (Guerrini, Cristiani, and Trasatti 2013).

An ML-MFC makes the process of treating wastewater economically viable by plummeting not only the capital speculation but also the overall operational cost for the membrane maintenance. In an ML-MFC, the basic requirements will be electrode material, bacterial culture used for electron transfer, and wastewater.

2.2.1 Electrode Material

This section gives a brief description of the electrode material and wastewater utilized by different researchers in an ML-MFC. As electrode is the major part of any fuel cell, different scientists and their co-workers have experimented with different types of electrode materials. They used various types of electrode for electron production with varying bacterial strain. Anodic material and cathodic material are the two types of electrode material present in a system. Table 2.1 shows the various types of electrodic material used by research community in the last few years.

2.2.1.1 Anode

In the ML-MFC, carbon is the basic anodic material due to its frequent availability and low cost. Various researchers have used carbon anode with metal coating such as Hu (2008), used carbon with platinum coating and Sun, Hu, Bi, et al. (2009), used 5 cm^2 carbon clothes as an anodic material. To minimize electrolytic losses, copper wires connected the anode with cathode. Aba, Marsili, Stante, et al. (2009) used anode material made of glassy carbon. In another study, Ghangrekar and Shinde (2007) created an ML-MFC of acrylic cylinder having internal diameter of 10 cm and total working volume of 4.6 L. Solid graphite was used for both anode

Table 2.1   List of electrode material used by different researchers in ML–MFC

S. No.	Wastewater	Electrode material		Voltage	Power density/ current density	Microorganism	References
		Anode	Cathode				
1.	Synthetic	Carbon felt	Carbon flake	44.4 ± 2.5 mW/m^2			Thung, Ong, Ho, et al. (2015)
2.	Wastewater from wastewater treatment plant	Carbon felt	Pt-loaded carbon paper	44.4 ± 2.5 mW/m^2			Thung, Ong, Ho, et al.(2015)
3.	Artificial	Graphite felt	Pt-coated graphite		1.3 mW/m^2		Jang, Phama, Changa, et al. (2004)
4.	Artificial wastewater containing glucose and glutamate	Graphite felt	Pt-coated graphite		2.60 ± 0.02 mA	Aerobic microbes	Jang, Phama, Changa, et al. (2004)
5.	Fresh washed anaerobic sludge	Carbon paper	Carbon cloth		184 mW/m^2		Zhiqiang (2008)
6.	Anaerobic sludge wastewater treatment plant	Carbon granules	Flexible carbon cloth with Pt coating		50.2 W/m^3		You, Zhao, Zhang, et al. (2007)
7.	Raw wastewater	Carbon cloth	Carbon cloth				Guerrini, Cristiani, and Trasatti (2013)
8.	Synthetic wastewater	Graphite rod	Graphite rod		10.13 mW/m^2		Ghangrekar and Shinde (2007)
9.	Municipal wastewater treatment plant	Bamboo charcoals	Carbon paper electrode containing 0.5 mg/cm2 of Pt as the catalyst on only one side		0.144 and 1.16 mW		Yang, Jia, and Liu (2009)

Contd...

Table 2.1 Contd...

S. No.	Wastewater	Electrode material		Voltage	Power density/ current density	Microorganism	References
		Anode	Cathode				
10.	Synthetic	Untreated glassy carbon	Untreated glassy carbon		70 mW/m²	Firmicutes, α-proteobacteria, β-proteobacteria, γ-proteobacteria, and Bacteroidetes groups	Aba, Marsili, Stante, et al. (2009)
11.	Anaerobic sludge collected from the wastewater treatment plant	Graphite rod	Two carbon cloth with Pt coating	450 mW/ mW/m²			Guo, Zhaoy, Wang, et al. (2010)
12.	Synthetic glucose	Graphite plates	Graphite plates	37.4 mW/ m²			Zhu, Wang, Zhang, et al. (2011)
13.	Domestic	Graphite matrix	Ferricyanide solution	90 W/m³			Korneel and Verstraete 2005
14.	Hospital	Graphite matrix	Ferricyanide solution	66 W/m³			Korneel and Verstraete 2005
15.	Brewery wastewater	Carbon cathode	Ni based and graphite with carbon cloths				Li, Zhang, Zeng, et al. (2009)
16.	Brewery wastewater	Graphite plates	Graphite plates		37.4 mW/m²		Zhu, Wang, Zhang, et al. (2011)

and cathode, separated by a distance of 20 cm. In this ML-MFC, they observed a maximum electricity generation of 10.13 mW/m^2. Yang, Jia, and Liu (2009) made a membrane-less single-chambered MFC consisting of bamboo charcoal anode electrodes. The anodic electrodes were made of "Jinhao" brand bamboo charcoals selected because of their resistance values less than 10 Ω by multimeter. Li, Zhang, Zeng, *et al.* (2009) developed a study where a membrane-less cloth cathode assembly was constructed by coating the cloth with nickel-based or graphite-based conductive paint and non-precious metal catalyst. The ML-MFC was treated with brewery wastewater and 10 mW/m^2 power output was produced (Li, Zhang, Zeng, *et al.* 2009).

Graphite is one of the traditional carbon allotrope which is used as anode material. Disk form graphite anode is used to treat artificial wastewater comprised of glucose and glutamate and gives an power output of 2.60 ± 0.02 mA (Jang, Phama, Changa, *et al.* 2004).

2.2.1.2   Cathode

Abiotic cathodes are most commonly used in MFC. Studies have used carbon paper, cloth, graphite, woven graphite, graphite granules, and brushe as cathode (Liu, Cheng, Huang, *et al.* 2008). When these material are used for the cathode, a catalyst is usually needed. In Yang, Jia and Liu's (2009) ML-MFC, the cathode electrode was a carbon paper containing 0.5 mg/cm^2 of Pt catalyst on only water-facing side with municipal wastewater and produced a maximal output power density of 0.144 mW and 1.16 mW. Pt is generally used as a catalyst in an MFC. At present, to reduce cost, non-precious metals are being explored as alternatives in hydrogen production. Stainless steel and nickel sheets can be also used as viable alternatives to Pt in cathode production. To generate hydrogen gas from organic matter, high surface area stainless steel brush cathodes were used. The rates and efficiencies were similar to those achieved with flat carbon cathodes containing nano-sized Pt. Rahimnejad, Adhami, Darvari, *et al.* (2015) helped in improving power density by using carbon paper as a cathode material in designed ML-MFC. In the same context Ni–Mo material was used as a cathode material and their output was found to be same as with Pt cathode used in microbial electrolytic cell for hydrogen production (Hu 2008). Tungsten carbide may also be useful material as cathode, although corrosion can be a problem in phosphate buffered, neutral pH solutions (Harnisch, Wirth, and Schröder 2009). The traditional and inorganic cathode material are compared together with biocathode, and the results showed that ideal biocathode material has large surface area and tough appearance. Due to these idealistic property, microbes found a large area to live and due to this there was a decrease in the power output of the material (Logan 2004). This was also confirmed by Sun, Hu, Bi, *et al.* (2009). However, the most suitable material has not yet been determined (Sun, Hu, Bi, *et al.* 2009). But, You, Zhao, Zhang, *et al.* (2007) showed and demonstrated that graphite fibre brush can be better utilized in MFC as biocathode material for sustainable electricity generation by utilizing the organic waste present in wastewater.

2.2.1.3 *Substrate/wastewater utilized in process*

In the microbial process of bioelectricity generation and wastewater treatment in MFC, selection of a substrate is an important parameter. For better utilization, proper growth of microbial community, and energy production by MFC, the substrate selected should be rich in organic content. Scientists have used both natural and artificial types of substrate/wastewater in this regard. To generate power and oversee the efficiency of the MFC system that removed wastewater pollutant, Gil, Seop, Byung, *et al.* (2003) created artificial wastewater (AW) that contained glucose and glutamate with BOD of approximately 300 mg/l in ML-MFC. Wastewater developed in the laboratory contained tap water with some modification. The solution per litre contained 3.95 g glucose, 0.38 g NH_4Cl, 0.015 g $FeSO_4$ × $7H_2O$, 0.026 g $MgSO_4$, 0.019 g $CaCl_2$, 0.76 mg $NiSO_4$ × $6H_2O$, and 9.36 g NaCl. The system was fed with the brewery wastewater collected from local industry with COD of around 2850 mg/l. Some amount of trace element and NaCl were adjusted as it is useful for microorganism growth (Zhu,Wang, Zhang, *et al.* 2011). This shows that to enable growth of microorganisms the substrate should contain proper composition of nutrients. Table 2.2 presents different types of wastewater used in ML-MFC for treatment application.

Table 2.2   Different types of wastewater used in ML-MFC for treatment

S. No.	Wastewater/substrate	References
1.	Synthetic	Aba, Marsili, Stante, *et al.* (2009)
2.	Anaerobic sludge collected from the wastewater treatment plant	Guo, Zhaoy, Wang, *et al.* (2010)
3.	Synthetic glucose	Zhu,Wang, Zhang, *et al.* (2011)
4.	Domestic	Rabaey and Verstraete (2005)
5.	Hospital	Rabaey and Verstraete (2005)
6.	Brewery wastewater	Li, Zhuang, Zeng, *et al.* (2009)
7.	Brewery wastewater	Zhu, Wang, Zhang, *et al.* (2011)

2.2.1.4   Inoculum/microorganisms

Microorganisms are the main driver of power production and degradation of organic substances, hence microbial inoculums are one of the major requirements for MFC system. A microorganism produces H^+ and generates electrons which help in production of electricity. The electrons present in the culture belong to anaerobic strains, such as *Geobacteraceae* (Bond, Holmes, Tender, *et al.* 2002). Different types of inoculums have been used by researchers to treat ML-MFC for power generation. In this regard Bond, Holmes, Tender, *et al.* (2002) used bacterial species belonging to *Geobacteraceae* family for substrate treatment in anodic chamber. To study different bacterial strains, Kim, Kim, Hyun, *et al.* (1999) and Kim, Park, Hyun, *et al.* (2002) studied the effect of mediators that help in electron transfer and work as inoculums for treatment process in MFC. They found that MFC can operate without using mediators. They also found metal reducing bacterial species, *Shewanella putrefaciens* as a novel discovery for MFC. With new research, new bacterial species such as *Geobacter sulfurreducence* also came into light. *Geobacter sulfurreducence* was used as an inoculum solution in ML-MFC to inoculate the anode chamber and accelerate the biofilm process (Du, Xie, Dong, *et al.* 2011). Tardast, Rahimnejad, Ghasem, *et al.* (2012) fabricated an ML-MFC system and inoculated the system with the bacterial species collected from the sludge of food and cheese whey industry. They tried to prove that the species collected can be utilized as substrate for bioelectricity generation and simultaneously wastewater treatment.

2.2.1.5   Electrolyte

In the MFC, an electrolyte is one of the major sources for electron transfer. In recent times, bioelectrolytes (microbes) are proving to be useful in mediator and ML-MFCs. Gil, Chang, Kim, *et al.* (2003) used NaCl as an electrolyte to transfer electron in a bulk/mass. They fed the design at a rate of 1.83 ml/min with an aeration rate of 200 ml/min. The liquid in the cathode chamber was removed and continuously fed with 1 M NaCl. It was found that when salt was added as an electrolyte, the production rate could reach up to 7.7 mA with the current generation density increasing from 3.5 mA to 4.7 mA. Du, Xie, Dong, *et al.* (2011) discovered that the anode and cathode chambers could be fed the same electrolyte that has a composition of KCl 0.13 g, NH_4Cl 0.31 g, $Na_2HPO_4 \times 12H_2O$ 6.1501 g, $NaH_2PO_4 \times 2H_2O$ 5.6182 g, and CH_3COONa 1.569 g. The solution is mixed with 1.25 mL of vitamin and mineral solutions to complete the electrolyte concentration and is used as a culture medium. When this same electrolyte is mixed with 4.643 g/L fumarate, it becomes an electron acceptor in the medium.

Hence, we can conclude that electrolyte is basically responsible for electron transfer and should contain the component which is responsible for electron transfer. These materials are now replaced by common bacterial strains for the same function in mediator-less MFC.

2.3 CLASSIFICATION OF MEMBRANE-LESS MICROBIAL FUEL CELL

Based on construction, structure, design, and development a ML-MFC is classified into different types. Nowadays, different types of ML-MFC are used by scientists as shown in Figure 2.4.

On the basis of work reported by different workers the ML-MFC can be described in two different ways, that is, for wastewater treatment and electricity generation. By using this type of MFC, the process becomes economically viable by utilizing the bacterial strain for electron transfer and for generating power without using any expensive membrane. Du, Xie, Dong, et al. (2011) avoided using expensive membrane by using continuous H$^+$ flow through electrolytic flow (see Figure 2.5). Both the electrode chambers are in cylindrical shape with a capacity of 785 mL volume. Using gravitational force, elctrolyte was flowed to the cathode chamber from the storage tank, which was kept at a height. A valve connects both the anode and the cathode chambers to control the connection status. Diameter of both the inlet and outlet is 6 mm. To minimize the cost of the whole system, a medical infusion set is used to control the flow rate in the system. Anode made up of ammonia-treated carbon cloth is used in the system. Plain carbon paper with a surface area of 20 cm^2 is used as cathode material. It contains 5% Pt in the form of catalyst. To facilitate the reaction faster, the cathode is kept close to the inlet point to enable the protons to meet the electron and catalyst as soon as they flow into the cathode chamber. To maintain sufficient supply of air in the cathode chamber, air is pumped artificially.

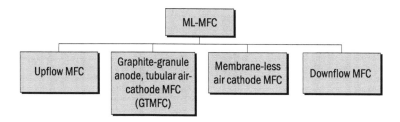

Fig. 2.4 *Types of membrane-less microbial fuel cell*

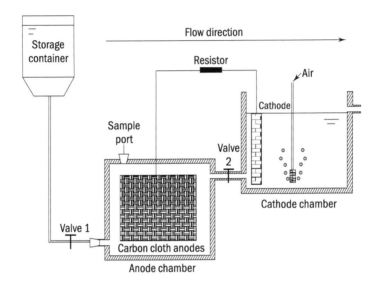

Fig. 2.5 *Design of ML-MFC*
Source   *Du, Xie, Dong, et al. (2011)*

Thung, Ong, Ho, *et al.* (2015) designed an innovative model of ML-MFC on the basis of upflow ML-MFC (Figure 2.6). This upflow ML-MFC had a diameter of 4.5 cm and height of 25 cm. A cylindrical hollow tube is filled with gravel which connects the anode and cathode chambers up

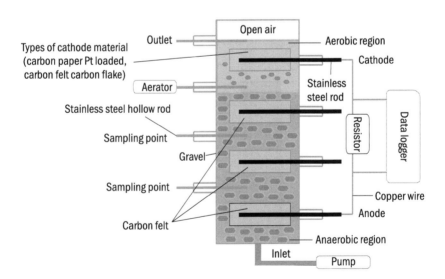

Fig. 2.6 *Upflow memebrane-less microbial fuel cell*
Source   *Thung, Ong, Ho, et al. (2015)*

to the height of 6 mm. The gravel, apart from acting as a membrane or separator between the anode and the cathode compartments, also functions as a medium for biofilm growth with the void volume of 0.11 L. It is kept submerged for a month before setting up of the UFML-MFC. The anode used in the UFML-MFC is made of carbon felt. In the system three anodes were used and kept at different distance from the cathode electrodes. Cathode of different materials and sizes were used in the system to compare the power generation efficiency. They were of (i) 3 cm × 2.5 cm × 2 cm (L × W × H) dimension, (ii) 2 cm × 5 cm (L × W) area made of Pt-loaded carbon paper, and (iii) carbon flake of 4.4 cm × 5 cm (D × H). Synthetic wastewater was used in the system and it flowed upwards from the anode region to the cathode region. A peristaltic pump was used to pump wastewater at a flow rate of 0.152 mL/min for 1h intervals on and off per day. A copper wire connected both the electrodes with an external resistance of 1000 Ω. In the cathode chamber air was transferred artificially and the system operated at room temperature.

Zhu, Wang, Zhang, *et al.* (2011) designed a downflow single-chamber MFC (Figure 2.7). The main part of the system comprised a cylindrical glass tube. The tube was 6 cm in inner diameter, 30 cm in height, and a working volume of about 850 mL. It was open at top. The rectangular anode material, 10 × 4 × 0.5 cm^3, was made of graphite. It was perpendicularly inserted into the anaerobic activated sludge from the bottom of the MFC. A circular graphite cathode of 5.6 cm diameter and 0.5 cm thickness was kept horizontally. It was half-submerged in the liquid. Both the anode material and the cathode material used did not have any coating. The electrodes of both the chambers were connected with copper wire to provide external circuit connections. Epoxy resin was used to seal all the connection points. The anode and cathode were connected

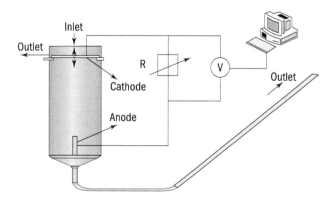

Fig. 2.7 *Downflow single-chamber MFC*
Source *Aba, Marsili, Stante, et al. (2009)*

with an external resistor of variable power to maintain the liquid level and to control the flow rate in the MFC. A plastic tube connected to the conical bottom of MFC was used. The designed MFC is inoculated with the anaerobic activated sludge, collected from the upflow anaerobic sludge bed reactor, used for treatment of synthetic glucose wastewater which is in working condition from last one year. The MFC chamber was filled with 200 mL of anaerobic sludge that has a concentration of ca. 10.0 g/L for volatile suspended solids. The downflow feeding was introduced once the MFC operated for at least 30 days in a batch operating mode. A peristaltic pump was used to maintain the flow of wastewater at a continuous flow rate of 4.0 mL/min from the bottom of the MFC. The system was maintained at a control temperature of $30 \pm 1°C$ with the help of water bath to obtain a maximum power output.

To detect electrochemical activity of microorganisms, Wang and Su (2013) fabricated a membrane-less microfluidic microbial fuel, as shown in Figure 2.8. They designed a Y-shaped µMFC channel, which is very common to microfluidic microbial fuel cells (Kjeang, Djilali, and Sinton 2009) because it generates laminar flow for well-separated anode and cathode electrodes. In microfluidic chemical fuel cells, the composition in both the anodal and the cathodal streams should be known and remain consistent for stable electricity generation.

Guo, Zhaoy, Wang, et al. (2010) developed a membrane-less fluidized-bed microbial fuel cell (MLFMFC) to treat wastewater and generate energy in the form of electricity. The system's volume was 1 L, diameter 40 mm, and maximum height 600 mm. The distributor was a porous glass plate with a thickness of 2 mm, pore size of 2 mm, and fractional perforated area of 20%. Graphite rod was used as anode. The fluidized bed was fed

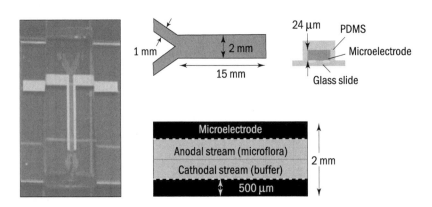

Fig. 2.8 *Microfluidic microbial fuel cell*
Source *Wang and Su (2013)*

with fresh and active carbon particles with an average of 0.2 mm–0.9 mm and was used in between cathode and anode as a carrier medium for biofilm. Two carbon clothes (3.14 cm^2, 0.35 mg/cm^2 Pt; 30% wet proofing, E-TEK, USA) were used as cathode to investigate the effect of electrode area and position on electricity generation (Figure 2.9). They were coated with four diffusion layers and fixed onto one side of the chamber wall (cathodes 1 and 2 were at 150 mm and 300 mm respectively, above the distributor). Wastewater used in the system was transferred from storage tank as fluidization liquid with a peristaltic pump. The reactor of the system was inoculated with 150 mL anaerobic sludge and particles of active carbon. All experimental data were gathered and recorded with the data logger for digital and analog I/O.

2.3.1 Applications of ML-MFC

2.3.1.1 Wastewater treatment

Earlier reported works showed that electricity consumption was high for wastewater treatment process and treatment of effluent with high COD. In both the cases, up to 95% of COD was removed. In the same context, Jang, Pham, Chang, et al. (2004) found that in an ML-MFC, COD gradually decreased from 300 mg/L to 30 mg/L and COD removal efficiency was 526.7 mg/m^3 day. It meant about 90% of COD was removed. Du, Xie, Dong, et al. (2011) designed an ML-MFC system of continuous flow and found that the efficiency of the system for soluble COD (sCOD) removal was about 90.45%. Moreover, it has been found that COD results in high biomass concentration which results in high COD removal (Tardast,

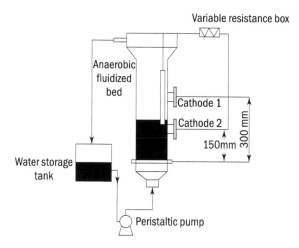

Fig. 2.9 *Design of membrane-less fluidized-bed microbial fuel cell*
Source *Guo, Zhaoy, Wang, et al. (2010)*

Rahimnejad, Ghasem, *et al.* 2012). This result showed that ML-MFC can be used for wastewater treatment. Several types of wastewater, including municipal, food processing, brewery, and animal wastewater have been successfully treated and electricity was also generated simultaneously as these wastewaters were found to be biocatalysts for direct power generation and waste treatment in MFCs.

2.3.1.2   *Electricity generation*

Production of electricity by MFC depends upon COD level in the wastewater and resistance used. Liu and Logan (2004) treated domestic wastewater in MFCs with PEM and without PEM. On comparing the power generating density in both types of MFCs, it was found that an MFC without PEM produces a maximum power density of about 146 mW/m^2. Zhu, Wang, Zhang, *et al.* (2011) found that at a maximum loading of 3500 mg/l COD, around 37.4 mW/m^2 of power density developed. In the same direction of research, Du, Xie, Dong, *et al.* (2011) found that 24.33 mW/m^3 of power density was generated from the designed system with up to 90.45% removal efficiency achieved. Other researchers also worked in increasing power density of the system and has been summarized in Table 2.3.

Table 2.3   Wastewater utilized in ML-MFC and power produced

S. No.	Wastewater	Power density/ current density	References
1.	Synthetic	44.4 ± 2.5 mW/m^2	Thung, Ong, Ho, *et al.* (2015)
2.	Wastewater from wastewater treatment plant	44.4 ± 2.5 mW/m^2	Thung, Ong, Ho, *et al.* (2015)
3.	Artificial	1.3 mW/m^2	Jang, Pham, Chang, *et al.* (2004)
4.	Artificial wastewater containing glucose and glutamate	2.60 ± 0.02 mA	Jang, Pham, Chang, *et al.* (2004)
5.	Fresh washed anaerobic sludge	184 (mW/m^2)	Hu (2008)
6.	Anaerobic sludge wastewater treatment plant	50.2 W/m^3	You, Zhao, Zhang, *et al.* (2007)
7.	Synthetic wastewater	10.13 mW/m^2	Ghangrekar and Shinde (2007)
8.	Municipal wastewater treatment plant	0.144 and 1.16 mW	Yang, Jia, and Liu (2009)
9.	Synthetic	70 mW/m^2	Aba, Marsili, Stante, *et al.* (2009)

Contd...

Table 2.3 *Contd...*

S. No.	Wastewater	Power density/ current density	References
10.	Anaerobic Sludge collected from the wastewater treatment plant	450 mW/m²	Guo, Zhaoy, Wang, *et al.* (2010)
11.	Synthetic glucose	37.4 mW/m²	Zhu,Wang, Zhang, *et al.* (2011)
12.	Domestic	90 W/m³	Rabaey and Verstraete (2005)
13.	Hospital	66 W/m³	Rabaey and Verstraete (2005)
14.	Brewery wastewater	37.4 mW/m²	Li, Zhang, Zeng, *et al.* (2009)

2.4 CONCLUSION

Renewable energy-based electricity generation and simultaneously wastewater treatment is a main key research for the scientist in the current scenario. New researches have proved that fuel cell will be the promising technology for upcoming future having the possibility of new dimension of research. MFC is one of the new technologies which utilizes different waste/wastewater as substrates for production of many types of energy in the form of electricity, hydrogen, biogas and many more, and in providing environmental clean approach for sustainable development. This technology is somewhat costly and is removed by development of ML-MFC which produces a maximum power of 32.3 mW/m³ and having a COD removal efficiency of 94%. This technology has the technical challenge of low-power generation that needs proper research strategies to improve the generation capacity for utilization at commercial scale.

ACKNOWLEDGEMENTS

Financial assistance in the form of Junior Research Fellowship to Atin Kumar Pathak under Inspire Fellowship Scheme by Department of Science and Technology, Ministry of Science and Technology, Government of India is gratefully acknowledged.

REFERENCES

Aba, Aldrovandi, Enrico Marsili, Loredana Stante, Patrizia Paganin, Silvia Tabacchioni, and Andrea Giordano. 2009. Sustainable power production in a membrane-less and mediator-less synthetic wastewater microbial fuel cell. *Bioresource Technology* 100: 3252–3260

Allen, R. M. and H. P. Bennetto. 1993. Microbial fuel-cells. *Applied Biochemistry Biotechnology* 39: 27–40

An, Junyeong, Bongkyu Kim, Jae Kyung Jang, Hyung-Sool Lee, and In Seop Chang. 2014. New architecture for modulization of membraneless and single-chambered microbial fuel cell using a bipolarplate – electrode assembly (BEA). *Biosensors and Bioelectronics* 59: 28–34

Bond, D. R., D. E. Holmes, L. M. Tender, and D. R. Lovley. 2002. Electrode-reducing microorganisms that harvest energy from marine sediments. *Science* 295: 483–485

Choi, H-K, T. L. Adams, R. W. Stahlhut, S-I Kim, J-H Yun, B-K Song, J-H Kim, J-S Song, and S-S Hong. 1999. Method for mass production of taxol by semi-continuous culture with Taxus chinensis cell culture. *US Patent* 5871979

Du, Fangzhou, Beizhen Xie, Wenbo Dong, Boyang Jia, Kun Dong, and Hong Liu. 2011. Continuous flowing membraneless microbial fuel cells with separated electrode chambers. *Bioresource Technology* 102: 8914–8920

Fornero, J. J., M. Rosenbaum, M. A. Cotta, and L. T. Angenent. 2008. Microbial Fuel Cell Performance with a Pressurized Cathode Chamber. *Environmental. Science Technology* 42: 8578–8584

Ghangrekar, M. M. and V. B. Shinde. 2007. Performance of membrane-less microbial fuel cell treating wastewater and effect of electrode distance and area on electricity production. *Bioresource Technology* 98: 2879–2885

Gil, Geun-Cheol, Chang In-Seop, Hong Kim Byung, Kim Mia, Jang Jae-Kyung, Park Hyung Soo, Kim Hyung Joo. 2003. Operational parameters affecting the performannce of a mediator-less microbial fuel cell. *Biosens and Bioelctro* 18: 327-334

Guerrini, Edoardo, Pierangela Cristiani, and Stefano Pierpaolo Marcello Trasatti. 2013. Relation of anodic and cathodic performance to pH variations in membraneless microbial fuel cells. *International Journal of Hydrogen Energy* 38:345–353

Guo, Qingjie, Shuju Zhaoy, Xuyun Wang, Xuehai Yue, and Liangyu Houyy. 2010. Electricity generation Characteristics of an Anaerobic fluidized bed Microbial fuel cell. *The 13th International Conference on Fluidization - New Paradigm in Fluidization Engineering.*Details available at http://dc.engconfintl.org/cgi/viewcontent.cgi?article=1037...fluidization_xiii

Harnisch, F., S. Wirth, and U. Schröder. 2009. Effects of substrate and metabolite crossover on the cathodic oxygen reduction reaction in microbial fuel cells: Platinum vs. iron(II) phthalocyanine based electrodes. *Electrochemistry Communications* 11: 2253–2256

Hu, Zhiqiang. 2008. Electricity generation by a baffle-chamber membraneless microbial fuel cell. *Journal of Power Sources* 179 (1): 27–33

Huang, Liping, Raymond J. Zeng, and Irini Angelidaki. 2008. Electricity production from xylose using a mediator-less microbial fuel cell. *Bioresource Technology* 99:4178–4184

Jadhav, G. S. and M. M. Ghangrekar. 2009. Performance of microbial fuel cell subjected to variation in pH, temperature, external load and substrate concentration. *Bioresource Technology* 100: 717–723

Jang, Jae Kyung, The Hai Phama, In Seop Changa, Kui Hyun Kanga, Hyunsoo Moon, Kyung Suk Cho, and Byung Hong Kim. 2004. Construction and operation of a novel mediator- and membrane-less microbial fuel cell. *Process Biochemistry* 39:1007–1012

Kaufmann, F. and R. Lovely. 2001. Isolation and characterization of a soluble NADPH-independent Fe (III)-reductase from *geobacteria* sulphur reducens. *Journal Bacteriology* 185 (15): 4468–4476

Kim, B. H., D. H. Park, P. K. Shin, I. S. Chang, and H. J. Kim. 1999. *Mediator-less biofuel cell*. Google Patents: 5976719

Kim, B. H., H. J. Kim, M. S. Hyun, and D. H. Park. 1999. Direct electrode reaction of Fe (III) reducing bacterium, *Shewanella putrefacience*. *Journal of Microbiology and Biotechnology* 9:127–31

Kim, H., M. Hyun, I. Chang, and B. H. Kim. 1999. A microbial fuel cell type lactate biosensor using a metal-reducing bacterium, *Shewanella putrefaciens*. *Journal of Microbiology and Biotechnology* 9: 365–367

Kim, Hyung Joo, Hyung Soo Park, Moon Sik Hyun, In Seop Chang, Mia Kim, and Byung Hong Kim. 2002. A mediator-less microbial fuel cell using a metal reducing bacterium, *Shewanella putrefaciens*. *Enzyme and Microbial Technology* 30: 145–152

Kim, Jung Rae, Giuliano C. Premier, Freda R. Hawkes, Richard M. Dinsdale, and Alan J. Guwy. 2009. Development of a tubular microbial fuel cell (MFC) employing a membrane electrode assembly cathode. *Journal of Power Sources* 187: 393–399

Kjeang, Erik, Ned Djilali, and David Sint.2009. Microfluidic fuel cells: A review. *Journal of Power Sources* 186: 353–369

Leong, Jun Xing, Ramli Wan Daud Wan, Ghasemi Mostafa, Ben Liew Kien, Ismail Manal. 2013. Ion exchange membranes as separators in microbial fuel cells for bioenergy conversion: A comprehensive review. *Renewable and Sustainable Energy Reviews* 24575–587

Lewis, K. 1966. Symposium on bioelectrochemistry of microorganisms. IV. Biochemical fuel cells. *Bacteriol. Rev.* 30: 101–113

Li, Z., X. Zhang, Y. Zeng, and L. Lei. 2009. Electricity production by an overflow-type wetted-wall microbial fuel cell. *Bioresource Technology* 100:2551–2555

Liu, H. and B. E. Logan. 2004. Electricity generation using an air-cathode single chamber microbial fuel cell in the presence and absence of a proton exchange membrane. *Environmental Science & Technology* 38: 4040–4046

Liu, H., S. A. Cheng, L. P. Huang, and B. E. Logan. 2008. Scale-up of membrane-free single chamber microbial fuel cells. *Journal Power Sources* 179: 274–279

Liu, Zhi-Dan and Hao-Ran Li. 2007. Effects of bio- and abio-factors on electricity production in a mediatorless microbial fuel cell. *Biochemical Engineering Journal* 36:209–214

Logan, B. E. 2004. Biologically extracting energy from wastewater: biohydrogen production and microbial fuel cells, *Environ. Sci. Technol.* 38: 160A–167A

Mokhtarian, N., D. Ramli, M. Rahimnejad, and G.D. Najafpour. 2012. Bioelectricity generation in biological fuel cell with and without mediators. *World Applied Science Journal* 18: 559–567

Pant, Deepak, Van Bogaert Gilbert, Ludo Diels, and Karolien Vanbroekhoven. 2010. A review of the substrates used in microbial fuel cells (MFCs) for sustainable energy production. *Bioresource Technology* 101: 1533–1543

Peighambardoust, S., S. Rowshanzamir, and M. Amjadi. 2010. Review of the proton exchange membranes for fuel cell application. *Int. J. Hydrogen Energy* 35: 9349–9384

Potter, M. C. 1911. Electrical effects accompanying the decomposition of organic compounds. *Proc. Royal Soc. London, Ser. B, Containing Pap. Biol. Charact.* 84:260–276

Rabaey, Korneel and Willy Verstraete. 2005. Microbial fuel cells: novel biotechnology for energy generation. *Trends in Biotechnology* 23: 291–298

Rahimnejad, M., A. A. Ghoreyshi, G. Najafpour, and T. Jafary. 2011. Power generation from organic substrate in batch and continuous flow microbial fuel cell operations. *Applied Energy* 88: 3999–4004

Rahimnejad, Mostafa, Arash Adhami, Soheil Darvari, Alireza Zirepour, and Sang-Eun Oh. 2015. Microbial fuel cell as new technology for bioelectricity generation: A review *Alexandria Engineering Journal* 54(3): 745–756

Sirinutsomboon, Bunpot. 2014. Modeling of a membrane-less single-chamber microbial fuel cell with molasses as an energy source. *International Journal of Energy and Environmental Engineering* 5:93

Stirling, J. L., H. P. Bennetto, G. M. Delaney, J. R. Mason, S. D. Roller, K. Tanaka, and C. F. Thurston. 1983. Microbial fuel cells. *Biochem. Soc. Trans.* 11:451–453

Sun, Jian, Yongyou Hu, Zhe Bi, and Yunqing Cao. 2009. Improved performance of air-cathode single-chamber microbial fuel cell for wastewater treatment using microfiltration membranes and multiple sludge inoculation. *Journal of Power Sources* 187: 471–479

Tardast, Ali, Mostafa Rahimnejad, Ghasem D. Najafpour, Ali Asghar Ghoreyshi, and Hossein Zare. 2012. Fabrication and operation of a novel membrane-less microbial fuel cell as a biotechnology generator. *Iranica Journal of Energy & Environment* 3 (Special Issue on Environmental Technology): 1-5. Details available at https://www.researchgate.net/file.PostFileLoader.html?id.

Thung, Wei-Eng, Soon-An Ong, Li-Ngee Ho, Yee-Shian Wong, Yoong-Ling Oon, Yoong-Sin Oon, and Lehl Harvinder Kaur. 2015. Simultaneous wastewater treatment and power generation with innovative design of an upflow membrane-less microbial fuel cell. *Water Air Soil Pollution* 226: 165

Virdis, B., K. Rabaey, Z. Yuan, and J. Keller. 2008. Microbial fuel cells for simultaneous carbon and nitrogen removal. *Water Research* 42(12):3013–3024

Wang, Hsiang-Yu and Jian-Yu Su. 2013. Membrane-less microfluidic microbial fuel cell for rapid detection of electrochemical activity of microorganism. *Bioresource Technology* 145: 271–274

Yang, Shaoqiang, Boyang Jia, and Hong Liu. 2009. Effects of the Pt loading side and cathode-biofilm on the performance of a membrane-less and single-chamber microbial fuel cell. *Bioresource Technology* 100: 1197–1202

You, Shijie, Qingliang Zhao, Jinna Zhang, Junqiu Jiang, Chunli Wan, Maoan Du, and Shiqi Zhao. 2007. A graphite-granule membrane-less tubular air-cathode microbial fuel cell for power generation under continuously operational conditions. *Journal of Power Sources* 173: 172–177

Zhu, Feng, Wancheng Wang, Xiaoyan Zhang, and Guanhong Tao. 2011. Electricity generation in a membrane-less microbial fuel cell with down-flow feeding onto the cathode. *Bioresource Technology* 102: 7324–7328

Zhuang, Li, Shungui Zhou, Yueqiang Wang, Chengshuai Liu, and Shu Geng. 2009. Membrane-less cloth cathode assembly (CCA) for scalable microbial fuel cells. *Biosensors and Bioelectronics* 24: 3652–3656

3

Hydrogen Energy: Present and Future

Jyoti Pandey

Department of Applied Chemistry, Babasaheb Bhimrao Ambedkar University,
Lucknow, Uttar Pradesh 226025
E-mail: drjyotibbau@gmail.com

3.1 INTRODUCTION

Hydrogen is the most abundantly found element in the universe. It is estimated that the universe is made up of 92% of hydrogen and 7% of helium. The remaining 1% comprise all other elements. However, hydrogen is not abundantly present in the earth's atmosphere. In contrast, hydrogen is the tenth most abundant element on the earth's crust. It occurs in vast quantities as water in oceans, rivers, lakes, and so on. Hydrogen is a natural resource and found in nature, such as solar, wind and bioenergy, hydropower, and fossil fuels. Hydrogen constitutes about 6% by weight of dry biomass (Balta, Dincer, and Hepbasli 2010). At present, fossil fuels are largely used as an energy source in the industrial, residential, and other sectors. These fuels release greenhouse gases (GHGs), such as CO_2, CH_4, and water vapour, owing to the presence of hydrocarbon content in them. These GHGs destroy the environment by polluting atmosphere, depleting ozone layer, creating global warming, and so on. These effects can be minimized by utilizing alternative green energy sources (Veziroglu 2007; Nema, Nema, and Rangnekar 2009). The green energy source includes those materials which generate energy without creating any kind of pollution. The green energy sources include water, wind, solar, bio-energy, hydrogen, geothermal, and so on. However, generation of green energy resources, such as wind and solar energies, are not technically feasible at all sites (Veziroglu 2007; Nema, Nema, and Rangnekar 2009; Thirugnanasambandam, Iniyan, and Goic 2010). Therefore, these energies have not been largely used for electricity generation (Georgiou, Tourkolias, and Diakoulaki 2008; Sherwani, Usmani, and Varun 2010).

Hydrogen and electricity are the secondary sources of energy and produced from primary natural renewable energies. Both these forms of energies are environmentally and climatically clean. Some bio-organisms such as algae and bacteria produce hydrogen using sunlight, which is called biohydrogen. Biohydrogen is a clean, green, renewable, and highly efficient source of energy. Wind hydrogen (Veziroglu 2000; Bendaikha, Larbi, and Bouziane 2011), solar hydrogen (Granovskii, Dincer, and Rosen 2006; Lutfi and Veziroglu 1991), and geothermal hydrogen (Yilanci, Dincer, and Ozturk 2009) are also included in the clean energy economy gambit. The secondary source of energies are exchangeable through electrolysis and fuel cells. Hydrogen is produced from natural gas, coal, or oil by electrolytic method. Hydrogen stores more energy than electricity. The hybrid energy system of hydrogen and fuel cell is based on the proton exchange membrane fuel cell technology. The hydrogen-fuelled fuel cells are small, compressed, clean, and more efficient than the Carnot heat engines (Lutz, Larson, and Keller 2002). The capacity of these fuel cells ranges from less than watt to more than megawatt.

Hydrogen-fuelled fuel cells are used in movable electronic devices, such as television, cameras, laptop computer, and mobiles. It also finds application in methanol and ammonia synthesis, food industry (for fat hardening), and as a cleaning agent in glass and electronics industry.

The use of hydrogen as a fuel results in a decrease in the carbon dioxide emission at the rate of 5% (1.4 Gt/year), in comparison to the efficiency measures with petrol-electric hybrid vehicles and other fuels like ethanol. Unlike hydrocarbons, hydrogen cannot be destroyed, and during ingestion, it changes only the state from water to hydrogen and is again converted back to water (Scheme 3.1).

Hydrogen use can be evaluated in terms of the issues of environmental emissions, sustainability, and energy security. The ultimate goal is the production of clean, natural, renewable, sustainable, and green energy. Hydrogen is one of the most promising energy haulers for the forthcoming generation because it can be used in two ways: materially (synthesis of NH_3 and CH_3OH and hardening of fat) and as an energy fuel (fuel cells and aircraft). It can play an important role in the world economy (Sorensen 2005).

Water + Energy $\longrightarrow$ Hydrogen + Oxygen

Hydrogen + Oxygen $\longrightarrow$ Water + Energy

Scheme 3.1 *Generation of hydrogen from water and regeneration of water*

Hydrogen is considered as the "future fuel" since it can be produced, like electricity, from any primary energy fuels, such as coal, oil, and natural gas. Hydrogen is an energy efficient and clean fuel. Whether hydrogen is utilized in fuel cell or burnt in air, the ultimate product is water (H_2O) and oxides of nitrogen (NO_x) in minor quantity.

Production of hydrogen and electricity is largely dependent on the same primary energy sources. Therefore, their production competes for the same primary energy sources. However, they are electrochemically exchangeable through fuel cell and electrolysis. In fuel cell vehicles, pure hydrogen is used as a fuel (this technique is largely used nowadays) or changed into synthetic liquid hydrogen fuels (Veziroglu 2008). With respect to its energy density per weight, high efficiency, and sustainability, hydrogen is considered as a future fuel. The global importance of hydrogen is because of its sustainability, cost-effectiveness, and eco-friendly nature. Hydrogen is sustainable climatically, societally, and economically.

The ultimate goal of production of hydrogen is to replace the anthropogenic energy cycle from the renewable energies and finally change into clean energy, such as hydrogen and electricity, which are used in industrial, residential, storage, transport, and other sectors. This secondary derived energy does not create any type of harmful environmental effects, and it is also used in fuel cells.

3.2 HYDROGEN PRODUCTION

Hydrogen does not exist alone in nature. It always exists in the form of differently substituted compounds. The direct production of hydrogen is not known, and its production needs consumption of high-value energy. All the known processes for hydrogen production are based on the isolation of hydrogen from hydrogen embracing compounds such as fossil fuel or water. Till the 1970s, hydrogen had received very little attention as an energy hauler. The concepts of "hydrogen energy", "hydrogen economy", and "hydrogen energy system" were not introduced at that time. Hydrogen energy system offers the best solution to the interrelated global energy problems elicited by the increasingly rapid depletion of fossil fuel sources, as well as the environmental problems caused by their current usage, for example, global warming, climate change, ozone layer depletion, acid rains, water and air pollution, and oil spills (Schmidt and Gunderson 2000). Different routes for hydrogen production are as follows:

- Steam methane reforming
- Partial oxidation of hydrocarbons
- Electrolysis

- Gasification of biomass
- Pyrolysis of biomass
- Thermochemical process
- Photochemical process

The production of hydrogen is generally carried out by separating hydrogen from hydrogen-rich compounds, such as coal, oil, natural gas, and biomass. The separation of hydrogen from hydrocarbons is carried out by the reforming process. At present, large amounts of hydrogen are prepared by reforming of natural gas. Natural gas is produced through the decay of organic compounds, releasing large amounts of methane in the process, which act as a fuel (Kreutz, Williams, Consonni, *et al.* 2005).

3.2.1   Steam Methane Reforming and Partial Oxidation of Hydrocarbons

This is the most common and least expensive method for commercial bulk production of hydrogen at 700–1100°C. In the presence of a nickel-based catalyst (Ni), steam reacts with methane to yield carbon monoxide and hydrogen. Additional hydrogen can be recovered by a lower temperature water–gas shift reaction, with the production of carbon monoxide. The reaction is known as water–gas shift reaction (Scheme 3.2).

$$CH_4 + H_2O \rightleftharpoons CO + 3H_2$$

Shift reaction

$$CO + H_2O \rightleftharpoons CO_2 + H_2$$

Scheme 3.2 *Steam methane reforming*

Partial oxidation of hydrocarbons, such as diesel fuel and residual oil, produces hydrogen. This technology involves the compression and pumping of hydrocarbon feedstock to produce hydrogen.

Natural gas, which is composed of methane and other hydrocarbons, undergoes partial oxidation in the presence of a limited amount of oxygen (typically from air and is not enough to completely oxidize hydrocarbons to carbon dioxide and water) to produce hydrogen (Scheme 3.3) (Curry Hydrocarbons Inc. 2008).

$$2CH_4 + O_2 \longrightarrow 2CO + 4H_2 + Heat$$

Scheme 3.3 *Partial oxidation of methane*

3.2.2   Electrolysis of Water

In this process, water is first distilled and then passed through several ion exchange columns to remove any mineral contaminants. Then it is pumped to electrolysis cells. During electrolysis, water is broken down into its constituents, that is, oxygen and hydrogen, by applying electrical current. Electrical current is produced from any renewable sources, such as photovoltaic, hydropower, and grid systems, which are used in the electrolysis of water to produce pure hydrogen (Scheme 3.4).

Reaction at cathode:

$$2H^+_{(aq)} + 2e^- \longrightarrow H_{2(g)}$$

Reaction at anode:

$$2H_2O_{(l)} \longrightarrow O2_{(g)} + 4H^+_{(aq)} + 4e^-$$

Over reaction:

$$2H_2O_{(l)} \longrightarrow 2H_{2(g)} + O_{2(g)}$$

Scheme 3.4   *Electrolysis of water*

In electrolysis, electrical current is supplied to the electrolyzer produced by solar, wind, or hydropower. If the source of electricity is solar, it is called solar hydrogen, and if it is wind, then it is called wind hydrogen. Biological production of hydrogen involves biological organisms, such as bacteria or algae, which use sunlight as the energy source (Ryazantsev and Chabak 2006).

3.2.3   Hydrogen from Solar Energy

In this technique, the production of hydrogen from solar energy is used for electricity generation by using photovoltaic or solar cells, and this electricity is used in the electrolyzer for electrolysis of water to produce hydrogen. Photovoltaic technology involves the direct conversion of solar radiation into electricity, and this technique is called indirect photoelectrolysis (Solarbuzz 2008).

Silicon-based Solar Cell

Silicon-based solar cells are currently the most significant energy source and they account for more than 90% of global solar energy sources (Lardic and Mignon 2008). These devices perform two functions: (i) photo generation of charge carriers (electrons and holes) in a light-absorbing material (semiconducting materials) and (ii) separation of charge carriers to a conductive contact that will transmit electricity. The following are the three generations of development of silicon-based solar cells:

(i) The **first-generation photovoltaic cells** are also known as silicon wafer-based solar cells and consist of a large area, single layer p-n junction diode, which is capable of generating electricity from light sources such as sunlight. This is the prevailing technology, and it has recorded more than 86% commercial production of solar cells in the solar cell market.

(ii) The **second-generation photovoltaic materials** are based on the use of thin-film deposits of semiconductors. Owing to the thin film, the mass of material is reduced. This contributed to a considerable reduction in the cost of thin-film solar cells. Recently, amorphous silicon, polycrystalline silicon, microcrystalline silicon, cadmium telluride, and copper indium selenide/sulphide materials have been used in the production of photovoltaic materials. The efficiencies and cost of thin-film solar cells are lower in comparison to silicon (wafer-based) solar cells. Owing to the reduced mass of thin-film solar cells, less support is required for placing solar panels on rooftops, and it allows fitting of panels made up of light and flexible materials, such as textiles.

(iii) The **third-generation photovoltaic cells** are broadly defined as semiconductor devices that do not depend on a traditional p-n junction to separate photo-generated charge carriers, thus differing from the other two types of photovoltaic cells. These types of photovoltaic cells include photoelectrochemical cells, polymer solar cells, and nanocrystal solar cells (Ramanathan, Contreras, Perkins, et al. 2003). Thin-film panels without silicon are required, and these cells are lower in efficiency compared to silicon-based photovoltaic cells.

$$\text{Solar cell + Electrolyzer} \longrightarrow \text{Hydrogen}$$

3.2.4   Hydrogen from Wind Energy

Wind energy is the clean, non-polluting, renewable energy and it cannot be generated at all sites. So this energy is feasible only in limited areas, such as coastal regions, where wind blows at high speed. Wind energy is a direct form of solar energy. Electricity from wind is produced by rotation of turbine driven by stream of wind in wind mills. The turbines are designed both horizontally and vertically. Currently, horizontal axis turbines are mostly used owing to their high efficiency (Bockris and Veziroglu 2007).

The production of hydrogen from wind energy is completely autonomous (self-dependent) on the grid system. Because of the independence of the grid system, electrical energy is used in the electrolysis of water for the production of hydrogen. This electrical energy comes from wind turbines.

Wind hydrogen is not continuously produced because no continuous output comes from the wind turbines, and so the capacity of wind turbines is low. Therefore, the capacity of electrolyzers will also be low. If we use electrolyzers for lower wind plants, the cost of electrolyzers will decrease, but we cannot use the high speed of wind in electrolyzers owing to the requirement of a constant supply of electricity in the grid system (AWEA 2008).

In the electrolysis of water, an electrolyzer cannot operate well because for the smooth operation of the electrolyzer (2V), electricity is required. The lower the voltage is supplied, the lower the amount of hydrogen and oxygen is produced. The problem of continuous production of hydrogen is very sensitive in alkaline electrolyzers, and so their efficiency decreases, whereas in proton exchange membrane (PEM) electrolyzers, this problem is less compared to that of alkaline electrolyzer (Sherif, Barbir, and Veziroglu 2005).

3.2.5 Photochemical Water Splitting

The direct solar water splitting involves the direct production of hydrogen from water by using solar energy, that is, without going through the intermediate electrolysis step. In this technique, semiconducting electrodes in a photoelectrochemical cell are used to convert light energy into chemical energy of hydrogen. Basically, two types of photoelectrochemical systems are known: those based on semiconductors or dyes and those based on dissolved metal complexes (Glatzmaier, Blake, and Showalter 1998).

3.2.6 Photobiological Process

In photobiological process, hydrogen is generated from the biological system by using sunlight. Certain algae and bacteria are known to produce hydrogen under suitable conditions. The pigments present in the algae absorb solar energy in the presence of sunlight, and enzymes in the cell act as catalysts to break down water into its components: hydrogen and oxygen. Hydrogen production is catalyzed by two enzymes, which are known as nitrogenase and reversible hydrogenase. Their functioning in cyanobacteria involves the utilization of photosynthetic products and the generation of hydrogen from water. Thus, it is possible to design bioreactors in which solar energy could be used to produce hydrogen from water and cyanobacteria may be used as a biocatalysts (Hall and Rao 1989).

3.2.7 Thermo-electrochemical Cycles

In these cycles, solar heat is utilized to produce hydrogen and oxygen by splitting water at a high temperature (2000°C), which is achieved by concentrating solar energy with the help of mirror (Richards, Shenoy, Schultz, *et al.* 2006). For example, sulphur–iodine thermochemical cycle.

In the sulphur–iodine thermochemical cycle, sulphur dioxide (SO_2) and iodine (I_2) act as a chemical catalyst, and this cycle includes the following three steps:

(i) $I_2 + SO_2 + 2H_2O \longrightarrow 2HI + H_2SO_4$ at 120°C

HI and H_2SO_4 may be separated by distillation.

(ii) $2H_2SO_4 \longrightarrow 2H_2O + 2SO_2 + O_2$ at 850°C

(iii) $2HI \longrightarrow H_2 + I_2$ at 450°C

In general, thermolysis of water needs high temperature, and these thermocycles can meet this demand and provide a mechanism for the separation of oxygen and hydrogen components from water. The low cost and higher efficiency of hydrogen have been achieved by a system that uses electrolysis. For example, the thermochemical system (Dincer 1998).

3.2.8 Hydrogen from Biomass

Biomass, a renewable organic resource, includes biological materials, such as plants, wood, animal waste, and industrial waste materials. Generally, wood is used in furnaces for combustion, and so these fuels are used as a source of heat energy. Biomass is classified into four categories (Ni, Leung, Leung, *et al.* 2006). They include energy crop, agriculture waste, forestry waste, and industrial and municipal waste (Figure 3.1).

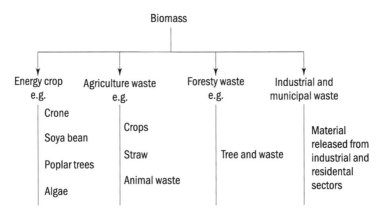

Fig. 3.1 *Classification of biomass*

According to Searchinger, Heimlich, Houghton, *et al.* (2008), the production of ethanol from corn and switchgrass increases GHG emission by 93% and 50%, respectively, in comparison to gasoline. Hydrogen production from biomass involves two processes: pyrolysis and gasification.

Pyrolysis

Pyrolysis at the temperature range of 650–800 K, pressure of 1–5 bar, and an inert environment converts biomass into liquid, solid, and gaseous products, such as oils (liquid), charcoal (solid), hydrogen (H_2), methane (CH_4), carbon monoxide (CO), and carbon dioxide (CO_2), and these are produced in a large amount. Slow pyrolysis at a low-temperature range yields charcoal as the major product (Baumlin, Broust, Bazer-Bachi, *et al.* 2006).

Gasification

Gasification converts biomass into a gaseous mixture of H_2, CO, CO_2 (synthesis gas or "syngas"), and other components when heat is supplied in the presence of steam and a controlled amount of oxygen under pressure. The CO then reacts with water to form CO_2 and more hydrogen. This reaction is known as the water-gas shift reaction (Albertazzi, Basile, Brandin, *et al.* 2005) (Scheme 3.5).

$$Biomass + O_2 + H_2O \longrightarrow CO + CO_2 + H_2$$

$$CO + H_2O \longrightarrow CO_2 + H_2 + Heat$$

Scheme 3.5   *Gasification of biomass*

3.2.9   Biological Process

Four methods are used to produce hydrogen by the biological process. These methods include direct and indirect biological water–gas shift reaction, bio-photolysis, photo-fermentation, and dark fermentation.

All these processes are based on the application of tailored biological organisms which produce hydrogen in place of oxygen in the presence of solar energy during photosynthesis. In another method, hydrogen is produced directly by fermentation of biomass or indirectly from biomass-derived liquid such as bio-oil. The biomass undergoes fast pyrolysis to generate bio-oil, followed by the catalytic steam reforming of the oil or its fractions to produce hydrogen (Kolb, Diver, and Siegel 2007). The direct fermentation of carbohydrate feedstock by microorganisms is one of a number of potential technologies for producing renewable hydrogen (Scheme 3.6).

$$C_6H_{12}O_6 \longrightarrow 2C_2H_5OH + 2CO_2$$

$$C_2H_5O_H \xrightarrow{\text{Catalytic steam reforming}} H_2$$

Scheme 3.6 *Bio-oil from boimass*

3.3 STORAGE TECHNOLOGIES FOR HYDROGEN

3.3.1 Compressed Gas

This is the economical method of storage. Compressed hydrogen gas is commonly stored in cylinder-shaped vessels at around 200 bar pressure. The large quantity of compressed gas with low storage density is stored either in tanks or underground at low pressure. Storage systems aboard vehicles involve storage in composite materials-based tanks at pressure up to 700 bar. The storage density can be increased by increasing pressure; however, it would also increase the cost of the technology and safety measurements.

3.3.2 Liquid Hydrogen

Liquid hydrogen exists in a liquid form at −253°C, and it has a very high energy density (three times in comparison to petrol). Therefore, highly insulated pressure vessels are needed for its storage, which make aboard liquid hydrogen storage relatively weighty.

3.3.3 Chemical Storage

Metal hydrides are well known to store hydrogen within their chemical structure and release hydrogen when heated. Carbon materials and liquid hydrides can also serve as chemical hydrogen storage.

3.4 HYDROGEN ENERGY AND ENVIRONMENT

The undesirable changes in the environment are a result of GHGs. These gases are produced by burning of fossil fuels which release CO_2, CH_4, N_2O, and H_2O vapour.

According to the report of Intergovernmental Panel on Climate Change (IPCC), much increase in the global average temperature since the mid-20th century has been owing to the observed increase in the man-made GHG concentration. As a result of GHG emissions, the temperature of the earth increases continuously, resulting in the melting of glaciers, and the rising of the sea level. Three major areas of environmental problems are as follows:

(i) Acid rain (Dincer 2000)

(ii) Ozone depletion (Ahrendts 1980)

(iii) Greenhouse effect (Holladay, Hu, King, *et al.* 2009)

These effects can be minimized by taking the following steps:

(i) Avoid clearing of trees or deforestation and change the procedures followed in industrial agriculture.

(ii) Issuance of worldwide CO_2 emission certificate which can act as an economic means of progress.

(iii) Elevate energy, particularly rational efficiencies, along all associates of energy conversion chains.

(iv) Generate electricity from renewable energy sources, limiting the use of fossil and nuclear sources.

(v) Introduce hydrogen technologies such as fuel cells or ICE (internal combustion engine).

(vi) Prevent other emissions such as CH_4, NO_x, and fluorine gases which largely contribute to the rise in CO_2 equivalents.

By following the above methods, we can generate clean and green energy, such as electricity and hydrogen. Clean energy is eco-friendly, sustainable, renewable, and safe, and also it is interchangeable through electrolyzer and fuel cell. From the environmental perspective, renewable energy is viewed as a source of clean, pollution-free, and green energy.

Hydrogen does not contain any elements that cause pollution. The combustion of hydrogen produces water as the final product. Sometimes it may produce NO_x at a high temperature from nitrogen and oxygen present in the air, which can be the case for other fuels as well. However, this production can be controlled. So hydrogen is almost a clean source of energy and it is unlikely to produce hazardous volatile chemicals, such as oxides of sulphur and carbon. Hence, the use of hydrogen causes no pollution.

The choice of suitable renewable energy technologies is based on several factors, climatological conditions, renewable energy potential, load factor, and the amount of hydrogen required by refuelling stations. The production of hydrogen is generally based on electricity and this electrical energy is generally produced by wind turbines or photovoltaic cells.

3.5 HYDROGEN AND OTHER ALTERNATIVE SECONDARY ENERGIES

The secondary energies are produced by or derived from primary energy sources (raw materials). Hydrogen and electricity are the most common energy sources. Sometimes they compete with each other and in other cases their combination can be beneficial. These secondary energies are environmentally and climatically clean. Electricity and hydrogen are switchable through electrolysis and fuel cell, and are prepared globally in different capacities.

3.5.1   Hydrogen and Electricity

We can store and transport electricity in the form of information, for example, in communicating devices. However, storage in large quantities in the form of energy is not possible. Hydrogen cannot store and transport information, but it has the potential to store and transport energy for long route. Hence, it can be used as a vehicle fuel for generation of heat and electricity. Hydrogen is utilized in the following two ways:

(i) **Materially:** Since the discovery of hydrogen in the 18th century, it has been used as an economic commodity. Approximately 50 million tonnes/year of hydrogen is used in various chemical and food manufacturing industries.

(ii) **Energy carrier:** Hydrogen does not exist in large quantities in nature in a useful form as oil, gas, and coal exist. Hydrogen is not only used as an energy carrier, but also finds application in energy conversion technologies. Hydrogen-based energy conversion, technologies, among all kinds of energy conversion chain technologies, are of overarching significance. Hydrogen-fuelled fuel cells having capacities of kilowatts to megawatts can substitute batteries in various portable electronic equipment, such as mobiles, TV, cameras, and laptops, as well as in transport vehicles on earth, sea, air, and space.

The following are the three features of hydrogen and electricity:

(i) They have their own spheres.

(ii) Both are cohorts.

(iii)   Both are competitors.

At present, aircraft and spacecraft engines are operated by the necessary sets of batteries which are too heavy and bulky, but in the near future it will undisputedly be the hydrogen domain. On the other hand, in the communication and power sectors (for light), production of electricity is necessary. So these zones will be the electricity zone.

Both hydrogen and electricity act as cohorts for all chemicals–electrical energy converters, for example, electrolytic cells and fuel cells, which convert hydrogen energy into heat and electricity in various heat and power applications in industry, households, and office buildings. Mobility or transportation sector can be governed by electricity or hydrogen (Scott 2008).

3.5.2   Hydrogen and Methanol

Methanol is an alternative source of hydrogen. It is produced from renewable sources or by the reduction of CO_2 (Dunwoody, Chung, Haverhals, *et al.* 2009). The source of CO_2 is atmosphere or the CO_2-rich exhaustive emissions of power stations. Hydrogen-derived methanol is generally termed as "liquid form" of hydrogen (Olah 2005). The chemical and physical properties of methanol are entirely different from oil, coal, and biomass. Its liquid form makes it easy to handle. All highly dense hydrocarbons in their liquid form have the tendency to store hydrogen and they can be directly used in DMFC (direct methanol fuel cell) without regeneration of hydrogen (Wall 1995). The only drawback of methanol is its toxicity.

Methanol produced from renewable sources cannot be maintained for a long time because it damages reserves; that is, it results in the depletion of geological resources and change of climate. Only 40% of CO_2 released from the combustion of a fossil fuel is available for storage. Therefore, in other way, it may be produced from biomass or by the conversion of hydrogen to methanol using CO_2 (Bockris 2010).

The hydrogen generation from methanol needs CO_2, which will be captured from atmosphere. But in this case the concentration of CO_2 is less compared to CO_2 emissions from a power station, and thus it increases the energy demand for CO_2 capture from atmosphere (Veziroglu and Sahin 2008).

In addition, this technique will phase out the production of hydrogen from electrolysis and fossil fuels.

3.6   HYDROGEN SAFETY

Hydrogen safety deals with the safe production, storage, and applications of hydrogen, particularly hydrogen gas and liquid hydrogen. The main concern about hydrogen safety is the inflammable nature of hydrogen. Therefore, hydrogen installation needs care and should be insulated enough to prevent leakage. For example, a collision of hydrogen car in an open area would be less risky compared to a petrol car because hydrogen is light and disperses quickly into the air, thus reducing the possibility of any hazard such as fire or explosion. As hydrogen is non-toxic in nature, its leakage would not cause any environmental destruction. In addition, hydrogen containers are significantly harder than petrol containers; therefore, there are less chances of leakage.

Every energy technology is relatively safe and has particular safety standards. According to the safety standards, different energy technologies are used for different purposes. The selected safety related data for hydrogen, methane, and gasoline are given in Table 3.1 (Specht 1992).

Table 3.1   Comparative safety data of hydrogen and other fuels

Properties	Gasoline	Methane	Hydrogen
Flammability limits in air (volume)	1.0–7.6	5.3–15	4.0–75
Ignition temperature (°C)	228–471	540	585
Flame temperature in air (°C)	2197	1875	2045
Explosion energy (g TNT/kJ)	0.25	0.19	0.17
Flame emissivity (%)	34–43	25–33	17–25

Source Lewis and von Elbe (1987)

3.7   FUTURE PROSPECTS OF HYDROGEN

A hydrogen economy includes production of hydrogen from various disseminated sources and utilization of this power in different sectors, such as transport, homes, and industry. Therefore, hydrogen technology has tremendous scope in the present and future energy systems. Hydrogen energy technology is well taken by industries and further commercialization of the fuel and other uses is expected in the future.

At present, the dependency on fossil fuels for primary energy is about 80% according to U.S. Energy Information Administration (EIA). Moriarty and Honnery (2009) stated that this dominance will be somewhat reduced in the coming decades.

Hydrogen can be produced from various sources by a number of different pathways, such as the following:

- From hydrocarbons, for example, coal, oil, and natural gas
- From biomass/wastes
- By electrolysis of water utilizing electricity produced from fossil fuels and nuclear or renewable energy sources

At present, hydrogen is mainly produced from fossil fuels (BP 2010; Veziroglu and Sahin 2008); however, it could also be produced from nuclear reactors, solar energy, biomass, or renewable energy sources in future (Marban and Valdes-Solis 2007). Direct photolysis of water is also the source of hydrogen generation (Miyake and Kawamura 1987). The combinations of various biohydrogen approaches, such as photo-biohydrogen production and artificial photosynthesis, result in the increased production rates of hydrogen (Abbott 2010). In the combined heat and power systems (CHP) applications where hydrogen is used directly, any of these methods of hydrogen production will become economically viable in the longer term, thus boosting the use of these applications.

For transportation, hydrogen can be burned in ICE, similar to burning of petrol or natural gas, and in the process it produces water as the main product along with small amounts of NO_x, which act as an air pollutant.

Hydrogen has the potential to bring about energy revolution, much like the way Internet has brought about the information revolution. Fuel cells are a "critical technology" that will bring a total energy revolution and change the course of history. The vital goal is to use the renewable solar energy for splitting of water into its elementary components: oxygen and hydrogen. The hydrogen economy has the potential to open the doors for ultimate changes in our socio-economic and political status.

Hydrogen economy envisions hydrogen as the power generation fuel of the future that will enable the world to reduce oil dependence, greenhouse gas effects, and poverty.

To conclude, we can say that the use of energy in future will be renewable, clean, safe, and eco-friendly. So the hydrogen will also join the electricity in future and also to the great development of fuel cell technology.

REFERENCES

Abbott, D. 2010. Keeping the energy debate clean: How do we supply the world's energy needs? *Proceedings of IEEE* 98(1): 42–66

Ahrendts, J. 1980. Reference states. *Energy* 5: 667–668

Albertazzi, S., F. Basile, J. Brandin, J. Einvall, C. Hulteberg, G. Fornasari, V. Rosetti, M. Sanati, F. Trifiro, and A. Vaccari. 2005. The technical feasibility of biomass gasification for hydrogen production. *Catalysis Today* 106: 297–300

AWEA (American Wind Energy Association). 2008. Top 20 States with Wind Energy Resource Potential. Details available at http://www.awea.org/pubs/factsheets.html, last accessed on 28 February 2008

Balta, M. T., I. Dincer, and A. Hepbasli. 2010. Geothermal-based hydrogen production using thermo-chemical and hybrid cycles: A review and analysis. *International Journal of Energy Research* 34(9): 757–775

Baumlin, S., F. Broust, F. Bazer-Bachi, T. Bourdeaux, O. Herbinet, F. T. Ndiaye, M. Ferrer, and J. Lédé. 2006. Production of hydrogen by lignins fast pyrolysis. *International Journal of Hydrogen Energy* 31: 2179–2192

Bendaikha, W., S. Larbi, and M. Bouziane. 2011. Feasibility study of hybrid fuel cell and geothermal heat pump used for air conditioning in Algeria. *International Journal of Hydrogen Energy* 36(6): 4253–4261

Bockris, J. O. M. and T. N. Veziroglu. 2007. Estimates of the price of hydrogen as a medium for wind and solar sources. *International Journal of Hydrogen Energy* 32: 1605–1610

Bockris, J. O. M. 2010. Would methanol formed from CO_2 from the atmosphere give the advantage of hydrogen at a lesser cost? *International Journal of Hydrogen Energy* 32: 5165–5172

BP. 2010. *Statistical Review of World Energy*. London: BP

Curry Hydrocarbons Inc. 2008. Chemical engineering plant cost index. Details available at http://ca.geocities.com/fhcurry@rogers.com/, last accessed on 12 July 2008

Dincer, I. 1998. Energy and environmental impacts: Present and future perspectives. *Energy Sources* 20(4–5): 427–453

Dincer, I. 2000. Renewable energy and sustainable development: A crucial review. *Renewable and Sustainable Energy Reviews* 4(2): 157–175

Dunwoody, D. C., H. Chung, L. Haverhals, and J. Leddy. 2009. Current status of direct methanol fuel-cell technology. In *Alcoholic Fuels*, edited by S. Minteer. London: Taylor and Francis

Georgiou, P., C. Tourkolias, and D. Diakoulaki. 2008. A roadmap for selecting host countries of wind energy projects in the framework of the clean development mechanism. *Renewable and Sustainable Energy Reviews* 12(3): 712–731

Glatzmaier, G., D. Blake, and S. Showalter. 1998. Assessment of methods for hydrogen production using concentrated solar energy. Golden, CO: NREL

Granovskii, M., I. Dincer, and M. A. Rosen. 2006. Life cycle assessment of hydrogen fuel cell and gasoline vehicles. *International Journal of Hydrogen Energy* 31(3): 337–352

Hall, D. O. and K. K. Rao. 1989. Immobilized photosynthetic membranes and cells for the production of fuel and chemicals. *Chimica Oggi* 7: 40

Holladay, J. D., J. Hu, D. L King, and Y. Wang. 2009. An overview of hydrogen production technologies. *Catalysis Today* 139: 244–260

Kolb, G. J., R. B. Diver, and N. Siegel. 2007. Central-station solar hydrogen power plant. *Journal of Solar Energy Engineering* 129: 179–183

Kreutz, T., R. Williams, S. Consonni, and P. Chiesa. 2005. Co-production of hydrogen, electricity and CO_2 from coal with commercially ready technology. Part B: Economic analysis. *International Journal of Hydrogen Energy* 30: 769–784

Lardic, S. and V. Mignon. 2008. Oil prices and economic activity: An asymmetric cointegration approach. *Energy Economic* 30(3): 847–855

Lewis and G. von Elbe. 1987. Combustion, flames and explosions of gases, 3rd Edn., Academic Press, Orlando, pp. 717

Lutfi, N. and T. N. Veziroglu. 1991. A clean and permanent energy infrastructure for Pakistan: Solar-hydrogen energy system. *International Journal of Hydrogen Energy* 16(3): 169–200

Lutz, A. E., R. S. Larson, and J. O. Keller. 2002. Thermodynamic comparison of fuel cells to the Carnot cycle. *International Journal of Hydrogen Energy* 27(10): 1103–1111

Marban, G. and T. Valdes-Solis. 2007. Towards the hydrogen economy? *International Journal of Hydrogen Energy* 32: 1625–1637

Miyake, J. and S. Kawamura. 1987. Efficiency of light energy conversion to hydrogen by the photosynthetic bacterium *Rhodobacter sphaeroides*. *International Journal of Hydrogen Energy* 12: 147–149

Moriarty, P. and D. Honnery. 2009. What energy levels can the earth sustain? *Energy Policy* 37: 2469–2474

Nema, P., R. K. Nema, and S. Rangnekar. 2009. A current and future state of art development of hybrid energy system using wind and PV solar: a review. *Renewable and Sustainable Energy Reviews* 13(8): 2096–2103

Ni, M., D. Y. Leung, M. K. Leung, and K. Sumathy. 2006. An overview of hydrogen production from biomass. *Fuel Process Technology* 87: 461–472

Olah, G. A. 2005. Beyond oil and gas: the methanol economy. *Angewandte Chemie International Edition* 44: 2636–2639

Ramanathan, K., M. A. Contreras, C. L. Perkins, S. Asher, F. S. Hasoon, J. Keane, D. Young, M. Romero, W. Metzger, R. Noufi, J. Ward, A. Duda. 2003. Properties of 19.2% efficiency ZnO/CdS/CuInGaSe$_2$ thin-film solar cells. *Progress in Photovoltaics: Research and Applications* 11: 225 - 230

Richards, M., A. Shenoy, K. Schultz, and K. Brown. 2006. H$_2$-MHR conceptual designs based on the sulphur–iodine process and high-temperature electrolysis. *International Journal of Nuclear Hydrogen Production and Applications* 1: 36–50

Ryazantsev, E. and A. Chabak. 2006. Hydrogen production, storage, and use at nuclear power plants. *Atomic Energy* 101: 876–881

Schmidt, D. D. and J. R. Gunderson. 2000. Opportunities for hydrogen: an analysis of the application of biomass gasification to farming operations using micro-turbines and fuel cells. *Proceedings of the 2000 Hydrogen Energy Program Review*, NREL/CP-570–28890. Golden, CO: National Renewable Energy Laboratory

Scott, D. S. 2008. *Smelling Land: the Hydrogen Defence against Climate Catastrophe*. Montreal, Canada: Canadian Hydrogen Association

Searchinger, T., R. Heimlich, R. A. Houghton, F. Dong, A. Elobeid, J. Fabiosa, S. Tokgoz, D. Hayes, and T.-H. Yu. 2008. Use of U.S. croplands for biofuels increases greenhouse gases through emissions from land use change. *Science* 319: 1238–1240

Sherif, S., F. Barbir, and T. Veziroglu. 2005. Wind energy and the hydrogen economy review of the technology. *Solar Energy* 78: 647–660

Sherwani, A. F., J. A. Usmani, and Varun. 2010. Life cycle assessment of solar PV based electricity generation systems: a review. *Renewable and Sustainable Energy Reviews* 14(1): 540–544

Solarbuzz. 2008. Portal to the world of solar energy. Details available at http://www.solarbuzz.com/index.asp, last accessed on 11 January 2008

Sorensen, B. 2005. *Hydrogen and Fuel Cells: Emerging Technologies and Applications*. San Diego, CA: Academic Press

Specht, M. 1992. Hydrogen energy progress IX. *Proceedings of the Ninth World Hydrogen Energy Conference*, Paris (France), p. 527

Thirugnanasambandam, M., S. Iniyan, and R. Goic. 2010. A review of solar thermal technologies. *Renewable and Sustainable Energy Reviews* 14(1): 312–322

Veziroglu, T. N. and S. Sahin. 2008. 21st Century's energy: Hydrogen energy system. *Energy Conversion and Management* 49: 1820–1831

Veziroglu, T. N. 2007. IJHE grows with hydrogen economy. *International Journal of Hydrogen Energy* 32(1): 1–2

Veziroglu, T. N. 2000. Quarter century of hydrogen movement 1974–2000. *International Journal of Hydrogen Energy* 25: 1143–1150

Veziroglu, T. N. 2008. International Association for Hydrogen Energy, Details available at http://www.iahe.org, last accessed on 6 December 2008

Wall, G. 1995. Exergy and morals. Paper presented at *Second-Law Analysis of Energy Systems: Towards the 21st Century*, edited by E. Sciubba and M. J. Moran, 5–7 July, Rome, Italy, pp. 21–29

Yilanci, A., I. Dincer, and H. K. Ozturk. 2009. A review on solar-hydrogen/fuel cell hybrid energy systems for stationary applications. *Progress in Energy and Combustion Science* 35(3): 231–244

4

Emerging Energy Alternatives for Sustainable Development in Malaysia

N.A. Rahim[a], Jeyraj Selvaraj[a], M.S. Hossain[a], and A.K. Pandey[a,*]

[a]University of Malaya Power Energy Dedicated Advanced Centre (UMPEDAC), University of Malaya, 59990 Kuala Lumpur, Malaysia
[*]E-mail: adarsh.889@gmail.com

4.1 INTRODUCTION

Renewable resources can generally be characterized as wellsprings that never run out. They can also be defined as "energy obtained from the continuous or repetitive currents of energy recurring in the nature" and also "energy flows which are replenished at the same rate as they are used" (Kyairul 2007). These resources are available in abundance around the globe and have been a part of our day-to-day lives for so long that they have been taken for granted and not exploited optimally. Electricity generation in Malaysia increased at an average rate of 8% per year from 69,280 GWh in 2000. In 2010 Malaysia's electricity generation totalled at 137,909 GWh. Malaysia, being close to the equator, receives between 4000 $Wh/m^2/day$ and 5000 $Wh/m^2/day$ of solar energy. This overwhelming amount of energy can be harnessed under Malaysian climatic condition and shows that there is a great potential for the smooth generation of solar energy in the country (Cristopher 2012; Hafiy 2014). At present, electricity is generated on a five-fuel combination (gasoline, coal, hydro, oil, and other sources). The use of gasoline to produce electrical energy has declined from 77.0% to 55.9%, hydro from 10.0% to 5.6%, and oil from 4.2% to 0.2%. On the other hand, the use of coal for electrical energy generation increased from 8.8% to 36.5% and the use of other sources from 0% to 1.8% (Cristopher 2012; Hafiy 2014).

Energy has a great impact on the socio-economic development of any country. Recently, there has been a total increase of 44.2% over the projection period for 2006–2030 in energy demand due to economic and technological developments around the world (Rahman and Lee 2006). An important portion of the energy is generated from petroleum, and it is a well-known and undeniable fact that burning of fossil fuels releases gaseous pollutants, which are responsible for causing global warming and greenhouse effects. For this reason, such activities will influence global climate change, stratospheric ozone depletion, loss of biodiversity, hydrological systems change and the supplies of freshwater, land degradation, and stress on food processing methods (Leo 1996). Malaysia, a member of the Association of Southeast Asian Nations (ASEAN), is one of the region's developing countries, with a gross domestic product (GDP) of US$ 19,789.20 per capita (PPP basis), and a steady GDP growth of between 4.5% and 5.5%, as recorded in 2015 (Watch 2013; Times 2015). The financial strength of Malaysia grew at the rate of 5% in 2005, and the total power demand is predicted to increase at a traditional rate of 6% every year. Parallel to Malaysia's speedy monetary development, the country's energy consumption grew at the rate of 5.6% million tonnes of oil equivalent (Mtoe) in 2000 and reached 38.9% Mtoe in 2005. Energy consumption in Malaysia is expected to reach 98.7% Mtoe by 2030, just about three times the 2002 growth rate (Oh, Pang, and Chua 2010). Renewable energy is being used in many countries around the world not only for harnessing energy but also for solving the environmental problems that result from the use of fossil fuels (Harikrishnan, Deepak, and Kalaiselvam 2014; Hossain, Pandey, Mohsin, *et al.* 2015).

4.2 RENEWABLE RESOURCES IN MALAYSIA

Renewable energy sources are abundantly available in Malaysia but it needs to be harnessed efficiently. The country has paid special attention to the need to find and use renewable resources and ease the energy demand. Other than apparent issues already mentioned, the country is also looking for ways to diversify its energy supply. Though the implementation of renewable energy can be a long and tedious undertaking, it is still a useful mission for the long term (Kyairul 2007).

Malaysia has an abundance of renewable energy assets in the form of biomass, biogas, wind, mini-hydro and solar energy; however, a lot of these renewable resources have not been properly exploited. Table 4.1 shows the renewable energy resource in Malaysia and their annual energy value in RM (million) (Fakhrur 2011).

Table 4.1   Renewable energy resource potential in Malaysia

Renewable energy resource	Energy value in RM million (annual)
Forest residues	11,984
Palm oil biomass	6,379
Solar thermal	3,023
Mill residues	836
Hydro	506
Solar PV	378
Municipal waste	190
Rice husk	77
Landfill gas	4

PV - photovoltaic
Source Fakhrur (2011)

Figure 4.1 indicates the national renewable energy goals in Malaysia, according to which solar photovoltaic (PV) is the leading source of power (Wei-Nee 2012). Table 4.2 gives the status of small renewable energy power (SREP) approved by Special Committee on Renewable Energy (SCORE) as of August 2004, and lists renewable energy resources and their capacities (Mazlina 2005).

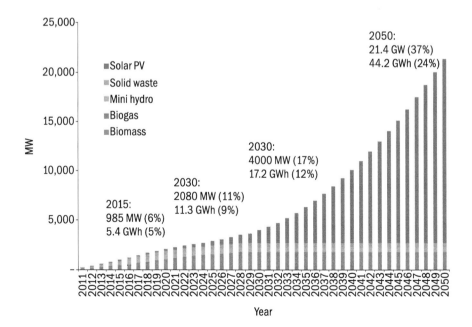

Fig. 4.1   *National renewable energy goals*
Source   *Wei-Nee (2012)*

Table 4.2   Status of renewable energy power programme projects

Type	Energy resource	Approved application	Generation capacity (MW)	Grid connected capacity (MW)
Biomass	Empty fruit bunches	22	200.5	165.9
	Wood residues	1	6.6	6.6
	Rice husk	2	12.0	12.0
	Municipal solid waste	1	5.0	5.0
	Mixed fuels	3	19.2	19.2
Landfill gas		5	10.2	10.0
Mini-hydro		26	99.2	97.4
Wind and solar		0	0	0.0
Total		60	352.70	316.1

Source Mazlina (2005)

4.2.1   Solar Photovoltaic Power

Malaysia's location within a tropical region and its exposure to endless and steady daylight, up to 8 hours a day, and an average radiation of 4500 kWh/m^2, makes it an ideal environment for the study and advancement of useful PV innovations. PV electricity generation is a renewable energy, which is clean and does not emanate any greenhouse gases (GHGs). It depends solely on solar energy that does not deplete as compared to fossil fuels. A sunlight-based PV set up in Malaysia would generate energy of around 900 kWh/kWp to 1400 kWh/kWp every year depending on the area. The Klang Valley has the lowest irradiance value, while Penang and Kota Kinabalu have the highest measured values. In any case, a set-up in Kuala Lumpur would yield around 1000 kWh/kWp to 1200 kWh/kWp annually which receives 30% more energy than an equivalent system in Germany. Despite the advantages of PV power, solar PV applications in Malaysia are often constrained to stand-alone PV systems, especially for rural electrification where the operation expenses are to a great degree supported. Other smaller PV systems are employed in parkways, solar street lights, information transfer, and solar water radiators (Fakhrur 2011; Hossain, Pandey, Mohsin, *et al.* 2015).

The endless availability of solar radiation differs by the region due to the change in climate by the region. The advantages of solar energy in Malaysia are indicated in Figure 4.2 (Seda 2009).

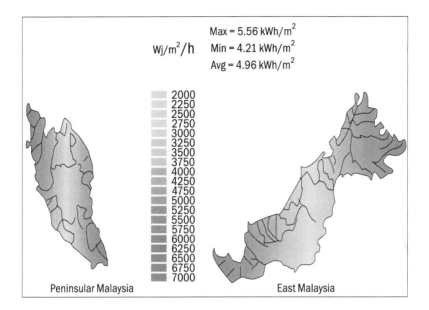

Fig. 4.2 *Annual average of daily solar irradiation in Malaysia*
Sources *Seda (2009); Mekhilefa, Safari, Mustaffaa, et al. (2012); Adawati (2011)*

Malaysia receives around 17 MJ/m^2 of solar radiation every day [Figure 4.3(a) and 4.3(b)]. In the period between 1989 and 2008, there is no indication that the consistency of solar radiation expanded or diminished, aside from urban communities such as Kuala Terengganu and Senai—where there is a weak linear trend showing a decline in solar radiation received by these two towns. Kota Kinabalu in Sabah also confirmed a decline in solar radiation from 1990 to 1999, after which radiation increased and balanced out at around 20 MJ/m^2 every day (Cristopher 2012).

From Figure 4.4(a) it can be seen that biomass, small hydropower, and solar power (non-individual) have received feed-in tariff (FIT) authorization in equal proportions. However, a look at all the renewable energy facilities installed, as shown in Figure 4.4(b), reveals that the installation of solar power, which is easily achieved, has been given first priority. Malaysia's FIT is structured such that the authorized installed capacity is determined for each recruitment period, and the purchase price is reduced according to the authorization period and the period in which operation begins. This indicates that it was designed as a system for promoting the steady generation of renewable energies and curbing the excessive demands of electricity consumers (Asia 2015).

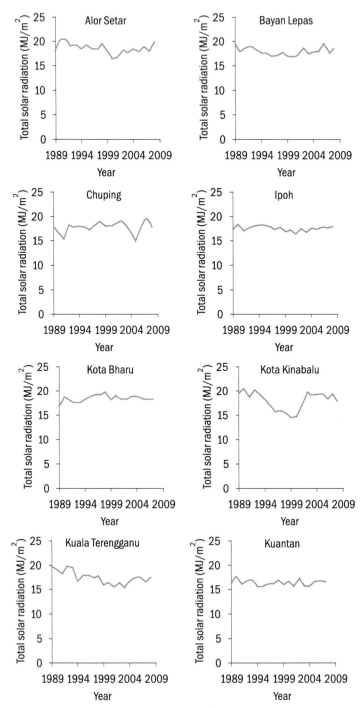

Fig. 4.3(a) *Average daily solar radiation (MJ/m²) for some towns in Malaysia from 1989 to 2009*

Source *Cristopher (2012)*

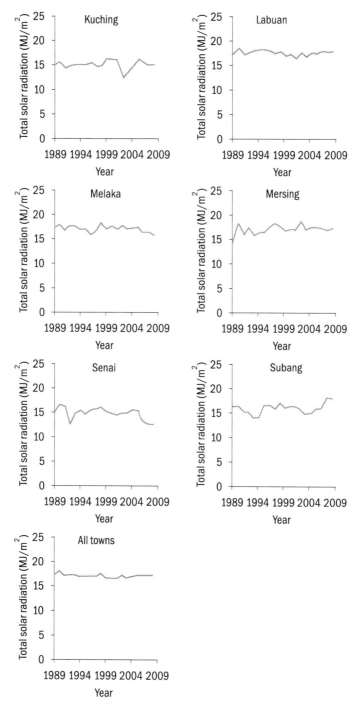

Fig. 4.3(b) *Average daily solar radiation (MJ/m²) for some towns in Malaysia from 1989 to 2009*

Source   *Cristopher (2012)*

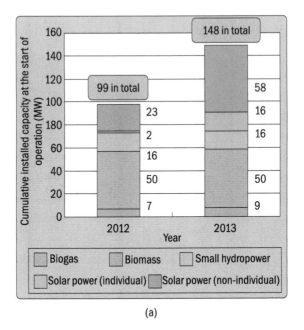

(a)

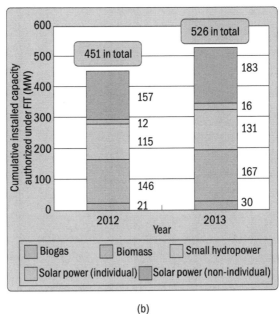

(b)

Fig. 4.4   (a) Cumulative installed capacity at the start of operation (power generation from renewable energies) and (b) FIT-authorized power generation

Sources   Asia (2015); Yee (2013)

Malaysia is the world's fourth largest manufacturer of PV modules. One reason for this is that the budget for installing PV systems in Malaysia is high. For instance, in 2005, the cost of PV system per kW cost was RM 31,410, tumbling to RM 24,970 in 2007, and to RM 20,439 in 2009. Today, the cost has diminished to about RM 15,000 for each kW peak, even though it is a rate unreasonably expensive or unrealistic to most Malaysians (Cristopher 2012).

4.2.2 Biomass in Malaysia

Biomass stands out as a critically important source of renewable energy in Malaysia. It is found in oil palm deposits and in timber remains. Currently, biomass contributes towards 16% of the total energy distribution in the country, 51% of it being from palm oil and 22% from wooden waste. Resources are also available from other rural sources and agro-based commercial ventures. Figure 4.5 shows the various sources of biomass. There are five main biomass sources in Malaysia: oil palm, wood, rice, sugarcane, and metropolitan waste. Each category has different classes as shown in Figure 4.5 (Shah Alam, Omar, Ahmed, *et al.* 2013).

Table 4.3 presents information regarding the amount of waste produced yearly from various sectors. The palm oil mills provided the highest

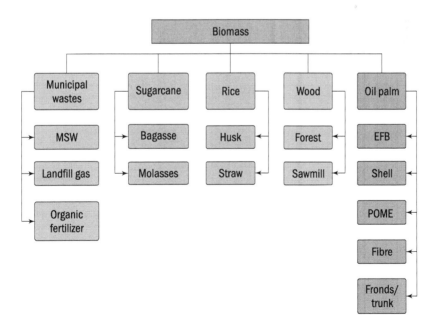

Fig. 4.5 *Biomass resources in Malaysia*
Sources *Elham (2014); Seda (2012)*

Table 4.3   Biomass resource potential

Sector	Quantity ktone/(yr)	Potential annual generation (GWh)	Potential capacity (MW)
Rice mills	424	263	30
Wood industries	2,177	598	68
Palm oil mills	17,980	3,197	365
Bagasse	300	218	25
POME	31,500	1,587	177
Total	72,962	5,863	665

POME - Palm Oil Mill Effluent
Source Kyairul (2007)

potential generating capacity of 365 MW and annual generation potential of 3197 GWh. Palm Oil Mill Effluent (POME) also can give significant contribution to the potential generation capacity with 177 MW. Bagasse is the waste produced after juice is extracted from sugarcane stalks, and rice mills also have a modest generation capacity of 25 MW and 30 MW, respectively (Kyairul 2007).

The top priority in Malaysia should be the use of palm oil as fuel to bring about energy security. Among the total biomass produce in Malaysia, palm oil has the largest percentage (85.5%, as demonstrated in Figure 4.6). After investigation and experiments, palm oil is now being utilized as biofuel. The biofuel strategy has triggered the creation of B5 ("ENVO") diesel

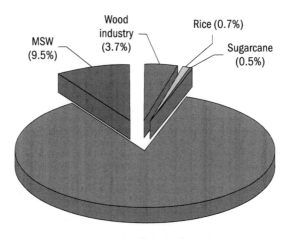

Fig. 4.6 *Total production of biomass in Malaysia*
Source *Shah Alam, Omar, Ahmed, et al. (2013)*

which is a blend of 5% processed palm oil and 95% petroleum diesel. This B5 fuel will be utilized for transport and mechanical purposes in Malaysia. However, for the production of B5, there needs to be a programme for technical development (of the palm oil component), testing and acceptance by engine manufacturers of the blended product, and the development of a supply infrastructure in Malaysia. Administrative organizations began using B5 in February. By this point, under Green Technology Financing Scheme (GTFS), funds worth RM 1.5 billion were earmarked for Tenth Malaysia Plan (10MP). In addition, under the Clean Development Mechanism biomass, biogas and landfill projects were included and the certified emission reduction issued to 10 companies, totalling to 1,231,655 tonnes of CO_2 equivalent. Table 4.4 shows the potential power generation from palm oil residues at Palm Oil Mills in Malaysia (Shah Alam, Omar, Ahmed, *et al.* 2013).

The potential power generations of palm oil residues are depicted in Figure 4.7. This also presents the comparisons between fibres, shells, and empty fruit bunches (EFB) at different levels of moisture content (MC). It can be seen that EFB with a low moisture content has the greatest potential for energy generation. The generation potential is expected to increase in the future due to expected expansion of the palm oil industry by 40% within the span of 20 years (Kyairul 2007).

Shells in palm oil industries refer to the fraction after the nut has been removed during the crushing process. These are fibrous materials which are easily handled in bulk directly from the product line to the end use. Their moisture content is low compared to other biomass residues; therefore, they have a slightly more electricity generation potential than EFB even though the amount of residue generated is much lower. Fibre

Table 4.4   Potential power generation using palm oil

Type of industry	Production (thousand tonne)	Residue	Residue product ratio (%)	Residue generated (thousand tonne)	Potential energy (PJ)	Potential electrical generation (MW)
Oil palm	59,800	EFB 65%	21.14	12,642	59	570
		MC	12.72	7,607	113	1,080
		Fibre shell	5.67	3,391	57	545
	Total solid			23,640	229	2,195
	POME (3.5m³ per tonne of CPO/65% of FFB)			41,860		346

POME - Palm Oil Mill Effluent; CPO - Crude Palm Oil
Source Baharuddin (2014)

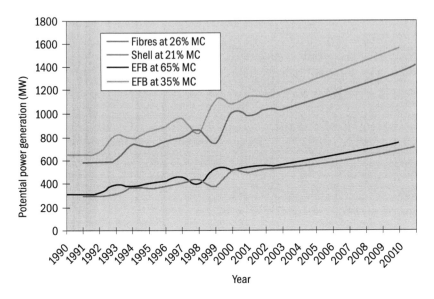

Fig. 4.7 *Oil palm residues' potential for power generation*
Source *Mazlina (2005)*

has the greatest potential for generating electricity at 1032 MW. Figure 4.8 and Table 4.5 show the potential energy that can be harnessed from paddy residues. Paddy residues are of two kinds—rice husk and paddy straws. It is clear from the chart that power generation was generally the same with a few minor fluctuations between 1991 and 2007. A slight increase is observed from 2007 to 2010, however, mainly the amount of residue will remain roughly the same because the paddy production output in Malaysia has reached its peak production capacity and land use. The only advancement would be in the biotechnological aspect of production. Because of its residue product ratio, rice straws can produce much more energy than rice husk. The total amount of residue generated according

Table 4.5   Residue product ratio and potential power generation of paddy residues

Industry	Production (thousand tonnes, 2000)	Residue	Residue product ratio (%)	Residue generation (thousand tonnes)	Potential energy (PJ)	Potential power (MW)
Rice	2140	Rice husk	22	471	7.536	72.07
		Rice straw	40	856	8.769	83.86
Total	2140			1327	16.305	155.93

Source Mazlina (2005)

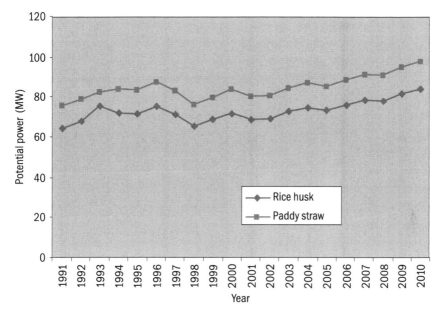

Fig. 4.8 *Potential power generation from paddy residues*
Source *Mazlina (2005)*

to data for year 2000 is 1,327,000 tonnes, which is equivalent to around 156 MW of electricity generation (Kyairul 2007).

In Malaysia, another available source of biomass is sugarcane. The residue product ratio of and potential power generation from bagasse is given is Table 4.6.

As mentioned earlier, wood residue also plays a major source of biomass energy. Three types of residues and their potential energy based on data available until 2002 are outlined in Figure 4.9. Plywood and Venner waste residue and moulding waste have almost no market value. Only sawn timber waste has considerable capacity to be utilized. The potential decreased from around 100 MW in 1997 to less than 50 MW in 1998 due to environmental issue, where the amount of trees reduced in conjunction

Table 4.6 Production from bagasse and molasses

Type of industry	Production (thousand tonnes)	Residue	Residue product ratio (%)	Residue generated (thousand tonnes)	Energy use factor	Amount for energy use (thousand tonnes)	Amount of surplus (thousand tonnes)
Sugar	1111	Bagasse	32	356	1.0	356	0
		Molasses	Not available				

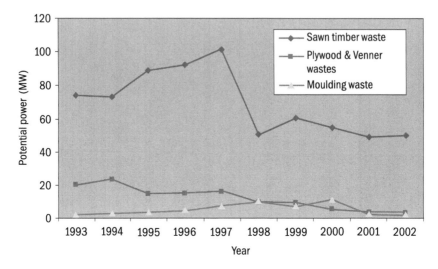

Fig. 4.9 *Potential for power generation from wood residues*
Source *Mazlina (2005)*

with the government's aspiration to preserve forests and slow down the process of lumbering (Kyairul 2007). Table 4.7 represents biomass potential in Malaysia.

4.2.3   Municipal Solid Waste

Malaysia has gone through tremendous changes in its economy in the last decade and the population growth has also brought together an increase in the amount of waste generated. In 2003, the amount of traditional waste came to weigh 0.5 kg/day – 0.8 kg/day, however, in urban areas the figure touched 1.7 kg/day. Municipal solid waste (MSW), more commonly known as trash or garbage, is a waste type that predominately includes household waste and also a portion from the industrial and commercial sector. Figure 4.10 shows sector-wise MSW, in terms of percentage.

Table 4.7   Biomass potential in Malaysia

Location	Hectares under oil palm cultivation (2005)	Number of palms (at 136/Ha) million	FFB delivered to mills (million tonnes)
Sabah	1,081,102	147	25
Sarawak	405,729	55	6
Peninsula Malaysia	1,956,129	266	43
Total	3,450,960	468	74

Source Baharuddin (2014)

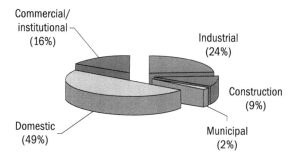

Fig. 4.10 *Sector-wise MSW in Malaysia*
Source *Mazlina (2005)*

Domestic sector dominates the waste portion with almost 50% (Kyairul 2007).

Municipal solid waste can be classified into five general groups: biodegradable waste, recyclable fabric, latent waste, composite waste, and unsafe waste. These types of wastes are usually deposited in specific landfills; however, the quick accumulation of waste and the limited number of landfills are making waste disposal difficult and inconvenient. To deal with this issue, technologies that reutilize waste need to be developed. Another way to deal with waste in an energy-efficient way is through incineration. Burning MSW can generate energy and in the meantime reduce the weight and volume of the waste amount by 75% and 90%, respectively (Kyairul 2007). Table 4.8 lists the amount of waste anticipated by 2020.

4.2.4 Biogas

Biogas, by and large, means biofuel that is generally made of methane and carbon dioxide. It can be derived from wastes through anaerobic conditions. In Malaysia, a part of the land that can be used to acquire biogas is in landfills. Natural gas is produced from organic wastes disposed on a landfill. The waste is covered and compressed mechanically. As

Table 4.8 Future projections of MSW in Malaysia

Year	Population	Estimated amount of waste (tonnes/yr)
1991	17,567,000	4,488,369
1994	18,917,739	5,048,804
2015	31,773,889	7,772,402
2020	35,949,239	9,092,611

Source Mazlina (2005)

conditions become anaerobic, organic waste is broken down and landfill gas is produced (Kyairul 2007).

Figure 4.11 shows the Jana Landfill undertaking, the first grid-connected renewable energy venture in Malaysia which was inaugurated in April 2004. It has 2 MW of mounted capacity and is fueled by biogas derived from landfills. A number of potential sites have been identified with projected capacity of approximately 20 MW (Adawati 2011).

4.2.5   Wind Energy Potential

Wind energy is a quickly developing industry; referred to as the new installation capital cost that is projected to grow from \$30.1 billion in 2007 to \$83.4 billion in 2017. The generation of wind energy has kept growing by 25%–30% every year since 2000 and to at least 93 GW in 2007 (Repn 2007). In 2007, the world recorded 20,000 MW of wind energy, which is corresponding to a total conventional power plant of 20 GW (Clean Edge 2010). Wind energy converts kinetic energy of wind into mechanical or electrical energy and then changes to mechanical energy through the rotor. Then this rotational energy may be utilized to replace electrical motors in countries and remote territories, and basically for pumping water. Wind energy can provide power for individual consumer as well as it can support the grid connected power supply (Adawati 2011).

An important investigation was made by Mekhilef and Chandrasegaran (2011) at some offshore wind farms in Malaysia. During the exploration, the HOMER system was used to confirm the potential of wind power

Fig. 4.11   *TNB Jana Landfill Project*
Source   *Adawati (2011)*

along the South China Sea coastline. They chose 16 locations and two exceptional turbine models V-47 and V-80, where 1 to 7 represent areas covering the east Peninsular Malaysia coastline that faces east of South China Sea. Locations 8 to 16 represent areas covering the north-west part of Borneo that forms pall of the Sarawak and Sabah coastline (Chiang, Zainal, Aswatha Narayana, *et al.* 2003; Mekhilef and Abdul Kadir 2010; Ustun and Mekhilef 2010; Mohamed and Lee 2006). Sites with numerical identification of 1, 2, 3, 4, 8 and 13 were selected. Figure 4.12 shows that power generation during the northeast rainstorm season is the

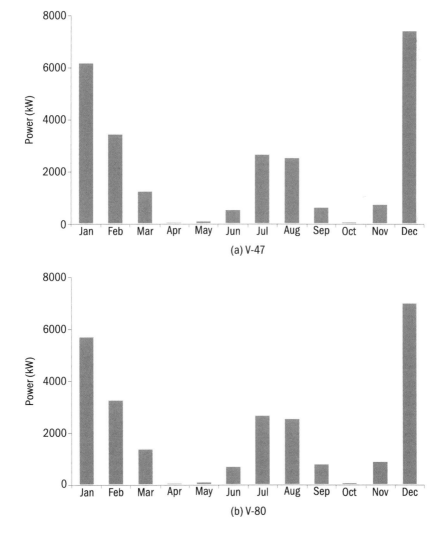

Fig. 4.12   *Monthly average energy production for two sites (a) V-47 and (b) V-80*
Source   *Mekhilef and Chandrasegaran (2011)*

Table 4.9   Techno-economic analysis for selected sites

Site	Model	Wind farm capacity (MW)	Initial capital (RM)	Operating cost (RM/ year)	Total NPC (RM)	COE (RM/ kWh)	E (kWh/yr)	Net specific production (MWh/MW)
1	V-47	12.54	162,828	614,232	154,480,404	0.85	13,445,249	1,072
1	V-80	14	132,750	871,438	83,294,371	0.64	13,451,246	961
2	V-47	12.54	162,828	1,439,335	97,132,399	0.55	18,286,012	1,458
2	V-80	14	132,750	1,856,770	69,903,402	0.40	18,225,138	1,302
3	V-47	12.54	162,828	628,527	108,151,545	0.77	14,357,679	1,145
3	V-80	14	132,750	1,043,950	80,949,874	0.58	14,287,065	1,021
4	V-47	12.54	162,828	472,174	123,110,393	1.40	9,024,842	720
4	V-80	14	132,750	20,300	94,861,629	1.04	9,327,512	666
8	V-47	12.54	162,828	806,762	127,657,583	1.77	7,403,764	590
8	V-80	14	132,750	309,527	99,344,070	1.32	7,729,525	552
13	V-47	12.54	162,828	599,816	108,541,727	0.79	14,218,581	1,134
13	V-80	14	132,750	1,030,091	81,137,637	0.59	14,220,128	1,016

Source Mekhilef and Chandrasegaran (2011)

most fruitful and is diminished for the rest of the year. Table 4.9 shows the techno-economic analysis for the selected sites in Malaysia (Mekhilef and Chandrasegaran 2011).

The sensitivity analysis confirms that the FIT is a significant criterion to determine the feasibility of offshore wind farm in Malaysia. FIT higher than the break-even point would attract private sectors to invest in this type of energy system. An attractive policy would determine the profitability of an investment in the offshore wind farms and encourage private sector to invest (Mekhilef and Chandrasegaran 2011).

4.2.6   Small Hydropower Potential

Hydroelectricity is a useful resource, and its generation is essential. It is presently a key contributor to the world energy supplies, and has been delivering power consistently and at reasonable costs for close to a century (Kyairul 2007). The real advantages of hydropower include its renewable nature, the fact that it does not produce any GHGs, and cost-effective way of storing electricity needed by consumers. It provides 17% of global generation capacity and 20% of energy produced annually (Othman 2005).

The mini-hydro potential of Malaysia was assessed and potential locations have been identified. Under the rural electrification programme, some of these sites were implemented with government funding based on keep run-of the river systems ranging from 500 kW to 1000 kW capacity. Currently there are 39 units with total capacity of 16.185 MW in Peninsular Malaysia, 7 models at a full capacity of 2.35 MW in Sabah and 5 models at a full capacity of 5 MW in Sarawak. There may be a need for significant involvement in building, running and maintaining mini-hydropower plants. Similar plants can be set up as and when needed.

Table 4.10   List of mini-hydro stations (TNB) in Peninsular Malaysia

Station	Town	State	Capacity (kW)
Sg. Ulu Langat	Kuala Lumpur	Kuala Lumpur	2200
Sg. Kerling	Rawang	Selangor	900
Sg. Benus	Bentong	Pahang	300
Sg. Perdak	Bentong	Pahang	364
Sg. Sempam	Raub	Pahang	1250
Sg. Sia	Raub	Pahang	548
Sg. Pertang	Raub	Pahang	492
Sg. Ulu Dong	Raub	Pahang	550
Sg. Rek	Kuala Krai	Kelantan	270
Sg. Sok	Kuala Krai	Kelantan	588
Sg. Lata Tunggil	Kuala Krai	Kelantan	700
Sg. Renyok	Jeli	Kelantan	1600
Sg. Kemia	Jerteh	Terengganu	526
Sg. Brang	Kuala Berang	Terengganu	422
Sg. Tersat	Kuala Berang	Terengganu	488
Sg. Cheralak	Dungun	Terengganu	500
Sg. Bil	Tanjung Malim	Perak	258
Sg. Kinjang	Tapah	Perak	349
Sg. Kenas	Kuala Kangsar	Perak	532
Sg. Asap	Kuala Kangsar	Perak	110
Sg. Gebul	Kuala Kangsar	Perak	120
Sg. Chempias	Kuala Kangsar	Perak	120
Sg. Lawin	Lenggong	Perak	270
Sg. Temelong	Lenggong	Perak	872
Sg. Tebing Tinggi	Selama	Perak	178
Sg. Mahang	Selama	Perak	483
Sg. Kupang	Baling	Kedah	216
Sg. Mempelam	Baling	Kedah	397
Sg. Tawar Besar	Baling	Kedah	540
Sg. Mentawak	Pulau Tioman	Pahang	500

Source Badrin (2012)

Mini-hydro is a mature technology based on proven equipment. Research and development is the creation and dynamic use of detailed hydrodynamic and computer models. It is a hope that these models can be used to predict problems by extreme conditions of weather or power demands (Shah Alam, Omar, Ahmad, *et al.* 2013). Table 4.10 lists the different mini-hydropower stations Tenaga Nasional Berhad (TNB) in Peninsular Malaysia.

There are also a few public licensed mini-hydropower installations owned by private sector such as AMDB Perting Hydro Sdn Bhd, Sg. Perting, Bentong, Pahang (4.2MW), Syarikat Esajadi power Sdn Bhd, Sg. Kaingaran, Tambunan, Sabah (2.5MW), Sg. Kadamaian, Kota Belud, Sabah (2MW) and Sg. Pangpuyan, Kota Belud, Sabah (4.5MW) (Status of Renewable Energy Power Programme 2009). Table 4.11 indicates the capacity of mini-hydro power stations in Malaysia (Badrin 2012).

At the national level, one of the chief difficulties faced by small hydropower ventures is attracting investment of clients and industry players. In addition, access to water and the use, control, and diversion of water flows is liable to government and state laws. There are other regulations that apply to physical alteration of a stream channel that may affect water quality or wildlife habitat. Hydropower methods should not lead to contamination when in operation. It is understood that the natural impact of small-scale hydropower plants is insignificant; however, the little ecological impact must be taken into consideration before construction begins. Proper caution must be taken in order to ensure that there will be no damaging impact on the surrounding environment or civil infrastructure. Small-scale hydropower plants can play a role in empowering sustainable watershed management. To ensure that these plans are set in motion, renewable energy engineers should be required

Table 4.11   Installed capacity mini-hydro power stations in Malaysia

Location		Capacity (MW)
Peninsular Malaysia	Kedah	1.556
	Perak	3.207
	Terengganu	1.936
	Kelantan	3.158
	Pahang	3.504
	Sub-total	13.361
East Malaysia	Sabah	8.335
	Sarawak	7.297
	Total	28.993

Source Badrin (2012)

to follow the frameworks and strategies of different organizations such as Department of Environment (DOE) for environmental impact assessments, state authorities for land conversions approval for water abstraction rights and permissions (in small-scale hydro plants), and local authorities for structural plan approval (Badrin 2012).

4.3   CONCLUSION

This chapter presented a potential and status of emerging energy alternatives in Malaysia which contributes towards the sustainable development of the nation. Some of the important conclusions obtained from the study are given as below:

- Malaysia is rich in renewable energies such as biomass, solar, wind, and hydropower; still it is mostly dependent on palm oil for energy generation. As far as renewable energy is concerned, the biomass energy followed by solar energy is having largest potential in Malaysia and available throughout the country and can be harnessed abundantly.
- However, a country like Malaysia needs special attention as the share of the total renewable energy is only around 5.5%. Increasing the share of renewable energy will attract the companies and ultimately clean energy will be produced which will not harm the environment and Malaysia will progress in sustainable manner.

Finally, from the study it can be concluded that in spite of abundant availability of renewable energies, it has not been harnessed efficiently till date and Government needs to focus on energy policies which can promote and encourage the clean and renewable energies among the industry players and users.

REFERENCES

Adawati, Y. 2011. Renewable energy potential in Malaysia: Masters thesis: 23-63. Universiti Malaya

Asia, B.O. 2015. Status for Renewable Energies in Malaysia. Details available at http://www.asiabiomass.jp/english/topics/1411_05.html

Badrin, M. A. M. 2012. Malaysia Report on Small Hydro Power (SHP): myForesight. *National Foresight Institute*

Baharuddin, B. A. 2014. Biomass as a Renewable Energy Source: The case of Converting Municipal Solid Waste (MSW) to Energy. *7th Asian School on Renewable Energy* 34

Chiang, E. P., Z. A. Zainal, P. A. Aswatha Narayana, and K. N. Seetharamu. 2003. The potential of wave and offshore wind energy in around the coastline of Malaysia that face the South China Sea. *Proceedings of the International*

Symposium on Renewable Energy: Environment Protection and Energy Solution for Sustainable Development. Kuala Lumpur, Malaysia

Clean Edge, I. 2010. *Clean Energy Trends 2008.* Details available at http://www.cleanedge.com/reports/charts-reports-trends2008.php

Cristopher, T. B. S. 2012. Electricity from solar energy in Malaysia: Clean, renewable, and abundant energy source, so what's the problem? Details available at http://christopherteh.com/blog/2012/05/solar-malaysia/

Elham, O. S. J. 2014. Biomass Energy. Details available at https://biomass energymicet.wordpress.com/

Fakhrur, R. 2011. Renewable Energy in Malaysia. 2010. *Cardas Research Consulting SDN.BHD.* Details available at https://www.academia.edu/3554682/Renewable_Energy_In_Malaysia_2010_

Hafiy, H. S. 2014. Status of Solar Energy in Malaysia. Details available at https://prezi.com/jec-nkrhuuob/status-of-solar-energy-in-malaysia/

Harikrishnan, S., K. Deepak, and S. Kalaiselvam. 2014. Thermal energy storage behavior of composite using hybrid nanomaterials as PCM for solar heating systems. *Journal of Thermal Analysis Calorimetry* 115(2): 1563–71

Hossain, M. S., A. K. Pandey, A. T. Mohsin, J. Selvaraj, K. E. Hoque, and N. A. Rahim. 2016. Thermal and economic analysis of low-cost modified flat-plate solar water heater with parallel two-side serpentine flow. *Journal of Thermal Analysis and Calorimetry* 123(1): 793–806

Kyairul, A. B. B. 2007. *Analysis of Renewable Energy Potential in Malaysia* (Thesis). Australia: University of New South Wales

Leo, M. A. 1996. Keynote address. Bakun Hydroelectric Project Seminar, Kuala Lumpur

Mazlina, H. 2005. Present Status and Problems in Biomass Utilization for Electricity Generation in Malaysia. Biomass-Asia Workshop. Asia-Pacific Economic Cooperation. Tokyo, Japan

Mekhilef, S and M. N. Abdul Kadir. 2010. Voltage Control of Three-Stage Hybrid Multilevel Inverter Using Vector Transformation. *IEEE Transactions on Power Electronics* 25: 2599–2606

Mekhilef, S. and D. Chandrasegaran. 2011. Assessment of off-shore wind farms in Malaysia. TENCON 2011-2011 IEEE Region 10 Conference held at Bali, Indonesia, 21-24 November, pp. 1351–55

Mekhilefa, S., A. Safari, W. E. S. Mustaffaa, R. Saidurb, R. Omara, and M. A. A. Younis. 2012. Solar energy in Malaysia: Current state and prospects. *Renewable and Sustainable Energy Reviews* 16: 386–396

Mohamed, A. R. and K. T. Lee. 2006. Energy for sustainable development in Malaysia: energy policy and alternative energy. *Energy Policy* 34: 2388–2397

Oh, T. H., S. Y. Pang, and S. C. Chua. 2010. Energy policy and alternative energy in Malaysia: Issues and challenges for sustainable growth. *Renewable and Sustainable Energy Reviews* 14(4): 1241–52

Othman, Z. A. 2005. The Future of Hydropower in Malaysiam, Water Resources Technical Division, Details available at-http://dspace.unimap.edu.my/dspace/bitstream/123456789/13823/1/The%20Future%20of%20Hydropower%20in%20Malaysia.pdf

Rahman, M. A. and K. T. Lee. 2006. Energy for sustainable development in Malaysia: Energy policy and alternative energy. *Energy Policy* 34(15): 2388–97

Repn. 2007. Renewable Energy Policy Network for 21st Century. *Renewable 2007 Global Status Report.*

Sustainable Energy Development Authority (SEDA). 2009. National Renewable Energy Policy and Action Plan. *KeTTHA* 33

Seda. 2012. Tapping biomass to the max. Details available at http://seda.gov.my/?omaneg=000101000000010101010001000010000000000000000000000&y=45&s=1613

Shah Alam, S., N. A. Omar, M. S. B. Ahmad, H. R. Siddiquei, and M. N. Sallehuddin. 2013. Renewable Energy in Malaysia: Strategies and Development. *Environmental Management and Sustainable Development* 2(1): 51–66

Status of Renewable Energy Power Programme (SREP). 2009. Status of SREP Projects in Malaysia 2009. *Energy Commission of Malaysia*

Times, T. 2015. Malaysia revises GDP growth for 2015 to 4.5%-5.5%, Details available at http://www.straitstimes.com/news/asia/south-east-asia/story/malaysia-says-weak-global-oil-prices-will-lead-deficit-more-5b-budge

Ustun, T. S. and S. Mekhilef. 2010. Effects of a Static Synchronous Series Compensator (SSSC) Based on Soft Switching 48-Pulse PWM Inverter on the Power Demand from the Grid. *Journal of Power Electronics* 10: 85–90

Watch, E. 2013. Malaysia GDP Per Capita (PPP), US Dollars Statistics. Details available at http://www.economywatch.com/economicstatistics/Malaysia/GDP_Per_Capita_PPP_US_Dollars/ *International Monetary Fund (IMF)*

Wei-Nee, C. 2012. Renewable Energy Status in Malaysia: Presentation. *Sustainable Energy Development Authority Malaysia* 8

Yee, M. C. 2013. Sustainable Energy Development Authority Malaysia. *SEDA, Annual Report*

5

Role and Initiatives of Indian Government Policies for Growth of Wind Energy Sector

Vijay K. Jayswal[a,*], V. V. Tyagi[b], Richa Kothari[a], D. P. Singh[a], and S. K. Samdarshi[c]

[a]Department of Environmental Science, Babasaheb Bhimrao Ambedkar University, Lucknow, Uttar Pradesh 226025

[b]School for Energy Management, Sri Mata Vaishno Devi University, Katra, Jammu and Kashmir 182320

[c]Centre for Energy Engineering, Central University of Jharkhand, Brambe, Ranchi, Jharkhand 835205

[*]E-mail: vkjay1991@gmail.com

5.1 INTRODUCTION

The world population is growing rapidly with time. To maintain and improve the quality of life, a growing energy demand is inevitable. Hence, the issues of both dramatic increase in the rate of consumption of fossil fuels with shrinking fossil fuel reserves and the consequent adverse impact on the environmental quality and global warming need immediate attention. Sustainable future can be ensured only through reduction in the greenhouse gas (GHG) emissions, which requires reduction in fossil fuel burning and generation of energy through renewable energy sources. The promotion of renewable energy generation can be the best alternative for fulfilling the growing energy demand without much increase in the amount of GHGs. The energy from renewable energy sources can be produced sustainably with zero or a very less amount of carbon emission. Renewable energy sources comprise small hydro, wind, tidal, solar, wave, municipal solid waste (MSW), geothermal, biomass, and so on.

Wind energy is defined as the kinetic energy associated with the flow of air. Wind electric power can be generated by converting wind energy into a useful form of energy with the help of wind turbines. Wind energy density is very high along the coastal line and its territorial waters. India has a coastal line of 7517 km and high wind energy density is extended up to 12 nautical miles into the sea water. According to Ministry of New and Renewable Energy (MNRE), the potential of Indian wind energy sector is estimated to be 49,130 MW at 50 m height and 2% of land availability. However, if the tower height is increased to 80 m, the potential shoots up to 102,778 MW (Khare, Nema, and Baredar 2013).

The current installed global wind capacity has crossed the mark of 300 GW. About 60% of the total installed capacity is in Europe and North America. Among the BRICS Brazil, Russia, India, China, and South Africa countries, India and China are the major destinations for wind installation. They account for almost 35% of the global wind installations. The total installed capacities of developed countries, as well as India plus China, account for 95% of global installed wind power capacity (Timilsina, Cooten, and Narbel 2013). It means a few major European, American, and BRICS countries contribute to the total installed wind capacity worldwide. The developing countries except BRICS are not able to match the growth rate of the global installed wind power. The entire Southern Africa, almost untouched by the wind energy revolution, is slowly becoming the major source of electricity generation. However, there has been some progress in recent times because as per reports 83 countries have installed wind turbines of varying capacities by 2010.

In the past 15 years, the global wind energy sector has grown at a rate of 27% per annum, while the Indian wind energy sector has grown at a decent growth rate of 19.5% in the past 5 years (Timilsina, Cooten, and Narbel 2013). Despite the high growth rates in recent years, the current share of wind power in global electricity generation is very small, accounting for 2%–3% of the global electricity generation. In India, the share of wind electric power was 7.4% of the total electricity generation.

International Energy Agency (IEA) has estimated that with the continuation of current policies, the global wind electricity generation is expected to increase from 342 TWh in 2010 to 2151 TWh by 2035, which will constitute 5% of the global electricity generation. If new and progressive policies are adopted, this figure would reach up to 4281 TWh/year by 2035, which will constitute about 13% of the total electricity generation. According to IEA projections for India, wind power generation will reach 81 TWh/year by 2020, and it might even reach 174 TWh/year by 2030 (Khare, Nema, and Baredar 2013). It is predicted that the cost of wind power and the cost of electricity generation may approach towards grid parity by 2022, and it may become cheaper than natural gas and coal in the course of time.

The wind electric generators (WEGs) can be installed in two modes: grid-tied and stand-alone modes. By installing stand-alone systems, the Indian rural population could be benefitted by exploiting the potential of distributed micro-generation technologies, which can provide energy under distributed generation or off-grid generation for rural and remote population. This will provide long-term sustainable solutions.

5.2　INDIA'S WIND ENERGY POTENTIAL AND INSTALLATION

The estimated gross wind energy potential of India is 49,130 MW and the total installation has reached 19,051 MW, which is 38.7% of the overall potential. India has doubled its installed wind capacity in the past 5 years, which is three times more than what it was 7 years ago.

Figure 5.1 shows the distribution of wind energy potential in different states. As shown in the figure, the maximum wind energy potential in India is concentrated in the coastal states, with 10.81% of the distribution potential in Kashmir, which is an exception from coastal areas because of high altitude, high wind velocity, and high wind density. Gujarat, with 21.6% wind energy potential, is the highest producer, followed by Karnataka. Andhra Pradesh, Tamil Nadu, Rajasthan, and Odisha have almost a similar potential distribution, but Madhya Pradesh, Kerala, and rest of India have a very small amount of wind energy potential. The rest of India represents a very small potential of less than 2%, which is scattered through several states with only a few units of turbines. Considering the potential distribution, we can say that the major focus should be on only the seven major states and regulations should be drafted to attract investment for installation of wind turbines. These seven states have the most potential for wind energy generation and if the government focuses on only these states for installation of wind power, they will be able to generate a very high amount of wind power in future.

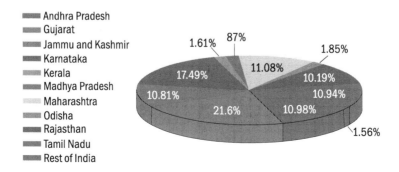

Fig. 5.1　*State-wise wind energy potential of India*

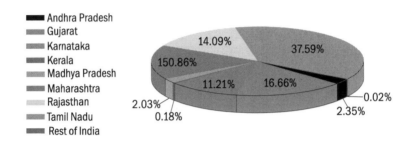

Fig. 5.2 *State-wise installation of wind turbines*

Figure 5.2 shows the distribution of current installed capacities of wind turbines in different states. We can see that Tamil Nadu is the leader in wind power installation programme with 37.5% share, while Gujarat, Karnataka, Rajasthan, and Maharashtra have collectively installed more than 55% of the total wind power installation. The wind power installation is concentrated in five major states, which have installed 95% of the total wind power installation in India. Almost 50% installations are concentrated in Gujarat and Tamil Nadu, indicating that they have crafted the wind energy policy very efficiently to attract chunk of investment for wind power installation. The other lagging states have started shaping their policies progressively to attract investors for wind energy installation.

Figure 5.3 shows the year-wise installation of wind turbines in India over a span of 13 years. As shown in the figure, more than 40% of

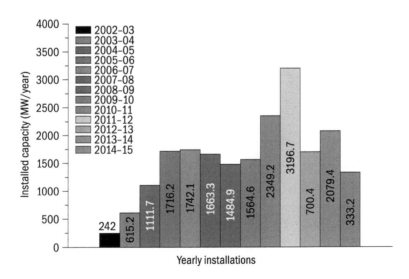

Fig. 5.3 *Installed capacity of wind energy projects in India*
Source *MNRE (2014)*

installations have taken place in the past 5 years. The increased wind turbine installation over this period in India reflects the success of the government efforts and policies.

5.3   INDIAN GOVERNMENT POLICIES FOR WIND ENERGY SECTOR

The Indian Government began making policies for the wind energy sector way back in 1994–95 through its Department of Non-conventional Energy Sources in the Ministry of Power. The Electricity Act, 2003 made specific provisions to promote renewable energy generation and it has changed the legal and regulatory framework for renewable energy sector. The act provides the way for policy formulation by Government of India (GoI) and mandates the State Electricity Regulatory Commission (SERC) to promote and implement the policy formulated for renewable energy generation.

The provisions of Electricity Act, 2003 prompted Central Electricity Regulatory Commission (CERC) and a large number of SERCs to announce regulations for cost plus tariff and set terms for generation and utilization of electricity produced by wind power projects.

Financial incentive policy announced by GoI included tax rebate through accelerated depreciation (AD) for procurement of WEGs. In 2010, CERC announced the regulations for renewable energy certificates (RECs). The RECs allowed another revenue mode for investors. The REC was dependent on renewable purchase obligation (RPO) by states. To deal with problems in financing of wind energy projects, GoI has established IREDA (International Renewable Energy Development Agency) that provides financial assistance for renewable energy projects, specially the independent power projects.

5.3.1   Direct Tax Benefits Policy by GoI

(i) Accelerated depreciation benefit at 80% of the project cost is available from the first year.

(ii) Exemptions in income tax on earnings from wind energy projects.

5.3.2   Generation-based Incentives by MNRE

Generation-based incentives (GBI) are provided for the project that cannot avail the direct tax benefits (DTB). GBI aims to broaden the investor base and incentivize actual generation/outcome-based incentive. GBI promotes the entry of large independent power producers (IPPs) and foreign direct investment in the wind energy sector.

Under the scheme, a generation-based incentive is provided to wind electricity producers at ₹0.50 per unit of electricity fed into the grid for a period of not less than 4 years and a maximum of 10 years with a cap of 100 lakh/MW. The total disbursement will not exceed one-fourth of the maximum limit of incentive, that is, ₹25 lakh/MW/year for the initial 4 years. The government has set target to cover 15,000 MW under GBI scheme for 12th Five-year Plan period (2012–17). GBI is over and above the tariff approved by various states. The disbursement of incentive is coordinated by IREDA.

5.3.3   National Action Plan on Climate Change for Grid Connected Systems

Through the action plan on climate change, India aims to adapt to climate change and enhance ecological sustainability. The Electricity Act, 2003 and National Tariff Policy, 2006 provide guidelines for regulating the purchase of a certain percentage of power from renewable-based sources. It also provides a mechanism for regulation of preferential power.

During 2009–10, through the establishment of dynamic minimum renewable purchase standard (DMRPS), it was suggested that 5% of national renewable energy may be set for total grid purchase, which should keep on increasing by 1% every year over a period of 10 years.

5.3.4   International Renewable Energy Development Agency

International Renewable Energy Development Agency is a non-banking financial institute under the control of MNRE. IREDA provides financial assistance to almost all projects. All types of applicants who have borrowing power and the power to take up new and renewable energy and energy efficiency projects are eligible for financial assistance from IREDA.

5.3.5   National Institute of Wind Energy

National Institute of Wind Energy (NIWE), formerly known as C-WET, is a national institution for wind energy research and development activities established in Chennai by MNRE. It has the responsibility to provide scientific and technical support for wind energy sector. It is involved in the development of a reliable and cost-effective technology, preparation of wind energy map, wind atlas, and reference wind data, certification/approval, and preparation of DPRs (detailed project reports).

5.3.6   Land Acquisition and Environmental Protection

There are several key provisions made in the land acquisition and environmental protection of wind farms. Thus, it should not be a matter of loss for the farmers lending their land and wind farms might not be destroying the environment of that particular place. The regulations for land acquisition and environmental protection of wind farms are as follows:

- Tips of wind turbine blades shall be painted orange colour to avoid bird hits.
- A distance of 300 m shall be maintained from highways and villages.
- Wind turbines of 1 MW capacity should be promoted for the optimum use of forest land.
- Lease period shall initially be for 30 years.
- Land area of circle of 100 m diameter shall be provided for installation of wind mast for every 500 hectares.
- From 65% to 70% leased area should be utilized for developing medicinal plant gardens and so on.

5.4   STATE GOVERNMENT POLICIES FOR WIND ENERGY SECTOR

5.4.1   Andhra Pradesh

Andhra Pradesh is a coastal state, and according to MNRE, Andhra Pradesh has a potential of 5394 MW of wind energy. The MNRE data indicates that Andhra Pradesh installed 912.5 MW of wind power until 31 December 2014. To encourage the optimum utilization of available wind power potential and installation of wind power projects in the state, the state government announced its policy for "Development of Wind Power in Andhra Pradesh" on 11 April 2008. Figure 5.4 shows the year-wise installation of wind turbines in Andhra Pradesh. We can see from the figure that the majority of wind turbines have been installed in the past 5 years. This implies that the installation of wind turbines was boosted after the announcement of policy for wind energy sector. The key features of the policy are as follows:

- The minimum capacity of WEGs shall not be less than 225 kW.
- WEGs should be approved by C-WET (NIWE), Chennai.
- Cost of electricity purchased from the wind power projects will be ₹3.50 per unit for the initial 10 years from the commercial operation date (COD). Tariff from 11th year onwards will be fixed by the Andhra Pradesh Electricity Regulatory Commission (APERC).
- District collectors are authorized for advance possession of government land for execution of wind projects. Developers can be allotted land for setting up wind power projects of 200 MW

capacity. Initially, land will be allotted for 100 MW capacity, and on its successful completion, land for another 100 MW project will be allotted.

- Developers are free to use the produced power for their captive consumption or they can sell the power to a third party or distribution companies (DISCOMs). A Power Purchase Agreement (PPA) will have to be signed between DISCOMs and the developer for a period of 20 years.

- Developers will have to prepare a power evacuation facility for interconnecting the wind farm with grid for delivery of power. The developer will have to bear the cost for erection of transmission line.

- Developers will be eligible for captive use or sale of power to a third party, but it will be governed by open access regulations of the Andhra Pradesh Government.

- Banking of energy will not be allowed.

- A 5% of concessional wheeling and transmission charge will be given by the state Government.

- Developers will be eligible for carbon credits under clean development mechanism (CDM). They will be allowed to retain 90% of CDM benefits and may pass on 10% CDM benefits to the distribution company.

5.4.2   Gujarat

Gujarat is the only state with abundance renewable power potential. According to the MNRE, Gujarat has a potential of 10,609 MW wind

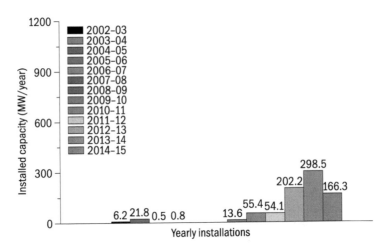

Fig. 5.4 *Year-wise installation of wind turbines in Andhra Pradesh*
Source *MNRE (2014)*

energy. The MNRE data indicates that the installed capacity of wind power in Gujarat was 3581.3 MW before 31 December 2014. To encourage the optimum utilization of available wind power potential and installation of wind power projects in the state, the government announced its wind power policy for the development of wind power projects in the state on 13 June 2007, which was titled "Wind Power Policy – 2007". Figure 5.5 shows the year-wise installation capacity of wind turbines in Gujarat. It can be seen from the figure that the wind turbine installed capacity in the state dramatically increased many times after the announcement of wind energy policy in 2007. The key points of the policy are as follows:

- The wind power projects installed and commissioned under the policy will be eligible to receive incentives for 20 years from the date of commissioning or lifespan of project.
- Wheeling charges at 66 kV and above will be at normal rates. However, wheeling at the voltage below 66 kV will be 10% of energy fed to the grid.
- For a third party sale, the developer will be allowed to charge electricity duty from the purchaser at applicable rates.
- The developer will be allowed to sell the electricity produced by wind turbine to state distribution companies at ₹3.65 per unit for 25 years.

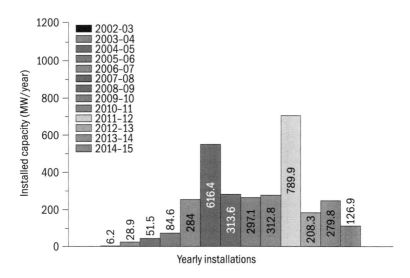

Fig. 5.5 *Year-wise installation of wind turbines in Gujarat*
Source *MNRE (2014)*

- Allotment of land will be done by the co-ordination committee of the government. The wind power projects can be set up on private land or revenue land/Gujarat Energy Development Agency (GEDA) land.
- The wind turbine generators (WTGs) should be new and approved by the MNRE/C-WET.
- The interconnection point for metering of electricity should be the delivery point at the relevant sub-station.
- Gujarat Government has made an RPO for distribution companies which states that a minimum of 5% of the total electricity consumption should be procured from wind-based projects. In the case of non-fulfilment of RPO by DISCOMs, they have to pay a penalty decided by the SERC.
- The power evacuation infrastructure to grid will be built by the developer at his cost up to 100 km. However, after 100 km, the Gujarat Energy Transmission Company will develop the transmission facility on its own cost. The voltage level should be 66 kV and above.
- The developer will deposit security money of ₹5 lakh/MW and will have to commission the project within a specified period of time or else the security money will be confiscated. The commissioning period is 1 year from the date of allotment of projects up to 100 MW capacity, 1.5 years for 101–200 MW capacity, 2 years for projects up to 201–400 MW, and 3 years for projects up to 600 MW.

5.4.3 Karnataka

Karnataka is a state located on the western coast of India. According to the MNRE, Karnataka has a potential of 8591 MW of wind energy. The MNRE reports indicate that Karnataka had the installed capacity of 2548.7 MW of wind power until 31 December 2014. To encourage the optimum utilization of available wind power potential and installation of wind power projects in the state, the government announced its wind power policy for the development of wind power projects in the state in 2009.

Figure 5.6 shows the year-wise capacity addition over the past 13 years, and there has been a gradual increase in wind turbine installations during these years. It indicates that the policies and support of the state government have always helped in maintaining consistency in the capacity addition.

The key features of the policy are as follows:

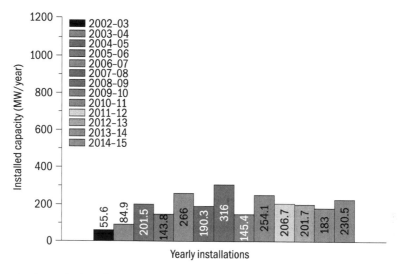

Fig. 5.6 *Year-wise installation of wind turbines in Karnataka*
Source *MNRE (2014)*

- Developers will be allotted specified capacity projects and they have to sell the generated electricity to the respective energy supply companies located in their geographical regions.
- The policy directs the government to establish a green energy fund for financing of renewable/wind energy projects.
- Unused land of government bodies will be made available to developers and the developers will be assisted by the government to acquire private land.
- Government agencies will acquire the forest land from the forest ministry and develop the land for erecting renewable energy infrastructure, and then it will sub-lease the land to developers for 30 years. The lease period will be renewed for 5 years at a time subject to the fulfilment of specific conditions.
- The SERC will direct DISCOMs to purchase a minimum 7%–10% of electricity from non-solar renewable energy sources of its total purchase in a year.
- Developers will sell the electricity at the tariff provided by the state ERC till the 11th year. After the 11th year, they have to sell power at tariff based on variable costs.
- Developers will be compensated for 5% of wheeling losses.
- Developers are allowed for banking of power.
- A PPA will be signed between the developer and DISCOMs.

- The payment of power purchased by DISCOMs must be made on a monthly basis, or else they will have to pay the interest after the month.
- Developers will be eligible for entry tax and other incentives available for renewable energy sector.
- Developers will be eligible for CDM benefits.
- All unskilled/skilled workers and other non-executives required for execution and maintenance will be local people.

5.4.4   Madhya Pradesh

According to the MNRE, Madhya Pradesh has wind energy potential of 920 MW. The MNRE data shows that Madhya Pradesh installed 567.3 MW of wind power capacity till 31 December 2014. To promote the optimum utilization of available wind power potential and installation of wind power projects in the state, the government announced its wind power policy for development of wind power projects in the state in 2012 and amended the same in 2013. Figure 5.7 shows the magnitude of capacity addition in the recent years. As discussed previously, Madhya Pradesh had already installed 62% of its total capacity before it enacted the wind energy policy in 2012, and this has contributed to its growth of capacity addition slowly. The key features of the policy are as follows:

- The projects can sell their electricity to DISCOMs or any other high tension (HT) line consumer for captive use within the state.

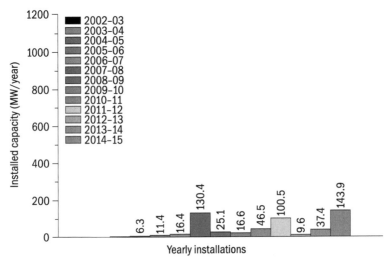

Fig. 5.7 *Year-wise installation of wind turbines in Madhya Pradesh*
Source *MNRE (2014)*

- Permission will be cancelled if the project execution is not commenced within 3 months from the date of sanctioning and if the project is not completed within 15 months.
- All the WEGs should be approved by the C-WET (NIWE).
- The preferred rating of WEGs should be 500 kW.
- Green energy fund will be created for financing.
- The rate of purchase of electricity and the quantity of electricity purchased by DISCOMs will be decided by SERC.
- Government will provide 4% of wheeling subsidy.
- All the expenses for erection of power plant will be borne by the developer.
- Third party purchasers will be allowed the facility to reduce the contract demand.
- Government land will be provided for 30 years or for the project life, whichever is less, on a token premium of ₹1 per annum.
- A non-refundable amount of ₹50,000 per MW will be paid along with the application. An extension of the date of completion may be provided for 3 months for justified reasons on the payment of a fine of ₹1 lakh/MW.
- Projects up to 5 MW shall be sanctioned by Madhya Pradesh Vidyut Utpadan Nigam (MPVUN) and those above 5 MW capacity by the government.
- Energy banking is permitted for a period of a financial year. Surplus energy at the end of the banking period will be purchased by DISCOMs and 2% of banked energy shall be payable as banking charges.

5.4.5 Maharashtra

According to the MNRE, Maharashtra has a potential of 5439 MW of wind energy. The MNRE data indicates that Maharashtra had the installed capacity of 4369.8 MW of wind power till 31 December 2014. To encourage the optimum utilization of available wind power potential and installation of wind power projects in the state, the government announced its wind power policy for development of wind power projects in the state in 2014. Figure 5.8 shows that the installation of wind turbines in the state maintained a good pace from 2005 onwards, but it showed a declining trend in the magnitude of capacity addition until the announcement of policy for wind energy in 2014. If the success of this policy is maintained, the state can add more capacity in the coming years. The important features of the policy are follows:

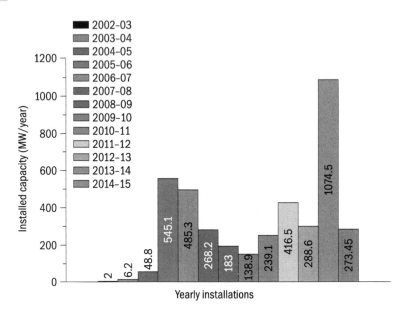

Fig. 5.8 *Year-wise installation of wind turbines in Maharashtra*
Source *MNRE (2014)*

- The policy makes it obligatory to sell 50% of electricity to DISCOMs under a long-term PPA, and the remaining 50% of electricity may be sold within the state.
- Power evacuation arrangements shall be made by the developer his own expense, and 50% of the expenditure will be reimbursed from the green energy fund.
- The approach roads will be constructed by the Maharashtra government.

- 11% of the total project cost will be paid from green energy fund for projects set up by the co-operative institutions.
- All the taxes paid by the developer will be reimbursed by the state Government from green energy fund.
- Government land reserved for industrial use and government barrier land will be provided on a rental basis for 30 years on a lease agreement.

5.4.6 Rajasthan

According to the MNRE, Rajasthan has a potential of 5005 MW of wind energy. The MNRE data indicates that Rajasthan had the installed capacity of 3052.7 MW of wind power until 31 December 2014. To promote the optimum utilization of available wind power potential and installation of wind power projects in the state, the government announced its wind power policy for development of wind power projects in the state in 2003. Rajasthan is the state which announced its policy for wind energy at the very beginning to tap its potential for wind energy. Figure 5.9 shows the year-wise capacity addition for Rajasthan, where the installation of wind turbines started right after its policy announcement and it has been able to speed-up the capacity addition after 2008. In the past 7 years from 2008, there has been 82% increase in its total capacity. The major features of the policy are follows:

- The electricity produced can be sold to DISCOMs or third party purchasers, or it can be reserved for captive use.
- The power should be transmitted to grid at 33 kV or above.
- A fee of ₹200 per MW will be paid to DISCOMs for creation of a proper facility to receive power.
- Transmission line from the project site to the grid sub-station will be constructed by the developer at his own cost.
- The transmission, wheeling, and other charges will be specified by the SERC.
- Banking of energy is permitted for 1 year.
- Wheeling and banking agreement will be assigned on one-time basis in each case on payment of ₹1 lakh per application.
- PPA should be signed for initial 20 years.

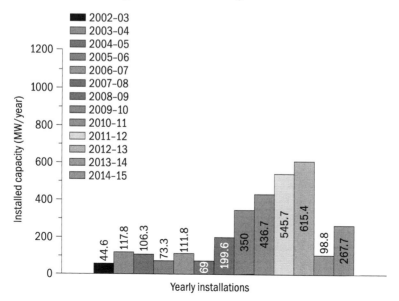

Fig. 5.9 *Year-wise installation of wind turbines in Rajasthan*
Source *MNRE (2014)*

- PPAs are allowed to be assigned to others on one-time basis on payment of ₹1 lakh per application to SERC.
- Energy sold to the third party will be exempted from the payment of electricity duty at 50% for a period of 7 years from COD.
- Registration of project can be done with a processing fee of ₹25,000/MW.
- A refundable security deposit of ₹5 lakh/MW will be forfeited if the project is not completed within the specified time period, which is from 6 months for 25 MW projects to 24 months for 100 MW projects.
- For the extension of time period, a fine will be charged, which is from ₹50,000/MW for 1 month to ₹5 lakhs/MW for a period of 3 months.

5.4.7   Tamil Nadu

Tamil Nadu is one of the states which has very high wind energy potential and has been able to add the wind capacity at the faster rate compared to other states in India. The MNRE data indicates that Tamil Nadu had the installed capacity of 7394 MW of wind power until 31 December 2014. As shown in Figure 5.10, the capacity addition in the state has always been very good from the start and it has attained a steady state in the past 3 years as the state has nearly achieved the threshold limit

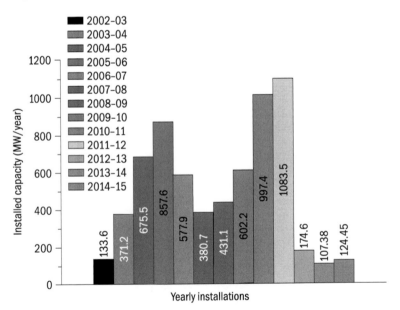

Fig. 5.10 *Year-wise installation of wind turbines in Tamil Nadu*
Source *MNRE (2014)*

of installation compared to its wind energy potential. To encourage the optimum utilization of available wind power potential and installation of wind power projects in the state, the government has announced its wind power policy for development of wind power projects in the state. The key features of the policy are as follows:

- Tamil Nadu Electricity Board (TNEB) accords sanction and grid connectivity approval for wind power projects to be set up in the state.
- A registration fee of ₹1000 per application is charged.
- A consultation charge of ₹10,000 is taken per application.
- An infrastructure development charge of ₹30 lakh/MW is levied.
- For operation and maintenance with 5% escalation per year for the life of project, ₹1.60 lakh/MW/year is charged, which is to be paid on a monthly basis.
- Sale of wind energy to third party is allowed.

5.5   CONCLUSION

In the past 5 years, India has increased 47% of its total installed capacity of wind energy generation capacity. Therefore, we can conclude that this period has witnessed the growth in installation of wind turbines, which has been very impressive compared to the overall installations in the past 13 years. The Indian Government through MNRE has announced a number of policies for the promotion of wind energy sector. All the states with abundance of wind power potential have also announced their policies for wind energy sector keeping pace with the GoI to promote the installation of wind turbines and the production of electricity from wind power plants. The state governments have also improved and amended their policies after consultation with the industries involved in the installation of wind turbines. This reflects the flexible approach of state governments to promote the wind energy sector. As already discussed, with an increase in the tower height, the potential of wind energy is doubled. The GoI has decided to increase its target of wind turbine installation in the next 7 years to enhance the share of electricity generated from renewable sources in order to match the increasing demand of electricity in the country. Now, the government has started the promotion of manufacturing of wind turbines in the country itself with its ambitious 'Make in India' initiative, which will see a significant increase in the domestic manufacturing sector for wind turbines. In the coming years, we can see a bigger wind energy market developing in the country, which will enable us to increase the capacity of wind turbine installations, create better opportunities of employment, and lead to bigger share of electricity produced from wind energy.

REFERENCES

CEA (Central Electricity Authority). 2013. *Growth of Electricity Sector in India from 1947–2013*. New Delhi: CEA

CSO (Central Statistics Office). 2014. *Energy Statistics*, Vol. 21. New Delhi: CSO

Government of Andhra Pradesh. 2008. Development of Wind Power in Andhra Pradesh, Government of Andhra Pradesh

Government of Gujarat. 2007. Wind Power Policy 2007, Government of Gujarat

GoI (Government of India). 2008. *National Action Plan on Climate Change*. New Delhi: Prime Minister's Council on Climate Change, GoI

Government of Karnataka. 2009. Karnataka Renewable Energy Policy 2009–14, Government of Karnataka

Government of Madhya Pradesh. 2013. Wind Power Project Policy 2012, Government of Madhya Pradesh

Government of Maharashtra. 2014. New Policy for Wind Power Projects, Government of Maharashtra

Government of Rajasthan. 2003. Policy for Promotion of Electricity Generation from Wind, Energy Department, Government of Rajasthan

IEA (International Energy Agency). 2012. *World Energy Outlook 2012*. Paris: OECD/IEA

IREDA (Indian Renewable Energy Development Agency). 2011. *Assessment of Investment Climate for Wind Power Development in India*. New Delhi: IREDA

IREDA (Indian Renewable Energy Development Agency). 2015a. *Direct Tax Benefits, Wind Energy Sector*. New Delhi: IREDA

IREDA (Indian Renewable Energy Development Agency). 2015b. *Extension Scheme for Generation Based Incentive for Grid Connected Wind Power Projects*. New Delhi: IREDA

IREDA (Indian Renewable Energy Development Agency Ltd). 2015c. Details available at http://www.ireda.gov.in/

Khare, V., S. Nema, and P. Baredar. 2013. Status of solar wind renewable energy in India. *Renewable and Sustainable Energy Reviews* 27: 1–10

Ministry of Power. 2015. *Electricity Act, 2003*. New Delhi: Government of India

MNRE. 2015. State-wise and year-wise installed wind power capacity. Details available at http://mnre.gov.in/file-manager/UserFiles/wp_installed.htm

National Institute for Wind Energy. 2015. Details available at http://niwe.res.in/

Tamil Nadu Electricity Regulatory Commission. 2014. Comprehensive Tariff Order on Wind Energy, Government of Tamil Nadu

Timilsina, G.R., G.C. Cooten, and P. A. Narbel. 2013. Global wind power development: Economics and policies. *Energy Policy* 61: 642–652

6

Improved Technology for Non-edible Seed Oils: Sources for Alternate Fuels

Rama Chandra Pradhan[a],* and Ashwani Kumar[b]

[a]*Department of Food Process Engineering, National Institute of Technology (NIT), Rourkela, Odisha 769008,*

[b]*Department of Botany, Dr Harisingh Gour University (Central University), Sagar, Madhya Pradesh 470003*

*E-mail: pradhanrc@nitrkl.ac.in

6.1 INTRODUCTION

The need for renewable, green, and non-polluting alternative sources of energy has assumed top priority throughout the world. At present, worldwide India's rank in the total energy consumption is sixth. Diesel is a very important fuel for the Indian economy because it is used in a wide range of applications such as trucks, road vehicles (bus), railways, agriculture, and power generation. The country's demand for diesel for transportation and mechanized agriculture sector has increased to 66.90 million tonnes in 2011/12, from 42 million tonnes in 2002/03. There is a need to substitute diesel with biomass-based products and vegetable oils. In the Indian context, there is a scope to produce vegetable oils by cultivating non-edible oilseeds on wastelands (Srivastava and Prasad 2000).

Due to the price of fossil fuels escalating and their increasing negative impact on the environment, some developed countries, particularly Germany and Australia, have started using vegetable oil as an alternative source of energy. Many developed countries have invested a large amount of funds in the research and development of generating biodiesel on a commercial scale. In India too, various organizations such as government agencies, research institutions, and automobile industry are taking up suitable steps to promote biodiesel as an alternate fuel (Nagar 2003). In the country, research is being undertaken to standardize the technique of esterification to convert oil into biodiesel. This has now encouraged scientists and farmers to grow oilseed-yielding species as an economically viable activity, particularly to develop marginal and large wastelands that are lying unutilized.

In India, emphasis is being laid on non-edible oilseeds (NEOs) for producing biodiesel. The country has a vast potential to grow NEO plants, such as karanja (*Pogamia pinnata* L.), simarouba (*Simarouba glauca* L.), and jatropha (*Jatropha curcas* L.). These NEOs are harvested, collected, processed (traditionally), and crushed for producing various kinds of oils. In the rural areas, oilseeds play an important role in generating employment. About 14 million farmers are engaged in the production of various oilseeds and 0.5 million in the processing sector. NEOs also have a huge employment potential for collection and post-harvest processing, which can be used for producing biodiesel.

Jatropha curcas L. belongs to family Euphorbiaceae and is found in many countries of South America and Africa (Bringi 1987). It is also found in India. Depending on the growing conditions, such as water and nutrient availability and the absence of pests and diseases, jatropha yield can vary from 4 tonnes to 8 tonnes of seeds per hectare. In India, jatropha bears fruits between September and December. Unlike most other species, which fruit in the monsoon and are accompanied with the difficulties of collection, drying, and storage, jatropha offers a natural advantage of the fruits being collected during the dry season (Bringi 1987). When the fruit dries, the hull or shell becomes hard and black. The dry fruits remain on the branches and contain two to three seeds. The fruits are either hand-picked or harvested by hitting the fruits with a long stick. At times, older trees are harvested by shaking the tree/branches. The collected fruits are then sun-dried for processing and decorticated manually to get the seeds (Lakshmikanthan 1978).

Karanja (*Pongamia pinnata* L.) belongs to the family Leguminosae and is found almost throughout India. The species is considered to be a native of Western Ghats and is chiefly found along the banks of streams and rivers or near sea coast and in tidal forests (Bringi 1987). The pods, collected from April to June, are dried in sun and the kernels are extracted by thrashing the fruits. The kernels comprise 95.5% of the dry fruits, while the shells comprise 4.5%. When the kernels are extracted in a *ghani*, the yield is 18% to 22% and in expeller, the yield is 24% to 27.5%. The yield of kernels per tree is reportedly between 8 kg and 24 kg and the oil content varies from 30% to 40% (Lakshmikanthan 1978; Bringi 1987).

Commonly known as paradize-tree or bitter wood, simarouba (*Simarouba glauca* L.) is an evergreen tree. It is medium size with a height of 7–15 m. It grows up to 1000 m above the sea level in all types of well-drained soil (pH 5.5–8.0) and has been found to be established in places with 250 mm to 2500 mm annual rainfall and temperatures going up to 45°C (Joshi and Hiremath 2000). In a hectare of land, approximately 200 trees can be planted. The simarouba fruits are similar to olives in their size, shape, and colour. The tree starts to yield fruit after four years but comes to full production at 6 years of age. The tree starts flowering during December and bear fruits in January and February. On an average, 6000 kg to 8000 kg of simarouba fruit per hectare can be obtained after 10 years of plantation (Joshi and Hiremath 2001).

The NEOs are cultivated and harvested to produce oil, which can be used in several technical applications, such as fuel, lubricants, or raw material for the production of chemical products. For this, it is necessary to separate the oil from the oil-bearing material. Post-harvest technology, out of many inputs, plays an important role in increasing the oil production by reducing the post-production losses and increasing the oil recovery from seed (Sahay and Singh 1996). In oil production, the post-harvest processes involved are drying, decortication, and oil expression with a mechanical oil expeller. Presently, decortication of seed/fruit is done by the rural people with the help of bricks or wooden plates. This process of decortication is a time-consuming and labour intensive process with low capacity. It was observed that the decorticated seeds yield better oil as compared to un-decorticated seeds. It was also observed that the cooking time, and moisture content and temperature of the seed at the time of oil expression have significant effects on oil yield of non-edible oilseeds.

The aim of this study was to exploit these as a potential source of much-needed non-edible oil in India. The separation of the seed-coat from the seed (decortication/shelling) is a requirement to obtain higher percentage of oil from seed, so that the oil can be processed into biodiesel. One of the major constraints in efficient utilization of NEOs is the non-availability of proper, improved, efficient processing technology and equipment such as decorticator and efficient oil expeller. For full commercial exploitation of NEOs, it becomes necessary to develop technologies for mechanical deshelling of fruits/seeds and extraction of oil from seeds. Very little is known about the physical and mechanical properties of these non-edible fruit/seeds and about the various processing conditions for deshelling and oil expression. Such information is necessary for the development of decorticator and oil expression technology.

6.2 MATERIALS AND METHODS

6.2.1 Sample

Non-edible seeds/fruits were collected from different Indian states for this study. Cleaning, sorting, and grading of seeds/fruits were done manually. All foreign materials such as dust, dirt, small branches, and immature fruits were removed. Then the seeds/fruits were sun-dried and stored in jute bags. They were kept under ambient room conditions (25°C–35°C, 75%–80% RH) to the equilibrium moisture. To obtain seeds/kernels, the fruits were manually decorticated and the seeds/kernels were stored for further analysis.

6.2.2 Moisture Content

The initial moisture content of the sample (fruit/seed) was determined by using the standard hot air oven method at 105°C ± 1°C for 24 h (Brusewitz 1975) and is expressed as percentage on a dry weight basis.

6.2.3 Oil Content

The oil content of the samples was determined by hexane extraction using a laboratory soxhlet apparatus (AOAC 1984) and the value reported are average of the three experiments.

6.2.4 Physical Properties of Non-edible Fruit and Seed

The bulk sample materials were divided into five groups. From each group, 20 samples were randomly selected to gather hundred samples each for the experiment. Therefore, measurements of all size and shape properties as well as the fruit and seed weight are replicated one hundred times. The fruit/seed size, in terms of major, intermediate and minor dimensions, that is length (L), width (W), and thickness (T) were measured using a digital vernier caliper (Mitutoyo, Japan). The fruit is composed of two parts: (i) the outer most part (shell fraction) and (ii) the inner portion (kernel fraction). These measurements of the two fractions were replicated 20 times to get the mean value. All the physical properties, namely, principal dimensions, sphericity, aspect ratio, one thousand unit weight (test weight), surface area, densities, porosity, angle of repose, and static coefficient of friction were determined as per the standard methods (Mohsenin 1980; Sacilik, Ozturk, Keskin 2003; and Singh, Mishra, Supradip 2010).

6.2.5 Decortication of Non-edible Oilseeds

A jatropha decorticator was designed and developed. To develop a process for decorticating the jatropha fruits and to optimize the factors affecting decortication, a jatropha fruit decorticator (40 kg/h) was conceived and tested.

6.2.6 Mechanical Expression of Jatropha Seed

There are several types of oil extraction units, namely, soxhlet extraction unit, screw press mechanical expeller, ghani, among others available for extracting oil from edible/non-edible oil seeds in different parts of the country. Experimental studies were initially conducted to evaluate the extraction efficiency of these units for jatropha seed. Figure 6.1 shows the different types of oil extraction units.

The optimization study for jatropha seed using the screw press oil expeller was also conducted for comparison. Studies were conducted to evaluate the effects of hydrothermal conditioning of the jatropha seed on oil recovery. A laboratory oven was used for conditioning the seeds at different hydrothermal treatments.

6.2.7 Seed Oil Characteristics

At optimum processing parameter, jatropha seed's oil was extracted and chemical properties of the oil was evaluated as per methods described by the Association of Official Analytical Chemists (AOAC 1984). Values are average of three replications.

6.3 RESULTS AND DISCUSSION

6.3.1 Moisture Content and Oil Content

The mean value of the moisture content and oil recovery of jatropha, karanja, and simarouba is given in Table 6.1. It is seen that the karanja fruit and kernel have maximum moisture content in comparison to the jatropha and

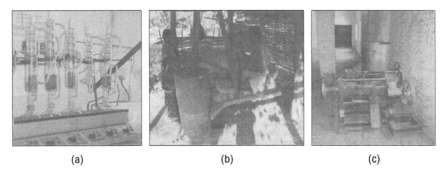

(a)	(b)	(c)

Fig. 6.1 *Different types of oil extraction units: (a) Soxhlet unit, (b) traditional ghani, and (c) screw press expeller*

Table 6.1   Average oil and moisture content of jatropha, karanja, and simarouba

Parameter	Jatropha			Karanja		Simarouba	
	Fruit	Seed	Kernel	Fruit	Kernel	Fruit	Kernel
Oil content (%)	20.12	38.32	45.03	18.45	34.67	15.28	61.04
	(2.11)	(4.61)	(7.86)	(1.79)	(3.51)	(2.35)	(5.67)
Moisture content (%, w.b.)	11.77	7.21	6.35	14.15	12.73	6.21	8.51
	(1.43)	(0.33)	(0.04)	(0.44)	(1.62)	(0.23)	(0.52)

Note *Figures in parenthesis are standard deviation.*

simarouba fruits and kernels. The moisture content (w.b.) of the jatropha fruit, seed, and kernel are 11.77%, 7.21%, and 6.35%, respectively. In the case of karanja, the moisture content of the fruit is 10.03% higher than that of its kernel whereas the moisture content of the simarouba fruit is 37.03% less than that of its kernel. From Table 6.1, it is seen that the kernel has higher oil content than the fruit. Hence, it is necessary to separate the shell/husk from the kernel before the oil separation process. The simarouba kernel has more oil content in comparison to the jatropha and karanja kernels.

6.3.2   Physical Properties of Non-edible Oilseeds

6.3.2.1   *Jatropha fruit, seed, and kernel*

Table 6.2 lists the physical properties of jatropha fruit, seed, and kernel. 1000 unit weight, seed and kernel fraction, arithmetic mean diameter, and geometric mean diameter are provided along with other physical parameters. The samples' shapes are described in terms of their sphericity and aspect ratio. The sphericity values indicate that the fruit's shape is close to a sphere while the seed and the kernels resemble an ellipsoid. The bulk density of the fruits was the lowest while that of the kernels was the highest. This indicates that the fruits need more space per unit mass than the nuts and kernels. The solid density of the fruit, nut, and kernel is less than the density of water (1000 kg/m^3). The porosity of the bulk of fruits was the highest and that of the kernels was the lowest. The fruit's surface area was larger than those of the seed and kernel by 73.45% and 87.90%, respectively.

6.3.2.2   *Karanja fruit and kernel*

Table 6.3 presents the average values of the three principal dimensions of the karanja fruit and kernel, namely length, width, and thickness as determined in this study. The average diameter calculated by the arithmetic mean and geometric mean are also presented and provided along with other physical parameters. Dutta, Nema, and Bharadwaj (1988) considered the grain as spherical when the sphericity value was more than 0.80. In this study, the karanja fruit and kernel are not treated as an equivalent sphere for calculation of the surface area. The surface area of the kernel is less than that of the fruits. The bulk densities of the fruit and kernel are 425.04 kg/m^3 and 651 kg/m^3 respectively, and the true densities are 601.35 kg/m^3 and 995.0 kg/m^3, respectively. The corresponding porosity of the fruit and kernel are 29.32% and 34.57%, respectively.

Table 6.2   Physical properties of jatropha fruit, seed, and kernel

Physical properties	n	Fruit	Seed	Kernel
Length (mm)	100	29.37 ± 0.87	18.83 ± 0.51	14.03 ± 0.74
Width (mm)	100	22.22 ± 0.62	11.47 ± 0.43	7.17 ± 0.35
Thickness (mm)	100	21.61 ± 0.65	9.01 ± 0.32	5.90 ± 0.12
1000 unit mass (g)	20	2280.35 ± 13.26	761.50 ± 3.25	476.17 ± 2.54
Seed fraction (%)	20	71.68 ± 7.35	100	NA
Kernel fraction (%)	20	44.73 ± 5.36	63.02 ± 5.78	100
Shell/husk fraction (%)	20	28.32 ± 7.35	37.13 ± 4.11	0
Arithmetic mean diameter (mm)	100	24.39 ± 0.51	13.10 ± 0.31	9.03 ± 0.67
Geometric mean diameter (mm)	100	24.15 ± 0.25	12.45 ± 0.30	8.41 ± 1.25
Sphericity (decimal)	100	0.82 ± 0.02	0.66 ± 0.01	0.59 ± 0.06
Surface area (mm^2)	100	1834.40 ± 77.73	486.94 ± 15.67	221.91 ± 12.63
Aspect ratio (%)	100	75.65 ± 8.23	60.91 ± 7.11	51.10 ± 5.89
Bulk density (kg/m^3)	20	275 ± 1.09	476 ± 1.97	588.26 ± 3.84
True density (kg/m^3)	20	520 ± 6.47	711 ± 7.97	865.87 ± 9.23
Porosity (%)	20	47.11 ± 0.11	33.05 ± 0.11	32.06 ± 2.67

NA - Not available
Note n is the number of samples. Data are mean values ± standard deviation.

Table 6.3   Physical properties of karanja fruit and kernel

Physical properties	n	Fruit	Kernel
Length (mm)	100	48.85 ± 4.54	25.68 ± 2.65
Width (mm)	100	19.55 ± 2.77	15.63 ± 1.86
Thickness (mm)	100	11.48 ± 1.46	8.16 ± 0.87

Contd...

Table 6.3 *contd...*

Physical properties	n	Fruit	Kernel
1000 unit mass (g)	20	3220 ± 19.18	1329.58 ± 13.24
Kernel fraction (%)	20	47.39 ± 6.67	100
Shell fraction (%)	20	49.61 ± 3.21	0
Arithmetic mean diameter (mm)	100	26.63 ± 1.74	16.49 ± 0.65
Geometric mean diameter (mm)	100	22.21 ± 1.04	14.81 ± 0.58
Sphericity (decimal)	100	0.45 ± 0.08	0.58 ± 0.05
Surface area (mm2)	100	1550.63 ± 23.32	689.64 ± 19.66
Aspect ratio (%)	100	40.02 ± 2.71	60.86 ± 4.11
Bulk density (kg/m³)	20	425.04 ± 4.33	651 ± 3.67
True density (kg/m³)	20	601.35 ± 7.81	995 ± 4.51
Porosity (%)	20	29.32 ± 0.89	34.57 ± 1.21

Note *n is the number of samples. Data are mean values ± standard deviation.*

6.3.2.3 Simarouba fruit and kernel

The mean values of the physical properties of simarouba fruit and kernel is summarized in Table 6.4. The 1000 sample weight, fraction of kernel and fruit parts, arithmetic mean diameter, and geometric mean diameter are evaluated along with other physical properties. The length, width, and thickness of the fruit are found to be 21.26 ± 2.01 mm, 13.81 ± 0.98 mm, and 11.03 ± 0.64 mm, respectively. The shape of the fruit is described in terms of its aspect ratio and sphericity. The mean value of the sphericity of fruit and kernel are observed to be 0.69 and 0.65, respectively. The surface area of the kernel is less than that of the fruit by 63.36%. The surface area of the kernel is less than that of the fruit which indicated that mass or rate of energy transfer through the surface of the fruit might be quicker than the rate for fruit. The bulk densities of the fruit and kernel are 622.27 kg/m^3 and 727.73 kg/m^3, respectively. This indicates that the bulk density of the kernel is 14.49% higher than that of the fruit. The simarouba fruit contains air pores between the shell and the kernel; therefore, the true density of the kernel is higher than the density of water. Hence, separation of fruit shells from kernels after decortication could be done by blowing air (winnowing) or floating in water. The porosity of the simarouba fruit and kernel are found to be 33.23% and 28.61%, respectively. The magnitude of variation in porosity depends on the value of bulk and true densities. The porosity of the bulk of the fruits is higher than that of the kernels. This indicates that the aeration of the bulk of fruit is easier than that of the bulk of kernel.

Table 6.4 Physical properties of simarouba fruit and kernel

Physical properties	n	Fruit	Kernel
Length (mm)	100	21.26 ± 2.01	13.78 ± 1.08
Width (mm)	100	13.81 ± 0.98	7.77 ± 0.86
Thickness (mm)	100	11.03 ± 0.64	6.71 ± 0.37
1000 unit mass (g)	20	1120.16 ± 52.34	330.26 ± 29.35
Kernel fraction (%)	20	29.01 ± 0.34	100
Shell fraction (%)	20	71.12 ± 1.04	0
Arithmetic mean diameter (mm)	100	15.37 ± 0.87	9.42 ± 0.58
Geometric mean diameter (mm)	100	14.78 ± 0.82	8.95 ± 0.57
Sphericity (decimal)	100	0.69 ± 0.03	0.65 ± 0.03
Surface area (mm²)	100	687.94 ± 37.68	252.08 ± 32.36
Aspect ratio (%)	100	64.95 ± 6.51	56.41 ± 5.54
Bulk density (kg/m³)	20	622.27 ± 15.64	727.73 ± 15.54
True density (kg/m³)	20	931.96 ± 33.08	1019.3 ± 19.65
Porosity (%)	20	33.23 ± 2.03	28.61 ± 2.861

Note n is the number of samples. Data are mean values ± standard deviation.

6.3.3 Decortication of Non-edible Oilseeds

A decorticator for jatropha fruit was fabricated (Figure 6.2). The equipment was developed based on engineering principle and importance was given on easy-to-adjust, easy-to-dismantle, low cost, and easy-to-run device for separating the jatropha fruit's shell from its seeds. The machine is

Fig. 6.2 *Developed jatropha fruit decorticator*

composed of a stand (frame), fruit hopper, decorticating chamber for fruits, a concave sieve, three rotating blades, outlet to discharge, and a separator which vibrates with sieve to separate the seed and shell.

6.3.3.1 Performance of the developed jatropha fruit decorticator

Four levels of moisture contents of jatropha fruit were taken for the performance evaluation of the developed machine. According to varieties, place, season, conditioning to fruit, and so on, the moisture content of fruit may vary at the time of post-harvest operations. Hence, four different levels of moisture content were chosen for experiments. After the decortication, various fruit parts were carefully collected at outlet points and grouped into various category, namely, fruits fully shelled (whole clean seed, N_1), broken seed and dust (N_2), fruits partially shelled (N_3), and unshelled fruit (N_4).

The effect of various moisture levels and concave clearance on different performance parameter for the developed decorticator is summarized in Table 6.5. Percentages of whole seed, broken seed and dust, partially shelled fruits, and unshelled fruits were determined at different moisture content and concave clearance. It was observed that at fruit initial moisture level of 7.97% (d.b.) and concave clearance of 21 mm, the percentages of whole seed is maximum (67.94% ± 2.48%) whereas the corresponding value is minimum (52.18% ± 3.09%), at a moisture content of 15.65% (d.b.) with a concave clearance of 27 mm. This was due to the fact that at low moisture content the fruit is more dry and becomes brittle and, therefore, susceptible to mechanical damage. At a moisture content of 15.65% (d.b.) and concave clearance of 27 mm, the percentage of broken seed and dust was lowest (0.02%) whereas this value was highest (15.09% ± 0.87%) at a fruit moisture content of 7.97% (d.b.) with a 18 mm concave clearance. The reason may be due to the fact that the fruit was dried at its initial moisture content as compared to other moisture level. Also, because of low concave clearance, that is, very less gap between sieve and blades with respect to their axial dimensions, the fruits get compressed and thereby the percentage of broken seed and dust is more. At a fruit moisture level of 7.97% (d.b.) and a concave clearance of 18 mm, the percentage of partially shelled fruits was lowest (1.21% ± 0.23%) whereas it was highest (12.61% ± 2.11%) at a moisture content of 15.65% (d.b.) with a concave clearance of 27 mm. Similarly, at fruit moisture content of 7.97% (d.b.) with a concave clearance of 18 mm, the percentage of unshelled fruit was lowest (0.26% ± 0.08%) whereas it was highest (14.11 ± 2.32) at a moisture content of 7.97% (d.b.) with a concave clearance of 27 mm.

Table 6.5 Performance of jatropha fruit decorticator at different variables

Concave clearance (mm)	Percentage of whole seed (η_s)	Percentage of broken seed and dust (η_b)	Percentage of partially shelled fruit (η_p)	Percentage of unshelled fruit (η_u)
Moisture content: 7.97% (d.b.)				
18	57.79 ± 1.65	15.09 ± 0.87	1.21 ± 0.23	0.26 ± 0.08
21	67.94 ± 2.48	2.11 ± 1.56	4.06 ± 0.11	2.16 ± 0.04
24	59.78 ± 1.72	1.78 ± 0.77	4.11 ± 0.23	3.81 ± 0.07
27	56.78 ± 2.34	0.08 ± 1.02	9.99 ± 0.51	13.17 ± 0.02
Moisture content: 10.53% (d.b.)				
18	57.11 ± 2.31	14.89 ± 1.05	2.76 ± 0.32	0.89 ± 0.06
21	62.04 ± 1.45	2.07 ± 0.86	5.33 ± 0.09	2.38 ± 0.1
24	60.87 ± 1.78	1.62 ± 0.81	4.41 ± 0.22	4.55 ± 0.05
27	55.65 ± 2.01	0.05 ± 0.27	11.13 ± 1.01	13.72 ± 1.12
Moisture content: 13.09% (d.b.)				
18	56.01 ± 1.35	12.66 ± 1.68	5.31 ± 0.67	1.01 ± 0.02
21	61.93 ± 2.56	1.51 ± 0.21	8.87 ± 0.28	3.34 ± 0.06
24	58.57 ± 1.09	1.32 ± 0.05	9.13 ± 0.22	4.62 ± 0.06
27	53.56 ± 1.33	1.03 ± 0.06	12.46 ± 0.91	13.99 ± 0.03
Moisture content: 15.65% (d.b.)				
18	54.71 ± 3.01	9.12 ± 1.2	6.23 ± 1.0	1.86 ± 0.01
21	60.18 ± 2.12	1.22 ± 0.07	9.64 ± 0.06	3.37 ± 0.04
24	57.66 ± 1.6	1.10 ± 0.02	10.13 ± 0.2	4.73 ± 0.11
27	52.18 ± 3.09	0.02 ± 0.001	12.61 ± 2.11	14.11 ± 2.32

The decortication efficiency (η_{de}) of the machine is calculated by using the following equation:

$$\eta_{de} = 1 - \left[\frac{(N_3 + N_4)}{N_0} \right] \times 100$$

where, N_0 = amount of fruit fed into the hopper (kg).

The decortication efficiency of the developed machine is shown in Figure 6.3. From Figure 6.3 can be seen that when the moisture level increases from 7.97% to 15.65% d.b., the decortication efficiency decreases linearly from 98.53% to 73.28% (statistically significant at $P < 0.05$), at any concave clearance between concave sieve and blades. At a higher moisture content, the fruit is wet and, thereby, more elastic than dry fruit. Hence, separation is not easy during decortication process. Therefore, it results

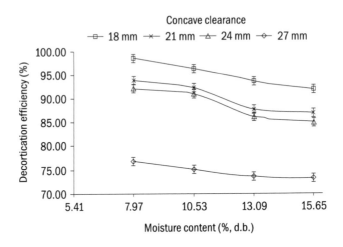

Fig. 6.3   *Effects of moisture content on decortication efficiency against different concave clearance*

in low efficiency. Figure 6.3 also concludes that at a moisture content of 7.97% d.b. and concave clearance of 18 mm, the developed equipment gives maximum decortication efficiency in comparison to other combinations. This is due to the fact that at a low moisture level, the fruit is dry and brittle and at a lower concave clearance, the machine developed more shearing force and, hence, more fruit detached.

The efficiency of the machine (η_{me}) was calculated by using the following equation:

$$\eta_{me} = 100 \times \left\{ 1 - \left(\frac{N_3 + N_4}{N_0} \right) \right\} \times \left(\frac{N_1}{N_1 + N_2} \right)$$

The effect of the different concave clearance and fruit moisture content on the machine efficiency is shown in Figure 6.4. It was observed that the machine efficiency decreases with an increase in the fruit moisture level (statistically significant at $P < 0.05$). The efficiency of the machine was highest (90.96% ± 0.74%) at fruit initial moisture content of 7.97% d.b. and a concave clearance of 21 mm, while at moisture content of 15.65% d.b. with a concave clearance of 27 mm, the machine efficiency was lowest (73.25% ± 0.64%) in comparison to other conditions. A similar result on efficiency of machine (85%–100%) has been reported by Adewumi and Fatusin (2006) for cocoa pod using an impact-type hand operated cocoa pod breaker. The efficiency of the machine decreases because at high moisture content, the seed coats or shells were sticky resulting in a high force of friction to

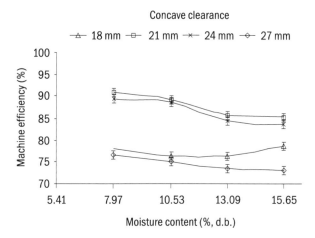

Fig. 6.4 *Effect of moisture content on efficiency of machine against different concave clearance*

separate the shell from the seeds. But at lower moisture content and at an optimum concave clearance, the fruits were less sticky and required less force to split and, therefore, able to separate much more easily.

6.3.4 Mechanical Expression of Jatropha Seed

At different operating parameters, the oil recovery from jatropha seed samples is summarized in Table 6.6. At the initial moisture level, the oil recovery from the cooked samples is higher than that of the uncooked samples. Because of cooking, the seed tissues become soft and viscosity of oil lowers. Due to softening of the tissues, the cellular structure weakens, making it highly susceptible to failure under pressure, whereas due to low viscosity, it enhances the flow rate of the oil (Indrasari, Koswara, Muchtadi, *et al.* 2001), and, hence, increases the oil recovery. Table 6.6 shows that at a moisture content of 9.69% (d.b.), the oil recovery was highest for cooked samples and at a moisture content of 12.16% (d.b.), the oil recovery was lower for the samples. It was also observed that oil recovery from samples conditioned to an initial moisture content of 12.16% (d.b.) were in general lower than the samples conditioned to other moisture contents used in this study (Table 6.6). Reduced oil recovery at higher temperatures could be due to oil degradations and a likely increase in the brittleness of the product. Sufficient heat treatment is needed for adequate coagulation of protein, breakdown of oil cells, and reduction in oil viscosity. Oil expression was highest (73.14%) at an optimum moisture content of 9.69% after cooking the sample at a temperature of 110°C and for 10 minutes (Table 6.6).

Table 6.6   Effects of initial moisture content, cooking temperature, and cooking time on percentage of oil recovery from jatropha seeds

Moisture (% d.b.)	Cooking temperature (°C)	Oil recovery (%) for corresponding cooking time (minutes)				Oil recovery (%) from uncooked sample
		5	10	15	20	
7.22	50	65.88 (3.67)	68.01 (4.56)	66.52 (3.71)	65.97 (2.67)	64.85 (1.92)
	70	65.92 (3.64)	68.11 (2.19)	66.76 (4.22)	66.33 (4.52)	
	90	66.03 (4.52)	68.23 (4.21)	67.09 (2.67)	66.48 (3.67)	
	110	66.42 (3.67)	69.15 (4.23)	67.78 (3.23)	66.76 (4.81)	
	130	65.13 (4.23)	67.27 (3.56)	66.88 (4.65)	66.11 (3.21)	
9.69	50	68.42 (1.23)	68.51 (2.16)	68.38 (3.45)	68.86 (4.12)	68.36 (1.23)
	70	69.76 (2.78)	69.98 (2.67)	68.51 (3.22)	69.44 (2.11)	
	90	70.98 (4.11)	71.22 (1.45)	69.46 (1.78)	69.83 (2.66)	
	110	71.69 (2.18)	73.14 (1.01)	71.81 (3.20)	70.66 (4.11)	
	130	69.02 (2.44)	68.89 (3.09)	68.55 (2.0)	68.41 (1.29)	
12.16	50	64.65 (3.04)	66.12 (3.11)	65.27 (2.19)	64.88 (2.22)	64.38 (1.55)
	70	64.89 (1.78)	66.65 (2.67)	65.72 (177)	65.32 (4.11)	
	90	65.31 (4.01)	67.01 (2.13)	66.42 (1.45)	65.69 (3.22)	
	110	65.54 (2.65)	67.45 (2.78)	66.88 (1.71)	65.87 (3.01)	
	130	64.93 (2.19)	66.24 (3.29)	65.36 (3.01)	69.32 (2.81)	

Note Figures in parenthesis represent standard deviation.

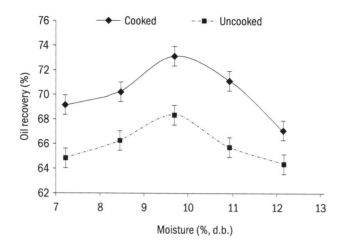

Fig. 6.5 *Effect of moisture content and cooking on oil recovery*

Oil recoveries on the basis of seed moisture content and cooking conditions are shown in Figure 6.5 (the standard deviations are shown as error bar). At seed moisture content of 8.19% (d.b.) for cooked seeds, it gives higher value of oil recovery whereas for uncooked seeds the highest oil recovery was obtained at 9.86% (d.b.) moisture content. Further oil recovery rapidly decreased with an increase in the seed moisture content. This can be related to mucilage development on oil cells (Vaughan 1970). There was swelling of mucilage at higher moisture level which in turn produced a cushioning effect on the seed. The swelled mucilage could have been an impediment to oil flow during expression while the cushioning effect on the seed reduced the rupturing of the particles and internal tissues during pressure application. A comparison of the oil recovery shows that the oil yields obtained from cooked samples were higher than those from uncooked samples, which indicated the necessity for adequate heat treatment. The increased oil recovery by cooking can be attributed to reduced viscosity of the oil in the capillaries of the seed tissues (Fasina and Ajibola 1989).

Table 6.7 describes the relationships between oil recovery and moisture content, which is represented by second-order polynomial equations. For both cooked and uncooked seeds, the relationships were statistically significant ($P \leq 0.01$). A similar result was reported for cooked and uncooked flaxseeds (in the moisture range of 5.3%–12.4% d.b.) (Singh and Bargale 1990) and rapeseeds (in the moisture range of 5.3%–12.4% d.b.) (Bargale and Singh 2000). However, other studies showed decrease in moisture content with increased oil recovery for soya bean and crambe seeds (Khan and Hanna 1983; and Singh, Wiesenborn, Tostenson, *et al.* 2002).

Table 6.7 Regression equations for oil recovery, residual oil, pressing rate, and sediment content as function of moisture content

Parameter	Equation	Correlation coefficient (r^2)
Oil recovery		
Cooked	$OR = 7.59 + 13.63M - 0.72M^2$	0.86
Uncooked	$OR = 23.70 + 9.13M - 0.48M^2$	0.79
Residual oil		
Cooked	$RO = 53.55 - 8.29M + 0.44M^2$	0.99
Uncooked	$RO = 49.43 - 7.03M + 0.36M^2$	0.89
Pressing rate		
Cooked	$PR = 33.15 - 0.94M + 0.06M^2$	0.94
Uncooked	$PR = 35.09 - 1.31M + 0.08M^2$	0.98
Sediment content		
Cooked	$SC = 9.41 + 0.08M - 0.04M^2$	0.99
Uncooked	$SC = 3.47 + 0.52M - 0.04M^2$	0.96

Table 6.8 Analysis of variance of oil recovery, cake residual oil, pressing rate, and sediment content

Parameter	Mean	Variance	df	F
Oil recovery				
Uncooked	65.83	3.19	4	0.288[*]
Cooked	70.54	11.07	4	-
Residual oil				
Uncooked	17.13	0.94	4	0.385[*]
Cooked	16.14	2.45	4	-
Pressing rate				
Uncooked	30.51	1.35	4	1.984[*]
Cooked	30.05	0.67	4	-
Sediment content				
Uncooked	5.03	0.61	4	0.225[*]
Cooked	6.54	2.73	4	-
Error	-	-	9	-

[*]Significant at $P < 0.01$

By comparing Figures 6.5 and 6.6, it is observed that increased oil recovery implies decreased cake residual oil (the standard deviations are shown as error bar). Thus with increase in moisture content, the residual oil decreases and reaches the lowest value at a moisture content of 9.69% (d.b.) for both cooked and uncooked seeds. Further there is an increase in residual oil content with increase in seed moisture levels from 9.69% to 12.16% (d.b.). This trend might be due to higher frictional resistance offered by low moisture seed in the barrel during pressing. Hoffmann (1989) also suggested that lower moisture content of the seed increases friction, whereas higher moisture acts as a lubricant during pressing. Singh, Wiesenborn, Tostenson, *et al.* (2002) observed a trend of decreased residual oil with decreased moisture content from 9.2% to 3.6% (d.b.) in a mechanical screw pressing of cooked and uncooked crambe seed. The residual oil in cooked seed was 2.8% – 5.7% lower than that of uncooked seeds, and this difference was statistically significant ($P \leq 0.01$).

As moisture content decreased from 12.16% to 7.22% d.b., the pressing rate decreased from 30.92 kg seed/h to 29.5 kg seed/h and 31.38 kg seed/h to 29.87 kg seed/h for cooked and uncooked seeds, respectively (Figure 6.7). (The standard deviations are shown as error bar.) These pressing rates were achieved at an optimum screw speed of 120 rpm. Bargale and Singh (2000) also reported decreased pressing rate with decrease in moisture content (in the range of 5.3% – 12.4% d.b.) in screw pressing of moisture-conditioned rapeseed but observed a peak at 10.1% d.b. in the screw pressing of hot-water-soaked and sun-dried rapeseed in the above-mentioned moisture-content range. From the above studies, it seems that the relationship between moisture content and pressing rate varies with the method of seed preparation and type of screw press.

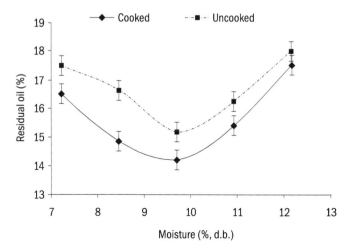

Fig. 6.6 *Effects of moisture content and cooking on residual oil content*

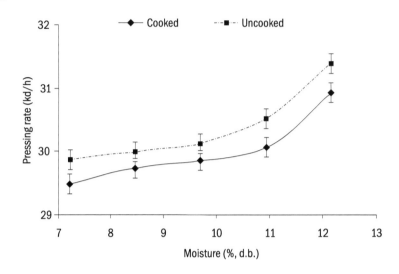

Fig. 6.7 *Effect of moisture content and cooking on pressing rate*

The relationships between moisture content and pressing rate were statistically significant at $P \leq 0.05$ in both the cases, and may be represented by second-order polynomial equations (Table 6.7). Moisture apparently behaves like a lubricant that reduces resistance during pressing, resulting in higher pressing rate. The higher pressing rate achieved with higher-moisture seed implied lower press cake residence time in the barrel, which probably contributed to the reduced oil recovery associated with this moisture content. The pressing rate of uncooked seed was slightly higher than that of cooked seed, and the difference was significant at $P \leq 0.05$. The reason for this difference could be that denatured protein in cooked seed caused higher frictional resistance in the barrel during pressing.

6.3.5   Some Chemical Properties of Jatropha Seed Oil

The chemical properties of any oil play an important role in the preparation of biodiesel. The average value of some chemical properties obtained for the oil expressed from jatropha seed are given in Table 6.9. The average values of the properties determined compare favourably with those obtained from the literature.

Table 6.9 Some physico-chemical properties of jatropha seed oil

Properties	Oil from screw press expeller			Literature values[a,b]
	Measured average	Measured standard deviation	Measured range	
Acid value (mgKOH/g)	3.21	0.95	3.90–12.2	3.00–38.00
Density (kg/m³)	870	260	870 – 910	800–910
Viscosity @ 40°C (m²s)	34.88×10^{-6}	2.59	34.30×10^{-6} 36.67×10^{-6}	31.0×10^{-6} 51.0×10^{-6}
Iodine value (gI2/100 g)	95.67	2.66	92.45–97.78	93.00–107.00
Saponification value (mgKOH/g)	190.01	8.78	189.20–195.02	188.00–198.00

[a,b]*Sources* Akintayo (2004); Meher, Naik, Naik, *et al.* (2009)

6.4 CONCLUSION

Alternative and renewable energy resources have recently become a high priority for many countries and this will play a major role in the agricultural industry in the near future. With the increasing demand of alternate energy, oilseeds are the major sources of raw materials for oil production. The high oil content for use in different industrial purposes including biofuels is an incentive for its post-harvest processing. Experimental finding proved that the physical properties, namely, length, width, thickness, one thousand unit mass, sphericity, angle of repose, surface area, densities and the static coefficient of friction at various surface for non-edible fruits/seeds are significantly different from each other. Hence, the physical properties of the fruit and seed were unique and different from other tree-borne oilseeds. The developments of equipment for jatropha processing such as decorticator and oil expeller are justified in the current situation. The fruit must be decorticated prior to oil expression processes. The developed manual decorticator which has an efficiency of 91% can be used to decorticate jatropha fruits at rural level as well as at hilly regions because of its easy handling, no electricity requirement, and low cost. Based on the study, it is strongly recommended that oilseeds must be conditioned and the rpm of screw expeller has to be set at its optimum for high oil yield. Among the considered variables, the seed moisture content, heating temperature, heating time, and the interactions among these variables were found to significantly influence the oil recovery from jatropha seeds. The highest oil recovery of 73.14% was obtained when jatropha seed was conditioned to 9.69% (d.b.) moisture and cooked at 110°C for 10 minutes. The oil recovery from screw press uncooked jatropha seed was lower than that from cooked seed, but the pressing rate from uncooked seed was higher than that from cooked seed. The values of chemical properties investigated were satisfying literature values. The oil expressed at optimum processing conditioned exhibited good chemical properties and could be useful as biodiesel feedstock and industrial application.

REFERENCES

Adewumi, B. A. and A. B. Fatusin. 2006. Design, fabrication and testing of an impact-type hand operated cocoa pod breaker. *Agricultural Engineering International: The CIGR Ejournal* 8(6): 1–12

Akintayo, E. T. 2004. Characteristics and composition of *Parkia biglobbossa* and *Jatropha curcas* oils and cakes. *Bioresource Technology* 92: 307–310

AOAC. 1984. *Official Methods of Analysis*, DC: Association of Official Analytical Chemists

Bargale, P. C. and J. Singh. 2000. Oil expression characteristics of rapeseed for a small-capacity screw press. *Journal of Food Science Technology* 37: 130–134

Bringi, N. V. 1987. *Non-Traditional Oilseeds and Oils in India*. New Delhi: Oxford and IBH Publishing Co Pvt. Ltd

Brusewitz, G. H. 1975. Density of rewetted high moisture grains. *Transaction ASAE* 18: 935–938

Dutta, S. K., V. K. Nema, and R. K. Bharadwaj. 1988. Physical properties of gram. *Journal of Agricultural Engineering Research* 39: 259–268

Fasina, O. O. and O. O. Ajibola. 1989. Mechanical expression of oil from conophor nut *(Tetracarpidium Conophorurn)*. *Journal of Agricultural Engineering Research* 44: 275–287

Hoffmann, G. 1989. *The Chemistry and Technology of Edible Oils and Fats and their High-fat Products*. New York: Academic Press

Indrasari, S. D., S. Koswara, D. Muchtadi, and L. M. Nagara. 2001. The effect of heating on the physicochemical characteristics of rice bran oil. *Indonesian Journal of Agricultural Science* 2: 1–5

Joshi, J. and S. Hiremath. 2000. Simarouba: A Potential Oilseed Tree. *Current Science* 78: 694–697

Joshi, S. and S. Hiremath. 2001. *Simarouba, Oil Tree*. Bangalore: University of Agricultural Sciences

Khan, L.M. and M.A. Hanna. 1983. Expression of Oil from Oilseeds: A Review. *Journal of Agricultural Engineering Research* 28: 495–503

Lakshmikanthan, V. 1978. *Tree Borne Oilseeds*. Mumbai: Directorate of Non-edible Oils and Soap Industry, Khadi and Village Industries Commission

Meher, L. C., S. N. Naik, M. K. Naik, and A. K. Dalai. 2009. Biodiesel production using Karanja (*Pongamia pinnata*) and Jatropha (*Jatropha curcas*) seed oil. In *Handbook of Plant-based Biofuels*, edited by A. Pandey, pp. 257–166. New York: CRC Press

Mohsenin, N. N. 1980. *Physical Properties of Plant and Animal Materials*. New York: Gordon and Breach Science Press

Nagar, M. D. 2003. *Jatropha and Other Tree-borne Oilseeds Species for Biodiesel Production*. BAIF Development Research Foundation. www.baif.com.

Sacilik, K., R. Ozturk, and R. Keskin. 2003. Some physical properties of hemp seed. *Biosystem Engineering* 86: 191–198

Sahay, K. M. and K. K. Singh. 1996. *Unit Operation of Agricultural Processing*. New Delhi: Vikas Publishing House Pvt. Ltd

Singh, J. and P. C. Bargale. 1990. Mechanical expression of oil from linseed (*Linum usitatissimum* L.). *Journal of Oilseeds Research* 7: 106–110

Singh, K. K., D. P. Wiesenborn, K. Tostenson, and N. Kangas. 2002. Influence of moisture content and cooking on screw pressing of crambe seed. *Journal of American Oil Chemistry Society* 79: 165–170

Singh, K. P., H. N. Mishra, and S. Supradip. 2010. Moisture-dependent properties of barnyard millet grain and kernel. *Journal of Food Engineering* 96: 598–606

Srivastava, A. and R. Prasad. 2000. Triglycerides-based diesel fuels. *Renewable and Sustainable Energy Review* 4: 111–133

Vaughan, J. G. 1970. *The Structure and Utilization of Oilseeds*. London: Chapman and Hall

7

Adsorption and Photodegradation of Sulfamethoxazole in a Three-phase Fluidized Bed Reactor

John Akach[a], Maurice S. Onyango[b], and Aoyi Ochieng[a,*]

[a]Centre for Renewable Energy and Water, Vaal University of Technology, Private Bag X021, Vanderbijlpark, 1900, South Africa

[b]Department of Chemical and Metallurgical Engineering, Tshwane University of Technology, Pretoria, Private Bag X680, Pretoria, 0001, South Africa

[*]E-mail: ochienga@vut.ac.za

7.1 INTRODUCTION

Antibiotic pollutants, such as sulfamethoxazole (SMX) in wastewater, have been linked to antibiotic resistance, highlighting the need to remove these antibiotics from wastewater (Kümmerer 2009). However, conventional wastewater treatment plants have not been efficient in removing antibiotics from wastewater due to their recalcitrant nature (Xekoukoulotakis, Drosou, Brebou, et al. 2011). Therefore, alternative removal methods, such as adsorption, photo-fenton, photolysis, ozonation, and semiconductor photocatalysis, have been developed to remove SMX from wastewater (Klavarioti, Mantzavinos, Kassinos 2009). Among these methods, adsorption and semiconductor photocatalysis have been shown to be highly effective in removing SMX. Semiconductor photocatalysis using nanophase titanium dioxide (TiO_2) and ultraviolet (UV) light from electric lamps have been widely used to photodegrade SMX (Xekoukoulotakis, Drosou, Brebou, et al. 2011; Nasuhoglu, Yargeau, Berk 2011; Abellán, Bayarri, Giménez, et al. 2007). However, the use of electricity to filter nanophase TiO_2 through membrane separation systems and to power UV lamps has been costly. In order to reduce cost, TiO_2 has been supported on various materials to improve its separation from wastewater after use

(Shan, Ghazi, Rashid 2010). Also, sunlight has been used as the source of irradiation instead of the costly UV lamps (Robert and Malato 2002).

One of the greatest challenges of solar photocatalysis has been the development of an efficient and cost-effective solar photoreactor. Several reactor designs such as water bell, thin film, compound parabolic concentrator, and fluidized bed reactors have been proposed. A fluidized bed photocatalytic reactor (FBPR) has been suggested as an efficient reactor due to its high mass transfer (Braham and Harris 2009). Although, air fluidized three-phase reactor has been found to be efficient and has been widely applied in other fields (Vunjak-Novakovic, Kim, Wu, *et al.* 2005; Schulz 1999), there has been very little application of the reactor in solar photocatalysis (Kimura, Yoshikawa, Matsumura, *et al.* 2004). In this work, a composite catalyst of TiO_2 and activated carbon (AC) bonded with silica xerogel was synthesized. The composite catalyst was then used to adsorb and photodegrade SMX in the three-phase FBPR. Sunlight was used as the source of UV light for the activation of TiO_2. The aim of the work was to investigate the hydrodynamic characteristics of the three-phase FBPRs and the effect of hydrodynamics on the adsorption and photodegradation of SMX.

7.2   METHODOLOGY

7.2.1   Reactor Equipment

The reactor set up (Figure 7.1) consisted of 10 FBPRs: (1) mounted on a frame that could be tilted both East–West and North–South to obtain the required azimuth and inclination angles to the sun. Air was supplied to the reactors from a compressor (2) through a common header (3) from where gas lines channelled the air into the respective reactors. Needle valves (4), rotameters (5), and non-return valves (6) were fitted to the gas lines from the header to control the air flow, metre the volumetric flow of air, and keep the liquid from entering the gas lines, respectively.

The reactor consisted of a glass section made of borosilicate glass screwed onto a bottom plastic section. The plastic section was connected to the non-return valves and also held the gas distributor which was made of sintered glass of pore size 10–16 μm. The porous glass distributor produced small-sized gas bubbles and prevented the catalyst from passing through the distributor. The glass section had an internal diameter of 32 mm, a thickness of 1.5 mm, and a column height of 480 mm. The gas disengaging section had a diameter of 67 mm and a height of 50 mm. Therefore, the glass section of the reactor had a total length of 569 mm with a maximum capacity of 629 mL and working solution volume of 450 mL. Four sampling ports were provided on the reactor body to be used for hydrodynamic

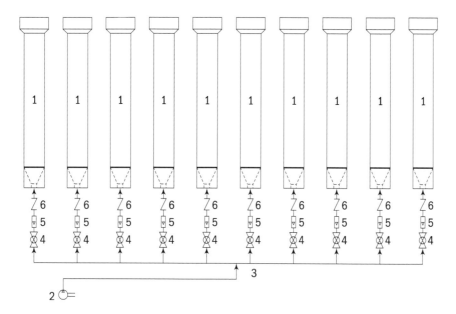

Fig. 7.1 *Experimental set up: (1) reactor, (2) air compressor, (3) common header, (4) ball valve, (5) rotameter, and (6) non-return valve*

experiments. The system was operated in a batch mode for the liquid and solid, with air being used to induce fluidization of the bed and also as a source of oxygen, which was used as an electron acceptor. The experimental rig was mounted on a rooftop where there were no obstructions to the rays of the sun.

7.2.2   Materials

Aeroxide P25 TiO_2 was purchased from Acros Organics while Ludox HS-30 colloidal silica and SMX were obtained from Sigma-Aldrich (South Africa). Commercial powdered activated charcoal (PAC) and HCl (32%) were purchased from Labchem (South Africa). All experiments were carried out with deionized water (resistivity 18.2 MΩcm) from a Millipore Direct Q reverse osmosis unit.

7.2.3   Preparation of PAC-TiO$_2$-silica Xerogel Composite Catalyst

Samples of PAC, TiO_2, and colloidal silica in the ratio 1:9:15 were mixed in a capped bottle and stirred magnetically until the mixture was homogeneous. Then 15 mL of the slurry mixture was spread on 30 cm × 30 cm glass plates. The glass plates were then put in an oven at 90°C for quick gelling of the colloidal silica sol. The resulting dry flakes of PAC and

TiO_2 bound by silica xerogel were crushed and sieved to a size range of 38–75 μm. This composite catalyst powder was then shaken in 100 ml of 0.05 M HCl solution to remove excess alkalinity due to NaOH in colloidal silica. The acid wash was followed by several washes with deionized water to remove the excess acid and loose particles of PAC, TiO_2, and silica xerogel until the pH rose to 6.5. Finally, the composite catalysts were dried at 50°C in the oven overnight to obtain the PAC–TiO_2–silica xerogel (CTS) composite catalyst.

7.2.4 Hydrodynamic Experiments

The solid concentration was determined by the method used by Matsumura, Noshiroya, Tokumura, *et al.* (2007). In this method, 5 g/L of the CTS composite was added to 400 mL of water in the reactors. The reactors were then inclined to the appropriate angle, and air sparging started and maintained at the required flow rate. After 10 minutes of fluidization, 10 mL samples containing water and CTS composite particles were taken from the 4 sampling ports on the side of the reactor. The solid concentration in the samples was then analysed gravimetrically by filtering the liquid through a filter paper to retain the solids and then drying the solid laden filter paper. The difference in weight of the filter paper before sample filtration and after drying was used to compute the solid concentration in the samples.

Global gas hold-up measurements were carried out using the quick stop method (Ochieng, Ogada, Sisenda, *et al.* 2002; Abraham, Khare, Sawant, *et al.* 1992). In this method, the reactor was inclined to the appropriate angle and then a certain volume of water corresponding to the required aspect ratio and 1.5 g/L of CTS composite was poured into the reactor. A mark was then made on the reactor wall corresponding to the gas-free level of the liquid. The calculation of gas hold-up was based on the difference between the static and expanded bed heights.

7.2.5 Adsorption and Photodegradation Experiments

The adsorption and photodegradation experiments were carried out between 8 am and 3 pm on sunny days in the FBPR. Dark adsorption experiments were carried out first by inclining the reactors at an appropriate angle and then covering the reactors with a black polythene sheet. Then, 1.5 g of the CTS composite catalyst was added into 450 mL of SMX in the FBPR to start adsorption. Air sparging was then started and the air flow was regulated and maintained at the required level using the rotameters. After reaching the adsorption equilibrium, the black polythene sheet covering the reactors was removed in order to start the photocatalysis. Periodically, 3 mL of the substrate was sampled and filtered using a 0.45 μm GHP syringe filter (Pall) for analysis using a UV-vis spectrophotometer (PG instruments T60) at λ_{max} of 256 nm. The azimuth angle of the reactors was adjusted every hour so that the reactors always faced the sun.

7.3   RESULTS AND DISCUSSION

7.3.1   Hydrodynamics

The hydrodynamic behaviour of the FBPR was such that the solid concentration distribution and gas hold-up of the reactor greatly influenced the adsorption and photodegradation of substrates in the reactor. Distribution of solid particles in liquid media is termed as solid concentration distribution. The solid concentration distribution determines the rate of substrate–catalyst mass transfer which influences the rate of adsorption and photodegradation of the substrate. Similarly, the gas holdup affects the rate of gas–liquid mass transfer which determines the amount of dissolved oxygen, an electron scavenger necessary for photodegradation. Hydrodynamic studies were, therefore, carried out to determine the solid concentration distribution and the global gas holdup at different superficial air velocities and reactor inclination angles.

Gas hold-up

The effect of the reactor inclination angle on the gas hold-up was investigated. The results (Figure 7.2) show a marked decrease in the gas hold-up with a decrease in the reactor inclination angle at high superficial air velocities. For example, the gas hold-up reduced from 0.2 to 0.12 when the inclination angle was reduced from 90° to 60°, where 90° represents the vertical position. This reduction is due to the fact that as the reactor was inclined, the air bubbles had a shorter distance to travel in the liquid before leaving the reactor. Also, when the reactor was inclined, the bubbles rose to the upper inclined wall of the reactor and crawled on the surface of the reactor until they reached the gas disengaging section. The high number of bubbles on the upper inclined wall of the reactor increased the bubble coalescence. The shorter bubble travel distance coupled with an increased bubble size due to coalescence resulted in a shorter bubble residence time in the reactor, thus reducing the gas hold-up. Similar results were reported by Ugwu, Ogbonna, and Tanaka (2002), who reported a decrease in the gas hold-up from 0.022 to 0.016 when they reduced the inclination angle from 45° to 8° in their inclined airlift reactor.

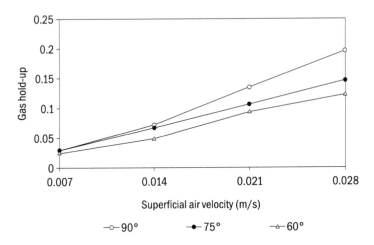

Fig. 7.2 *The effect of superficial air velocity and inclination angle (°) on gas holdup, aspect*
ratio = 6, solid loading = 2 g/L.

7.3.2 Solid Concentration Distribution

The solid concentration distribution gives an indication of the dispersion
of the solid catalyst particles in the reactor. Experiments were carried
out to determine the effect of superficial air velocity and reactor
inclination angle on the solid concentration distribution. Results (Figure
7.3a – c) show a general decrease in solid concentration with an increase
in distance from the distributor at different superficial air velocities.
This decrease in solid concentration with increasing distance from the
distributor was sharper when the reactor was vertical (inclination angle
of 90°). When the reactor was at this position, the solid concentration
dropped by 99% from 14.3 g/L to 0.16 g/L between 5.7 cm and 48.6 cm
from the distributor, respectively. Also, at 90° inclination angle, the
gradient of decrease in solid concentration with increasing distance from
the distributor did not change significantly with an increase in superficial
air velocity.

When the reactor inclination angle was reduced below 90°,
there was a significant improvement in solid concentration distribution.
For example, at an inclination angle of 60° and air superficial velocity
of 0.014 m/s, the solid concentration decreased by only 35% from 4.8 g/L
to 3.1 g/L between 5.7 cm and 48.6 cm from the distributor, respectively.
At reactor inclination angles below 90°, the superficial air velocity
significantly affected the solid concentration distribution with reactor
height. Figure 7.3a–c shows a sudden increase in the solid concentration
distribution with a decrease in the reactor inclination angle from 90°
to 60°. However, when the angle was reduced further to 45°, the solid

concentration distribution worsened. When the reactor was vertical, at an inclination angle of 90°, only air was responsible for fluidization. When the reactor was inclined, air bubbles moved along the upper surface of the reactor creating a low density and a high fluid flow region. The liquid in this region moved up the reactor carrying solid particles along to the top of the reactor. The space left by the upward moving liquid was filled by the liquid occupying the lower denser region of the reactor. This downwards moving liquid carried the solid particles back to the bottom of the reactor which were in turn picked and moved up by the upwards moving liquid. In this way, air induced a bulk circulation of the liquid in an inclined reactor which carried the solids along, thus fluidizing them. A similar

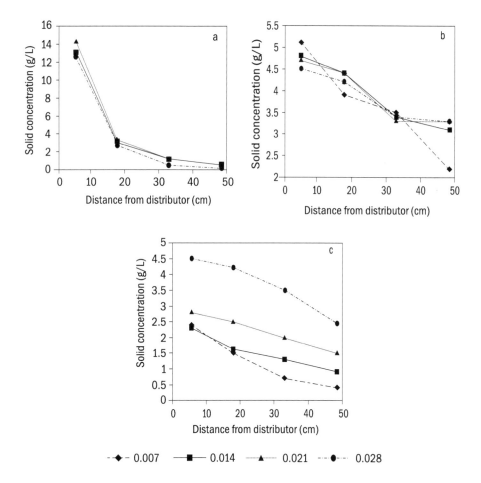

Fig. 7.3 *Variation of solid concentration with distance from the distributor at different superficial air velocities (m/s) and inclination angles: (a) 90°, (b) 60°, and (c) 45°; solid loading = 5 g/L, liquid volume = 400 mL*

behaviour has been observed for airlift reactors (Christi and Moo-Young 1993; Vunjak-Novakovic, Kim, Wu, *et al.* 2005). The inclined reactor essentially behaved as an airlift reactor with the inclined upper section of the reactor being analogous to the airlift riser while the inclined lower section acting like an airlift downcomer. For the reactor used in this work, there was no physical separation between the riser and the downcomer as found in airlift reactors. However, the region occupied by air bubbles was small enough for there to be sufficient hydrodynamic separation between the high and low density regions in the reactor.

The lower part of the inclined reactor was not as turbulent as the upper part and, therefore, particles tended to settle on the lower wall as they were dragged towards the distributor. Particles also tended to settle on the distributor since at least half the distributor was not sparged when the fluidized bed reactors were inclined. This is due to the fact that in the inclined reactor, bubbles on the distributor tended to rise upwards towards the upper reactor wall leaving the lower section of the distributor unsparged. The tendency of the solids to settle on the reactor walls and distributor was especially pronounced at an inclination angle of 45°. This settling of the solids on the walls and distributor increased with decreasing superficial air velocity due the reduction in turbulence at low superficial air velocities. However, very little settling was observed at 90° inclination angle due to the fact that the reactor being vertical, there could be no solid settling on the reactor walls. Instead, all the solid particles tended to settle on the distributor where air passing through the entire surface of the distributor lifted the solids, thus preventing the solids from settling on the distributor.

7.3.3 Effect of Hydrodynamics on Adsorption and Photodegradation

High solid–liquid and liquid–gas mass transfer rates are essential to achieve a high rate of adsorption and photodegradation. For photodegradation, the substrates have to come into contact with the photocatalyst and also oxygen in air has to dissolve in the liquid for efficient photodegradation to take place. Therefore, both the liquid–solid and gas–liquid mass transfer rates influence photodegradation. For adsorption, only the liquid–solid mass transfer rate determines the rate of adsorption. These liquid–solid and gas–liquid mass transfer rates depend on the solid concentration distribution and the gas holdup, respectively. Hydrodynamic studies have shown that the reactor inclination angle and superficial gas velocity affect the solid concentration distribution and the gas holdup. It is imperative to determine whether the hydrodynamic factors: reactor inclination angle and superficial velocity also affect the rate of adsorption and photodegradation.

The effect of superficial air velocity on the adsorption and photocatalytic degradation of SMX was investigated using 1.5 g/l of CTS composite. The superficial air velocity was varied from 0 to 0.028 m/s while maintaining the reactor inclination angle constant. The results (Figure 7.4a–c) show that the adsorption and photodegradation increased with increasing superficial air velocity up to an optimum of 0.007 m/s beyond which no further increase was observed. When the reactors were not sparged with air (0 m/s), 12% of SMX was adsorbed during the turbulent mixing of CTS composite and the substrate at the start of the experiment. After the turbulent mixing, some adsorption continued to take place as the CTS composite started settling in the reactor. By the time photocatalysis started, almost all the CTS composite particles had settled onto the distributor. Only those CTS particles at the very top of the distributor were exposed to sunlight and could, therefore, degrade the substrates.

When the superficial air velocity was increased from 0 m/s to 0.003 m/s, some fluidization of the catalyst was observed. There was also considerable settling of the CTS catalyst on the lower wall and on the distributor of the reactor. Due to the settling of the CTS composite, only a portion of the optimum amount of catalyst was fluidized and was, therefore, available for adsorption and photodegradation. At a superficial air velocity of 0.007 m/s, settling of the CTS on the reactor walls and distributor reduced, and almost all the CTS composite particles remained suspended in solution. The fluidized CTS composite particles had enough surface area exposed to the substrate and sunlight for adsorption and photodegradation. Increasing the superficial air velocity above 0.007 m/s, resulted in further reduction in the settling of the catalyst on the reactor walls and the distributor, but did not increase the adsorption and photodegradation. Matsumura, Noshiroya, Tokumura, *et al.* (2007) found similar results to those in this work by fluidizing larger (500 µm) but lighter (570 kg/m^3) particles in a larger (96 mm ID) reactor. It was reported (Matsumura, Noshiroya, Tokumura, *et al.* 2007) that the rate of photodegradation of *o*-cresol increased with increasing superficial gas velocity up to a 'critical gas velocity' beyond which no further increase in degradation was observed. They found an optimum superficial air velocity of 0.006 m/s which is comparable to the optimum superficial velocity observed in this work (0.007 m/s).

Fluidization is costly due to the electric power needed to run the compressor. A superficial air velocity of 0.007 m/s was, therefore, chosen as the optimum superficial velocity. It is worth noting that the

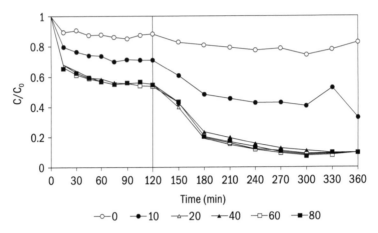

Fig. 7.4 *Effect of superficial air velocity (m/s) on the rate of adsorption and photocatalysis at 1.5 g/L CTS loading for: (a) SMX; (b) DCF; (c) CBZ*

optimum superficial air velocity of 0.007 m/s found during adsorption and photocatalysis experiments is lower than the optimum superficial velocity of 0.014 m/s found during hydrodynamic experiments. These findings suggest that a sub-optimum superficial air velocity is sufficient for adsorption and photocatalysis. Obtaining an optimum rate of adsorption and degradation with sub-optimal superficial air velocity is desirable from an economical point of view to reduce the cost of operation of the wastewater treatment system.

7.4   CONCLUSION

In this work, a composite catalyst of AC and TiO_2 bonded with silica xerogel (CTS composite) was synthesized. The CTS composite was used to adsorb and photodegrade SMX in a three-phase FBPR using sunlight as the source of UV light. The hydrodynamic behaviour of the reactor was characterized by the global gas hold-up and the solid concentration distribution. The gas hold-up was found to increase with increase in the superficial air velocity and reactor inclination angle. The optimum hydrodynamic condition was obtained at a superficial air velocity of 0.014 m/s. The effect of hydrodynamics on adsorption and photodegradation of SMX was also investigated at various superficial air velocities. It was found that the optimum adsorption and photodegradation was obtained when the superficial air velocity was 0.007 m/s. This shows that the solid concentration distribution affected the adsorption and photodegradation of SMX more than the gas hold-up. This implies that oxygen mass transfer was not a limiting factor. Air fluidization (rather than oxygen) is adequate for the reaction. The optimum superficial air velocity found during adsorption and photodegradation was lower than the superficial air velocity obtained during hydrodynamic experiments. This shows that a high removal of SMX by adsorption and photodegradation was achieved at sub-optimal hydrodynamic conditions. These experiments show the great potential of using a three-phase FBPR for the adsorption and solar photodegradation of SMX using CTS composite.

ACKNOWLEDGEMENTS

This work was supported by the Water Research Commission of South Africa (Project K5/2105).

REFERENCES

Abellán, M., B. Bayarri, J. Giménez, and J. Costa. 2007. Photocatalytic degradation of sulfamethoxazole in aqueous suspension of TiO$_2$. *Applied Catalysis B: Environmental* 74(3): 233–241

Abraham, M., A.S. Khare, S.B. Sawant, and J.B. Joshi. 1992. Critical gas velocity for suspension of solid particles in three-phase bubble columns. *Industrial & Engineering Chemistry Research* 31(4): 1136–1147

Braham, R.J. and A.T. Harris. 2009. Review of major design and scale-up considerations for solar photocatalytic reactors. *Industrial & Engineering Chemistry Research* 48(19): 8890–8905

Christi, Y. and Moo-Young. 1993. Improve the performance of airlift reactors. *Chemical Engineering Progress* pp. 39

Kimura, T., N. Yoshikawa, N. Matsumura, and Y. Kawase. 2004. Photocatalytic degradation of nonionic surfactants with immobilized TiO$_2$ in an airlift reactor. *Journal of Environmental Science and Health, Part A* 39(11–12): 2867–2881

Klavarioti, M., D. Mantzavinos, and D. Kassinos. 2009. Removal of residual pharmaceuticals from aqueous systems by advanced oxidation processes. *Environment International* 35(2): 402–417

Kümmerer, K. 2009. Antibiotics in the aquatic environment–a review–part II. *Chemosphere* 75(4): 435–441

Matsumura, T., D. Noshiroya, M. Tokumura, H. T. Znad, and Y. Kawase. 2007. Simplified model for the hydrodynamics and reaction kinetics in a gas-liquid-solid three-phase fluidized-bed photocatalytic reactor: degradation of o-cresol with immobilized TiO$_2$. *Industrial and Engineering Chemistry Research* 46(8): 2637–2647

Nasuhoglu, D., V. Yargeau, and D. Berk. 2011. Photo-removal of sulfamethoxazole (SMX) by photolytic and photocatalytic processes in a batch reactor under UV-C radiation (λ_{max} = 254 nm). *Journal of Hazardous Materials* 186(1): 67–75

Ochieng, A., T. Ogada, W. Sisenda, and P. Wambua. 2002. Brewery wastewater treatment in a fluidised bed bioreactor. *Journal of Hazardous Materials* 90(3): 311–321

Robert, D. and S. Malato. 2002. Solar photocatalysis: a clean process for water detoxification. *Science of the Total Environment* 291(1–3): 85–97

Schulz, H. 1999. Short history and present trends of Fischer–Tropsch synthesis. *Applied Catalysis A: General* 186(1): 3–12

Shan, A.Y., T.I.M. Ghazi, and S.A. Rashid. 2010. Immobilisation of titanium dioxide onto supporting materials in heterogeneous photocatalysis: a review. *Applied Catalysis A: General* 389(1): 1–8

Sheintuch, M. and Y.I. Matatov-Meytal. 1999. Comparison of catalytic processes with other regeneration methods of activated carbon. *Catalysis Today* 53(1): 73–80

Ugwu, C., J. Ogbonna, and H. Tanaka. 2002. Improvement of mass transfer characteristics and productivities of inclined tubular photobioreactors by installation of internal static mixers. *Applied Microbiology and Biotechnology* 58(5): 600–607

Vunjak-Novakovic, G., Y. Kim, X. Wu, I. Berzin, and J. C. Merchuk. 2005. Air-lift bioreactors for algal growth on flue gas: mathematical modeling and pilot-plant studies. *Industrial & Engineering Chemistry Research* 44(16): 6154–6163

Xekoukoulotakis, N.P., C. Drosou, Brebou, E. Chatzisymeon, E. Hapeshi, D. Fatta-Kassinos, and D. Mantzavinos. 2011. Kinetics of UV-A/TiO_2 photocatalytic degradation and mineralization of the antibiotic sulfamethoxazole in aqueous matrices. *Catalysis Today* 161(1): 163–168

8

Application of Cellulose Nitrate Membrane for Pervaporative Separation of Organics from Water

Shraddha Awasthi[a],[*], Dhanesh Tiwari[a], and
Pradeep Kumar Mishra[b]

[a]*Department of Chemistry, IIT (BHU), Varanasi, Uttar Pradesh 221005*
[b]*Department of Chemical Engineering and Technology, IIT (BHU), Varanasi,
Uttar Pradesh 221005*
[*]*E-mail: shraddhaawasthi02@gmail.com*

8.1 INTRODUCTION

Membrane technologies have become established in recent years as an alternative for environmental applications to conventional mass exchange technologies, such as absorption, adsorption, or extraction. Pervaporation is a separation process in which a liquid mixture is in direct contact with one side of a membrane and the permeated product called "pervaporate" is removed from the other side in a vapour state. Pervaporation is a pressure-driven process that differs from other membrane processes in that it involves a phase change requiring heat of vaporization. Maintaining the downstream partial pressure lower than the saturation pressure at that temperature brings about pervaporation. This is achieved by creating vacuum or sweeping an inert carrier gas (Koros, Ma, and Shimdidzu 1996).

Pervaporation can still be treated as a new technology, depending on the permeating components. It can be classified as (i) hydrophilic pervaporation and (ii) organophilic pervaporation (Lipnizki, Hausmanns, Ten, *et al.* 1999). In the hydrophilic pervaporation, water, the target compound, is separated from an aqueous–organic mixture. In this process, water permeates preferentially through the membrane (Smitha,

Suhanya, Sridhan, *et al.* 2004). The membranes used in hydrophilic pervaporation include polyvinyl alcohol (PVC), polyacrylonitrile (PAN), and cesium polyacrylate. Pharmaceutical waste streams that contain high concentration of water and solvents are separated using pervaporation, also used for volume reduction of pharmaceutical wastes (Shah, Ghorpade, Mangum, *et al.* 1999). Pervaporation is less familiar to the industries (Meuleman, Bosch, Mulder, *et al.* 1999). Some of the products separated successfully using pervaporation are methanol, ethanol, propanol, butanol, cyclohexanol, benzyl alcohol, acetone, butanone (MEK), methyl isobutyl ketone (MIBK), acetonitrile, benzene, toluene, phenol, methylacetate, ethylacetate, triethylamine, pyridine, methyl tert-butylether (MTBE), dioxane and tetrahydrofuran (Acharya and Stern 1988; Rautenbach, Franke, and Klatt 1991; Lee, Iwamoto, Serinoto, *et al.* 1989). It can also be used to recover chlorinated hydrocarbons from industrial waste water (Yong, Sang, Un, *et al.* 1990).

8.2 EXPERIMENTAL SET-UP

The set-up used for the experiment is shown in Figure 8.1. The experimental set-up consists of a feed tank, a stop valve, a tubular pipe made of Perspex, condenser, collector, and a vacuum pump arranged in series. The feed tank is made up of a small steel vessel that can hold around 1.2 L of the feed solution. It is placed at a height of 58 inches above the ground level so that the feed solution can pass through the membrane arrangement easily with the help of gravity. The feed solution is prepared based on the experimental run and fed into this steel vessel. From the feed tank, the solution slides down through the glass stop valve and reaches the process vessel. The glass stop valve is placed at 48 inches above the ground level. A 33-inch long mild steel pipe acts as an intermediary between the stop valve and the process vessel. The process vessel is fabricated by joining together two half pieces of Perspex tube, each measuring 6 inches long, with the

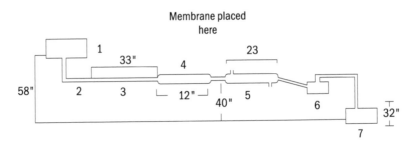

Fig. 8.1 *Experimental set-up: (1) feed tank (2) glass stop valve (3) mild steel pipe (4) process vessel made of perspex material (5) straight tube condenser (6) collector flask (7) vacuum pump*

help of flanges. The tube is 1.5 inches in diameter. The cellulose nitrate membrane, which is 47 mm diameter, is placed between the flanges. One side of the membrane remains in constant contact with the feed solution, whereas the other side of the membrane remains dry owing to the presence of vacuum, which immediately pulls away the liquid/vapour coming out of the membrane. The membrane is selective in nature; that is, it may allow the organic compound or water to pass through it. Then the vapour passes through a straight tube condenser, which is 23 inches long, and finally gets collected in a collecting vessel, which is placed at a height of 32 inches above the ground level. The vacuum pump, having a capacity of 100 L/min, is placed on the ground. The objective of this study is to examine the use of cellulose nitrate membrane in separation and treatment of various organic–water systems that are encountered in industrial usage.

8.3 RESULTS AND DISCUSSION

The reason for choosing cellulose nitrate is that it is a hydrophilic membrane, whose cost is lower than other expensive membranes. It is also readily available in the market and has a good shell life. Although experiments have been carried out using many expensive membranes and scientists are focusing on the development of newer types of membranes with superior properties, experimental work using cellulose nitrate can be conducted till the temperature of 145°C and with the pore size as low as 0.25 μm. The diameter of the membrane was 47 mm and it came in a pack of 100. Time lag (measured in seconds) refers to the time difference between the opening up of the feed valve and the collection of the permeate in the flask.

The system taken up was 1,4-dioxane–water system, whose standard graph for the plot of refractive index versus mixture composition is shown in Figure 8.2. The plot of permeate concentration versus feed concentration for dioxane–water system is shown in Figure 8.3. Series 2 indicates the feed concentration (mole fraction of dioxane), whereas Series 1 indicates the permeate concentration (mole fraction of dioxane). No separation is observed at low concentrations and separation starts to take place only at concentrations higher than 15%–18%. The ideal range for operation of 1,4-dioxane–water system is 20%–40% of organic, and this range is used for removing only a small percentage of the organic. The organic phase too cannot be removed beyond a stage of 10%. The upper limit for the use of this membrane is also fixed owing to the constraint of using the membrane only up to a limit of 60% organic since the membrane gets dissolved in this particular organic–water mixture. The pervaporation selectivity ranges from 1 to 0.833 (Figure 8.4), reaching a minimum of

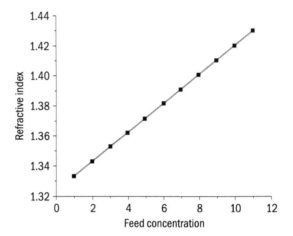

Fig. 8.2 *Refractive index versus mixture composition*
Note *Standard graph for dioxane–water system feed concentration is indicated in numbers with "1" indicating 0% dioxane and 11 indicating 100% dioxane in increasing steps of 10%.*

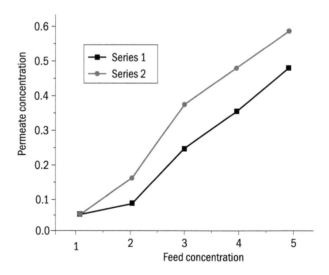

Fig. 8.3 *Plot of permeate concentration versus feed concentration*
Note *Series 1 indicates the feed concentration while series 2 indicates permeate concentration.*

0.75 for concentration of dioxane that ranges between 20% and 40%. The minimum the pervaporation selectivity, the better the outcome because the membrane used is a hydrophilic one. Cellulose nitrate is used only for separation in higher concentration ranges for this system.

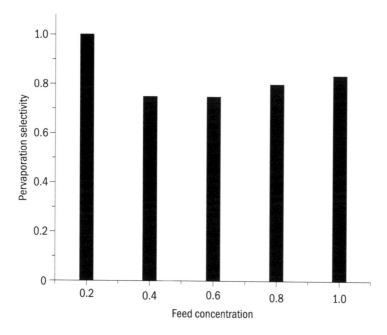

Fig. 8.4 *Variation of pervaporation selectivity with feed concentration*
Note *The first bar indicates pervaporation selectivity and the second one indicates the corresponding feed concentration.*

8.4 CONCLUSION

In recent years, membranes and membrane-related processes have gained much attention in several areas of science and technology. Among several membrane-related processes, the pervaporation technique is well known and has been used to separate mixtures of low molecular weight, close boiling point organic liquids, aqueous–alcohol mixtures, and so on (Shah, Ghorpade, Mangum, *et al.* 1999). The reason for choosing cellulose nitrate is that it is a hydrophilic membrane, whose cost is low compared to the other expensive ones. Cellulose nitrate can be used till the temperature of 145°C, and its pore size is as low as 0.25 μm. Cellulose nitrate can also be used in the treatment of wastewater containing particular organics, with the result that a cleaner technology can be employed. In this chapter, the system used for investigation is 1,4-dioxane–water (boiling point of 1,4-dioxane is 101°C and that of water is 100°C). The ideal range for operation of 1,4-dioxane–water system is 20%–40% of organic. The process can also be used for separation, with the result that it can be substituted for the conventional processes such as extraction and distillation.

ACKNOWLEDGEMENTS

It is my proud privilege to express deep sense of reverence and heartfelt thanks to Mr Surendra Kumar Verma (lab attendant), Department of Chemical Engineering and Technology, IIT (BHU) Varanasi for providing me the necessary laboratory facilities.

REFERENCES

Acharya, H. R. and S. A. Stern. 1988. Separation of liquid benzene/cyclohexane mixtures by perstraction and pervaporation. *J. Membrane Science* 37(1): 205–232

Koros, W. J., Y. H. Ma, and T. Shimidzu. 1996. Terminology for membranes and membrane processes (IUPAC Recommendations 1996). *Journal of Membrane Science* 120: 149–159

Lee, Yong Taek, Kazutoshi Iwamoto, Hideyuki Serinoto, and Manabu Seno. 1989. Pervaporation of water-dioxane mixtures with poly (dimethylsiloxane-CO-siloxane) membranes prepared by a sol-gel process. J. Membrane Science 42(1–2): 169–182

Lipnizki, F., S. Hausmanns, P.-K. Ten, R. W. Field, and G. Laufenberg. 1999. Organophilic pervaporation: Prospects and performance. *Chemical Engineering Journal* 73: 113–129

Meuleman, E. E. B., B. Bosch, M. H. V. Mulder, and H. Strathmann. 1999. Modeling of liquid/liquid separation by pervaporation: Toluene from water. *AIChE Journal* 45: 2153–2160

Rautenbach, R., M. Franke, and S. T. Klatt. 1991. Dehydration of organic mixtures by pervaporation. Experimental and theoretical studies on ternary-and multicomponent systems. *Journal of Membrane Science* 6: 31–48

Shah, D., A. Ghorpade, W. Mangum, and D. Bhattacharyya . 1999. Pervaporation of pharmaceutical waste streams and synthetic mixtures using water selective membranes. *Environmental Progress* 18 (1): 21–29

Smitha, B., D. Suhanya, S. Sridhar, and M. Ramakrishna. 2004. Separation of organic–organic mixtures by pervaporation—A review. *Journal of Membrane Science* 241(1): 1–21

Yong, Soo Kang, Sang Wook Lee Sang, Un Young Kim Un, and Jyong Sup Shim Jyong. 1990. Pervaporation of water–ethanol mixtures through crosslinked and surface-modified poly(vinyl alcohol) membrane. *Journal of Membrane Science* 51(1–2): 215–226

9

Green Chemistry: Mitigatory Measure for Environmental Pollution

Krishna Giri[a,*], **Gaurav Mishra**[a], **Shailesh Pandey**[a],
Rajesh Kumar[a], **and Deepika Rohatgi**[b]

[a]*Rain Forest Research Institute, Jorhat, Assam 785001,*
(Indian Council of Forestry Research and Education, Dehradun, an Autonomous
Body under the Ministry of Environment, Forest, and Climate Change,
Government of India)
[b]*Department of Scientific and Industrial Research, Technology Bhawan,*
Ministry of Science and Technology, New Delhi 110016
[*]*E-mail: krishna.goswami87@gmail.com*

9.1 INTRODUCTION

Green chemistry developed in response to the concerns of increasing chemical pollution in the environment. Although awareness of chemical contamination and pollution has increased in the last two decades, the environmental campaigns of today date in the year 1962 from the Rachel Carson's classic publication, *Silent Spring*. The book highlights the increasing concentration of DDT and other recalcitrant pesticides in the environment and their accumulation in animals through the food chain, which resulted in birds laying eggs that had soft shells and affected viable baby birds. This is also called as hatching problem in the birds. The message of this book was that substances harming birds' population might harm human population as well (Manahan 2000). Earlier, environmental issues were correlated with economic system and exploitation of natural resources. Green chemistry was then only a preliminary concept without any definition, principles, or proper practical relevance. At present, it has been defined as the design, development, and utilization of chemical processes and products to reduce or eliminate the formation of toxic by-products that cause serious environmental problems and health effects.

Chemistry, sustainability, and innovations are the three components that contribute in the well-being of any society. Chemistry is the key driver for economic growth, development, and quality of life. It has been considered as an important tool in protecting diversity, natural resources, and the environment. In the coming years, the society may face crisis in terms of energy, food, clean water, and medicine due to overexploitation of natural resources and degradation of environment. The great challenge for the chemical industry is to protect the environment and cultural heritage, while at the same time maintain the economic development of the country. At present, environment protection is the biggest concern for a society. Increasing chemical contamination, radioactive pollution, greenhouse gas concentration, global warming, ozone layer depletion, climate change, and subsequent natural disasters have exhibited the negative effects of anthropogenic interference with nature. These human endeavours have caused not only environmental problems but also negative effects in living organisms on planet earth. Therefore, the aim of sustainable chemistry is not only to minimize hazardous waste generation and environmental pollution caused by chemical manufacturing but also as an essential component for the well-being of the human society.

In the 21st century, a revolutionary movement is going on towards the design and development of green and eco-friendly technologies. This is a paradigm shift from traditional practices of the process efficiency focused on chemical yield to a new concept of economic value, waste elimination, minimum use of hazardous materials, and environmentally benign processes. Keeping in view the concept of green chemistry, several industries are making efforts in manufacturing the target compounds by following the principles and processes of green chemistry (Wardencki, Cury, and Namieoenik 2005).

9.2 GREEN CHEMISTRY CONCEPT

The basic idea of green chemistry arose in the United States of America as a multidisciplinary research programme of universities, R&D sector, industries, scientific organizations, and government agencies. These different organizations had their own programmes devoted to minimizing environmental pollution. The idea behind this interdisciplinary branch of chemistry was to introduce a novel approach of chemical synthesis and processing that stressed upon waste minimization and protection of environmental health. This concept is considered as environmentally benign, clean technology, atom economy, and benign-by-design.

The major focus of green chemistry is the prevention of waste generation using innocuous chemicals, processes, and necessary modifications in the hazard segment of a process. Green chemistry is an immensely flourishing branch of science that reduces chemical pollution through synthetic efficiency, catalysis, and changes in solvents. Various synthetic methods have been used to minimize energy consumption in the industrial sector. The use of bio-based feedstock is decreasing dependency on rapidly depleting conventional fossil fuel resources (EPA 1999).

9.3 HISTORY

In 1991, Professor Paul. T. Anastas of Yale University, United States coined the term "green chemistry" with the vision to ensure sustainable development in chemical industry, academia, and government organizations across the world. Thereafter, United States Presidential Green Chemistry Challenge was announced in 1995 and several other awards/honours were declared in the European countries also. A Working Party on Green Chemistry was formed in 1996 to act within the framework of International Union of Pure and Applied Chemistry. However, in 1997 Green Chemistry Institutes (GCI) were initiated in 20 countries to establish coordination between governments and industries in collaboration with universities and research organizations for designing and implementation of green technologies (Wardencki, Cury, and Namieoenik 2005). In 1997, a conference on green chemistry was held in Washington. Since that time, several scientific forums are regularly organized to highlight the concept and significance of green chemistry in order to achieve environmental sustainability. The first journal and book on this aspect were *Journal of Clean Products and Processes* and *Green Chemistry, respectively,* introduced in 1990. *Green Chemistry* is sponsored by the Royal Society of Chemistry. *Environmental Science & Technology* and *Journal of Chemical Education* have particular sections dedicated to dissemination of green chemistry research across the scientific community (Wardencki, Cury, and Namieoenik 2005).

9.4 A TIMELINE OF GREEN CHEMISTRY HIGHLIGHTS

- **1962:** *Silent Spring* published by Rachel Carson, which initiated the environmental movement. This publication created awareness about the hazards of pesticide usage and subsequent consequences on the environment.
- **1969:** Citizen's Advisory Committee on Environmental Quality and a cabinet-level Environmental Quality Council were established by Richard Nixon, President of the United States.
- **1970:** The Environmental Protection Agency of United States (US Environmental Protection Agency) was founded.

- **1980/1988:** Pollution Prevention and Toxics, 1988 led the paradigm shift from end-of-pipeline control to prevention of environmental pollution.
- **1990:** Under the chairmanship and administrative control of George H.W. Bush, the Pollution Prevention Act was approved and came into existence.
- **1993:** The concept of Green Chemistry was implemented by US Environmental Protection Agency to serve as a model for design and processing of chemicals without generating hazardous and toxic substances.
- **1995 and 1996:** Presidential Green Chemical Challenge Award was initiated by the US President Bill Clinton to encourage the peoples involved in developing environmentally sustainable design and processes in the chemical sector.[1]
- **1997:** An institute was established to implement the green chemistry principles into practice with the vision to benefit the earth and its people.[2]
- **1998:** Professor Paul T. Anastas and Professor John Warner developed a comprehensive set of 12 principles which are called principles of "Green Chemistry".
- **2001:** Professor Paul T. Anastas of Yale University was nominated as Head of Research and Development at the US Environmental Protection Agency.
- **2008:** A bill which serves to develop policy framework for green chemistry was signed by Governor Arnold Schwarzenegger.
- **2012:** International Green Chemistry World (IGCW 2012) was the largest and most comprehensive event showcasing contemporary developments of green chemistry and environmental sustainability.

9.5 SCOPE

There are many challenges in implementing green chemistry principles into practice. Only modification in the synthetic chemical processes and alternative solvent use without the integration of engineering, physics, and biology is not a sufficient solution for pollution control. In order to achieve the desired goals, proper exercise of comprehensive set of these principles and establishment of interdisciplinary collaboration is a necessary requirement. This can be accomplished through improvement in some

---

[1] Details available at http://portal.acs.org/

[2] Details available at http://portal.acs.org/

essential attributes, such as toxicity, persistence, or energy utilization. Exploitation of green chemistry principles in an integrated manner, rather than thinking of the principles as isolated parameters can ensure the sustainable future of chemical industries (Anastas and Williamson 1998). Integration of these interconnected issues at a stage where they all intersect at a molecular level will be one of the important strategies for implementing the comprehensive set of green chemistry principles.[3]

9.6 PRINCIPLES

Based on a diverse array of practices and contemporary research activities, a comprehensive set of 12 principles have been framed to implement green chemistry approaches into practice. These principles act as substitutes of starting and target materials, reagents, solvents, catalysts, modified processes, and process control (Anastas and Warner 1998). The implication of each principle is described below:

1. **Prevention is better than cure:** Prevention of waste generation is always better than treatment or disposal. Therefore, this principle deals with minimization of waste generation during chemical conversions and manufacturing rather than treatment after its entry into the environment.

2. **Maximization of atom economy:** Design of alternative synthetic methods which can incorporate maximum starting materials into fine products.

3. **Synthesis of less hazardous chemical reactions:** The synthetic methodologies should be designed in such a way that they produce little or no toxic substances. Proper exercise of this principle can minimize the hazards to human health and environment.

4. **Benign chemical design:** This principle deals with the design of less toxic chemicals in comparison to performance of desired functions.

5. **Use of safer solvent and auxiliaries:** During organic chemical conversions, auxiliary substances (such as solvents and separation agents) should be used wherever necessary or efforts should be made to maximize the use of innocuous auxiliary solvent, such as water and supercritical fluid carbon dioxide.

6. **Design and development of minimum energy consuming processes:** Considering environmental and economic impacts of energy resources, chemical conversions should be conducted with

---

[3] Details available at http://greenchemistry.yale.edu/green-chemistry-future, last accessed on 28 July 2015

minimum energy use. Whenever feasible, chemical synthesis should be carried out at ambient environmental conditions.

7. **Utilization of renewable raw materials:** Based on the technical and economical feasibility of synthesis and manufacturing of chemical products, renewable feedstock should be encouraged rather than the depleting one.

8. **Reduction of superfluous derivatization:** In order to prevent generation of toxic and hazardous pollutants, derivatization should be minimized or avoided. This is because such steps require additional reagents that may be toxic in nature and can produce environmentally undesirable toxic waste by-products.

9. **Promotion of catalytic reagents:** Catalytic reagents are better than stoichiometric reagents. Therefore, such reagents should be used as much as possible.

10. **Design of degradable products:** Design and synthesis of biodegradable chemical products minimize the generation of persistent organic pollutants and contribute substantially towards environmental pollution control.

11. **Real-time monitoring and analysis:** Some sophisticated real-time monitoring and analysis methodologies need to be developed that can prevent the formation and entry of pollutants in the environment.

12. **Inherently safer chemistry:** To minimize fires, explosions, or chemical accidents, safer substances should be used during chemical synthesis and manufacturing. Use of inherently non-toxic or non-hazardous materials can ensure such unforeseen events.

9.7 ROLE OF GREEN CHEMISTRY IN PREVENTION AND CONTROL OF POLLUTION

Modern technologies have contributed enormous changes in the environment due to increasing pollution levels. However, technologies useful for creating environmental responsiveness have also provided ways to deal with environmental pollution. Some important aspects in which contemporary technologies have resulted in environmental pollution and deterioration are described below:

- Pesticide and chemical fertilizer intensive agricultural practices have caused water, air, and soil pollution, and land degradation.
- Manufacturing of industrial products that consume large quantities of raw material produces large amount of hazardous waste by-products and causes environmental pollution.

- Exploitation and processing of minerals and other raw materials has contributed to large-scale land degradation, environmental pollution, and biodiversity loss.
- Coal mining, petroleum extraction, and use of fossil fuel as an energy source caused degradation of soil by mining as well as emission of acid rain forming gaseous air pollutants.
- Increase in the number of vehicles and transportation has led to an escalation in the consumption of fossil fuel and emission of large quantities of air pollutants in the atmosphere (Manahan 2000). To construct roads and highways, lands have been acquired in large scales resulting in the degradation of fertile lands.

Green chemistry aims to design environmentally benign processes and goods, and reflects an understanding about environmental policy of pollution control and disposal that are prevailing since the 1970s and were not successful at controlling the pollution caused by chemical industry. The negative consequence of this policy is the transport and spread of chemicals from one environmental medium to the other, making their way into various ecosystems. Even incinerators as a technological centrepiece of waste disposal and pollution control are now prime sources for a variety of metallic pollutants and the largest source of dioxins in the environment. Dioxins are formed due to incomplete combustion of organochlorine wastes (USEPA 1998; USATSDR 1990). The minimum use of toxic raw material prevents generation of hazardous wastes and, thereby, minimizes emissions and exposures in the environment. Though, green chemistry has been considered as a global solution of chemical contamination, this approach is not a sufficient measure, unless benign technologies are implemented at a large scale. Presently, alternatives of hazardous chemicals and substances have been developed through extensive research, but in certain circumstances their widespread application is not taking place. For instance, efficient chlorine-free pulp bleaching does not produce any persistent organochlorine pollutants associated with chlorine-based bleaches. This method is being used in many pulp mills of Europe but only in limited mills of the United States. Eco-friendly substitute to perchloroethylene in clothes cleaning, polyvinyl chloride (PVC) plastic as a construction material, and least pesticide application in agriculture have been implemented to only a limited extent (Thornton 2001). Proper adoption of environmentally benign processes, technologies complying with the green chemistry principles will definitely pave the way for implementing these principles into practices which ultimately helps in prevention and control of pollution.

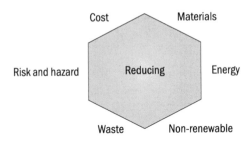

Fig. 9.1 *Green chemistry components of reduction*
Source *Clark (1999)*

Green chemistry has been recognized as a means of minimizations (Figure 9.1). These reductions provide benefits in improvement of economic, environmental, and social standards (USEPA 1998). Reduction of waste generation also minimizes disposal cost, energy use as well as less raw material consumption by making the processes more efficient. These reductions provide net environmental advantages in terms of feedstock utilization *vis-à-vis* waste disposal. In addition to this, the maximum use of renewable feedstock will make the chemical industry more sound, eco-friendly, and sustainable (USATSDR 1990). Safe handling of hazardous chemicals minimizes the risk of accidents and provides an additional social benefit to the workers and end users of the chemical industry.

9.8   ENVIRONMENTAL SUSTAINABILITY

Sustainable development has been considered as a tool for achieving social, economic, and environmental stability. Within these spheres of sustainability, chemistry plays a major role in improving the quality of life. This function of chemistry is not recognized by the government or the public because the three components, namely chemicals, chemistry, and chemists are seen as the root cause of problems related to pollution (Sharma 2010). Therefore, design of chemicals to conserve functional effectiveness and toxicity reduction is the prime role of the chemists because they are the designers of novel molecules and materials. A green chemistry expert ensures functions, safety, and eco-friendly nature of the chemical products synthesized using the green chemistry principles. In chemistry, manufacturing of safe and non-toxic chemical products is the ultimate goal than generation of toxic pollutants (Umakant 2014).

The green chemistry principles have been categorized into two major sectors: (i) reduction of energy use and waste generation, and (ii) manufacturing or utilization of safe goods and processes. Proper exercise of these rules will result in less energy use and minimum waste generation. These principles also ensure the safety and health of the peoples involved (Hughes, LeGrande, Zimmerman, *et al.* 2009; Ustailieva, Starren, Eeckelaert, *et al.* 2012). Similarly, policies framed to uphold green chemistry rules can ensure safety and health of the workers by adopting occupational safety and health criteria (Cunningham, Galloway-Williams, and Geller 2010). Many scientists have proposed safety within a safe-sustainability continuum, where commitment to safety serves in achieving sustainable business practices (Cunningham, Galloway-Williams, and Geller 2010; Gilding, Hogarth, and Humphries 2002). Green chemistry can move forward environmental sustainability by improvement in the processes and products, minimum energy use, conservation of resources, prevention of toxic waste generation, and protection of human health. This philosophical interdisciplinary idea has been articulated under the initiative 'Green'. The activities which utilize the green chemistry principles have been endorsed by the occupational health and safety of the society (Cunningham, Galloway-Williams, and Geller 2010). However, the prospect to integrate health and safety into the sustainability paradigm has not been realized yet (Schulte, McKernan, Heidel, *et al.* 2013).

9.9 CONCLUSION

Green chemistry is not a novel branch of science; it is an idea to implement the principles of chemistry into the practice for environmental sustainability. Several remarkable examples of green chemistry-based practices can be found in the published literatures which are relevant in synthesis, processing, and exploitation of chemical compounds. Based on green chemistry rules, several novel analytical methods have also been developed, which can be helpful in conducting chemical reactions and evaluation of their environmental effectiveness. Efforts have also been made to develop inherently non-polluting processes for chemical manufacturing which can prevent secondary by-product formation during the isolation and purification of final chemical products. However, green technologies developed in laboratory stage do not guarantee their performance with similar pace in commercial scale. Proper implementation of environment-friendly processes and technologies based on green chemistry principles definitely paves the way for practical utility of green chemistry principles into practice. This approach will ultimately help in prevention and control of pollution. Furthermore, accomplishments of green chemistry rules depend on the education and its dissemination to the chemists of this generation. Introduction of philosophy and practices of this new concept to the academician and researchers is an essential step in achieving sustainable development in the R&D sector.

REFERENCES

Anastas, P. T. and J. C. Warner. 1998. *Green Chemistry: Theory and Practice.* Oxford: Oxford Science Publications

Clark, J. H. 1999. Green Chemistry: challenges and opportunities. *Green Chemistry* 1(1): 1–8

Cunningham, T. R., N. Galloway-Williams, and E. S. Geller. 2010. Protecting the planet and its people: how do interventions to promote environmental sustainability and occupational safety and health overlap? *Journal of Safety Research* 41: 407–416

USEPA. 1999. *Green Chemistry Program.* Washington, DC: United States Environmental Protection Agency (EPA), Office of Pollution Prevention and Toxics. Details available at http://(www.epa.gov/gcc

Gilding, P., M. Hogarth, and R. Humphries. 2002. Safe companies: an alternative approach to operationalizing sustainability. *Corporation Environmental Strategy* 9: 390–397

Hughes, J. C., D. LeGrande, J. Zimmerman, M. Wilson, and S. Beard. 2009. Green chemistry and workers. New solutions. *Journal of Environmental and Occupational Health Policy* (19): 239–253

John, W. C., S. C. Amy, and M. D. Kevin. 2004. Green chemistry. *Environmental Impact Assessment Review* 24: 775–799

Manahan, Stanley E. 2000. Environmental science technology and chemistry. *Environmental Chemistry.* Boca Raton: CRC Press LLC

Paul, A. S., L. T. McKernan, S. H. Donna, A. H. Okun, S. D. Gary, J. L. Thomas, C. L. Geraci, E. H. Pamela, and M. B. Christine. 2013. Occupational safety and health, green chemistry, and sustainability: A review of areas of convergence. *Environmental Health* 12: 31

Sharma, S. K. 2010. Green chemistry for environmental sustainability. *Advancing Sustainability through Green Chemistry and Engineering.* Boca Raton, FL: CRC Press

Schulte, P. A., L. T. McKernan, D. S. Heidel, A. H. Okun, G. S. Dotson, T. J. Lentz, C. L. Geraci, P. E. Heckel, and C. M. Branche 2013. Occupational safety and health, green chemistry, and sustainability: a review of areas of convergence. *Environmental Health* 12: 2–9

Thornton, J. 2001. Implementing green chemistry, an environmental policy for sustainability. *Pure and Applied Chemistry* 73(8): 1231–1236

Umakant, C. 2014. Green chemistry: environmentally benign chemistry. *International Journal of Advanced Research in Chemical Science* 1 (1): 110–115

Ustailieva, E., A. Starren, L. Eeckelaert, and I. L. Nunes. 2012. *Promoting Occupational Safety and Health through the Supply Chain.* Bilboa, Spain: European Agency for Safety and Health at Work

Wardencki, W., J. Cury, and J. Namieoenik. 2005. Green chemistry: current and future issues. *Polish Journal of Environmental Studies* 14(4): 389–395

USTSDR. 1990. United States Agency for Toxic Substances and Disease Registry, U.S. Public Health Service

WEBSITES

http://www.cleanwater.org/

http://greenchemistry.yale.edu/green-chemistry-future

10

Building Energy Simulation for Improved Thermal Performance: A CFD Approach

Y. Anand[a], A. Gupta[a], M. Palliwal[b], S. Anand[a,*], and S. K. Tyagi[c]

[a]School of Energy Management, Shri Mata Vaishno Devi University, Katra, Jammu and Kashmir 182320

[b]School of Mechanical Engineering, Shri Mata Vaishno Devi University, Katra, Jammu and Kashmir 182320

[c]Sardar Swaran Singh National Institute of Renewable Energy, Kapurthala, Punjab 144601

*E-mail: anandsanjeev12@gmail.com

10.1 INTRODUCTION

The world's major primary energy-consuming sectors are transportation, buildings, and industry. Among these, buildings are the most dominant sector in terms of share of energy consumption throughout the world. Therefore, the sector needs to be examined carefully for its effective operations. There are various ways to cater to the need of energy developments in these sectors. These approaches may vary from the changes in the energy resources to the price differences with variations in their production. This will account for lesser energy usage and also the production of lesser energy-consuming systems. The industrial sector has already started to transcend from the heavily mechanized industry to a fully or semi-automated industry, aiding in the creation of a better quality product with higher yield. The transportation sector is focused on adopting newer energy-efficient technologies to produce automobiles and freight systems for better performance which includes research in the field of engines that use alternative fuels, highly efficient batteries, efficient electric vehicles, and so on. Indeed, strategies have been formulated based on public reaction resulting from the exorbitant increase in fuel prices, vehicular maintenance charges, and environmental impacts. These strategies include mass commute, car pools, and moving close to the workplace. In the building sector, efforts for designing energy-efficient buildings are underway. There are simplified manual methods for detailed hourly computerized programs to perform the building energy analysis (Al-Homoud 2001). However, the analysis also has to deal with the building's operations during its lifetime and, thus, requires attention. Energy conservation can be achieved by decreasing the air conditioning load, designing climate responsive buildings, constructing closely packed building envelopes, and developing and using energy-efficient equipment. Hence, opting for energy check prior to the construction can help a designer find good design solutions and operations.

10.2   THERMAL MASS AND BUILDING DESIGN

Thermal mass is a pivotal constituent in the success of some passive cooling strategies, active heating and cooling strategies, and most passive solar heating strategies, which are often applied to optimize the performance of energy-conserving buildings, relying heavily on the mechanical cooling and heating strategies (Haglund and Rathmann 1996). Most of us are bewildered by the difference between visible and thermal radiation. The common belief is that thermal energy is converted into mass, rather than being stored in the mass. Albert Einstein stated that mass and energy are interchangeable, just missing the daily workings of thermal mass, where the mass simply stores the excess thermal energy to be released during the required time. Thermal mass facilitates the regulation of energy flow of a building to the advantage of its occupants without the consumption of large amounts of high-grade fuels.

Also, nowadays almost everybody spends his or her time indoors. Thus, most of the commercial and institutional buildings have mechanical air-handling systems that are designed to provide proper ventilation and to ensure that the indoor sources of pollutants are quickly vented out. It is equally necessary to design systems such that the incoming outdoor air is free from odour or contamination. The positioning of air-handling units and exhaust stacks in buildings has also to be carefully analysed to ensure that building inhabitants are not exposed to irritants originating from outdoor sources or indoors. Studies of airflow around buildings generally involve the positioning of cooling towers, exhausts, and air-handling units. These projects have aided in the placement of large cooling towers and air-handling units to address odour complaints owing to exhaust sources near air intakes (McAlpine and Ruby 2004).

As comforters, buildings are designed to provide a serene shelter to occupants in the limited space, even though it might be backed by mechanical cooling and heating systems, and, as already stated, a proper design and operation of a building can have significant energy savings.

Thus, energy management and awareness are the key measures in the lifespan of buildings. A thorough building design can lessen the reliance on additional systems to achieve thermal comfort. The necessity for such systems depends upon climatic parameters and the function and schedule of their operations, over which designers have little control. Thus, the thermal comfort can be modified only through proper selection of building's physical components and their adequate integration throughout the design process. Also, some far-sighted decisions may significantly reduce energy consumption and improve thermal performance. However, the problem associated with modern buildings is that the factors and their dependencies that link them cannot be completely solved by a series of implicit evaluation studies. Such approaches tend to produce defective buildings in many aspects (Handler 1970).

To cope with these problems and improve the effectiveness of solutions, a lot of research is underway (Handler 1970). It has been found that with advanced technology, evolupion of living standards, and human knowledge, such factors are now being edded.

10.3   BUILDING ENERGY ANALYSIS

As buildings are long lasting, the need for energy analysis of buildings becomes crucial. The decision for improving the thermal effectiveness of a building has to be taken at the time of the building construction or a little later after the structure has been built because its implementation later will not be very effective. For having a controlled thermal comfort, most built structures require mechanical ventilating systems. The assessment of energy to operate them is necessary as this can also facilitate early investment returns (Al-Homoud 2001).

The advances in all the extensions of technology have led to the construction of complex building architectures. The increased use of equipment inside buildings and the harsh climatic conditions in many parts of the world have accounted for a lot of internal heat generation, thus making the use of air conditioning system an indispensable need of time. This dependency (that used to be a luxury) requires a lot of energy which needs to be optimized by a judicious application for providing thermal comfort in such buildings. Energy analysis of buildings is carried out for the following cases (Al-Homoud 2001):

  (i) Evaluation of alternative designs, systems, subsystems, components
 (ii) Allocation of annual energy budgets
(iii) Compliance with energy standards
 (iv) Economic optimization

10.4 NECESSITY FOR BUILDING SIMULATION

Design of buildings and their evaluations can be credited to the advancements in computer technology. This has made computer-aided designing (CAD) the most acceptable and frequently used tool as it provides convenience in availability, ease of use and reuse, flexibility, economy, speed, accuracy, and interactive display. This also allows designers to get the colour rendered and the three-dimensional (3D) animation view of the problem. Calculations can be made very rapidly and accurately, and the respective graphical representations can also be easily seen. Hence, CAD is widely used in designing and analysing a building. Energy simulation and optimization models are used for the prediction of thermal performance and for decision-making in structural and space layout aspects of a building design. But the complex and ill-defined nature of the building and the dynamism in the thermal behaviour limit the use of optimization techniques for designing a building. However, today's computers make it feasible and practical to use these models easily to achieve the comprehensive thermal design of buildings (Hong, Chou, and Bong 2000).

To achieve energy conservation, the understanding of heating, ventilation, and air conditioning (HVAC) and lighting systems is of great importance as they account for the major part of a building's energy use. A building's energy requirements depend on the overall performance of the building's walls, windows, and roofs and HVAC and lighting systems. Also, for a large commercial building, the interactions with its environments (complex and dynamic) need to be modelled and simulated for analysis using building simulation tools (Clark and Irving 1988; Shaw 1995; Shaw 1996).

Earlier, the building service engineers relied mostly on manual calculation methods which included standard design conditions. They extrapolated the results to the reviewed conditions, which led to the oversized system and plant and, thus, poor energy performance. So, there was no question of dealing with large buildings and complex designs without the proper building simulation programme (BSP).

Many BSPs have been developed and are in use. BSPs provide users with key indicators, such as temperature, cost, energy demand and use, and humidity (Crawley, Hand, Kummert, et al. 2008). A number of comparative surveys of energy programs have been published, which include the following:

(i) An evaluation procedure for simulation tools and microcomputer energy programs (Building Design Tool Council 1984, 1985; Willman 1985; Lawrie, Klock, and Levernz 1984).

(ii) Cross-examination of analysis tools and passive hybrid design tools for solar low-energy buildings and specific tools for Japan and Asia (Jorgensen 1983; Rittelmann and Ahmed 1985; Matsuo 1985).

(iii) Catalogue of the different programs in the areas of HVAC for the European Commission and for the lighting design software and daylight prediction models (Degelman and Guillermo 1986; Kenny and Lewis 1995; Lighting Design and Application 1996; Aizlewood and Littlefair 1996; de Boer and Hans 1999).

(iv) Comparison of various tools and the energy software (Wiltshire and Wright 1987, 1989; Corson 1990).

(v) Comparative study of thermal tools in Australia and the empirical validation (based on test room data) of thermal building simulation programs (Ahmad and Szokolay 1993).

(vi) Reviews in Engineered Systems Magazine (Amistadi, 1993, 1995; Lomas, Eppel, Martin, *et al.* 1994).

(vii) Cataloguing of DOE-based building energy tools and tools for energy auditing (Crawley 1996; Khemani 1997).

(viii) Comparison assessment of two HVAC simulation programs (Underwood 1997).

(ix) Evaluation of a wide range of simulation engines (Zmeureanu 1998; Haltrecht, Zmeureanu, and Beausoleil-Morrison 1999).

(x) Information of building energy, life cycle costing, and utility tools (Waltz 2000).

(xi) Cataloguing the existing simulation and design tools relative to the user requirements and providing recommendations for further tool development requirements pertaining to the entire building and its envelope, HVAC component, and simulation and design tools (Jacobs and Henderson 2002).

Simulation is used to reflect the performance of a system over a set of selected measures. Here, input and output variables are implicitly related, which are logically linked to each other. In a building design, models are used to evaluate the performance of a building under given conditions, with the provided values for the associated variables. There are two modelling methods, namely, the sequential method [referred to as the load, system, plant, and economics (LSPE) sequence] and the simultaneous solution method, which is used in the building energy simulation programs.

(i) **Sequential method:** In this method, the annual (hourly incremented) thermal (that is cooling and heating) loads are initially found for individual space based on the occupancy. This is followed by the secondary system simulation, where the calculations are made of energy flows at air-handling units or other equipment supplied by the central plant. Then calculations of the source energy requirements are made. These results are then used to deduce the cost or life cycle cost of the source energy. This method is advantageous for the computer as it consumes lesser time and memory. The disadvantage of this method is that it reports only to the unmet loads as it is not capable of interaction between system, load, and plant, which could produce challenging results.

(ii) **Simultaneous modeling method:** This method was developed to address the shortcomings of the sequential method. Here, the heat balance and the weighting factor methods are used for load modelling.

- In the **weighting factor method**, heat gains to building spaces are converted to cooling and heating loads using pre-calculated weighting factors. This procedure indeed improves the accuracy, but requires more computation time and computer memory (Sowell and Hittle 1995; ASHRAE Handbook 1997; Solar Energy Laboratory 1988).

- In the **heat balance method**, fewer assumptions are required as loads are calculated using a detailed heat model of the thermal transfer processes in rooms. Unknown surface and air temperatures are used to solve the equations for each enclosing surface and space air. Based on this, calculations for the convective heat flow to/from the space air mass are made. The packages based on this method are transient system simulation tool (TRNSYS) and building load analysis and system thermodynamics (BLAST).

10.5 APPLICATION OF BUILDING SIMULATION AND SYSTEMS APPROACH

A building has many components associated with it, namely, its structure, construction material, HVAC and other systems, lightings. Also, it is a common tendency to keep changing the internal arrangements and structure. There are several applications, such as the building heating/cooling load calculation, energy performance analysis (design and retrofitting), design of building energy management and control system, compliance with building regulations, cost analysis.

All the components interact with each other to account for the overall performance of the building. So, while dealing with such problems, an interdependent approach should be followed to reach to the solution of many performance problems. However, to get efficient solutions, there is a need for a proper guidance mechanism for building designers. Also, as the problem involves many components, the domain is divided into smaller and manageable components linked logically and systematically. Such an approach aids in finding the best solution for the system.

Thus, a systems approach implies the implementation of optimization techniques. Optimization techniques systematically model and analyse decision problems based on some mathematical functions. However, extreme care should be taken while formulating a problem. Therefore, in optimization, the best solution is sought that satisfies objectives from among a field of feasible solutions under certain constraints.

10.6   SOFTWARE

Software is used for many applications, namely, calculation of the predictive mean vote, wind and indoor air circulation, structural design work, pipe work design, drainage system design, and energy consumption calculations and analysis. Users of software include HVAC engineers, manufacturers, energy consultants, energy policy makers, and so on. Two of the most commonly used software are discussed in the following sections.

10.6.1   TRNSYS

Transient system simulation program (TRNSYS) is a FORTRAN-based program developed in 1970 at Solar Energy Laboratory, University of Wisconsin, Madison (Solar Energy Laboratory 1988). Its aim was to conduct transient analysis of active solar heating and cooling systems on an hourly time step. TRNSYS is a component-based simulator (Fireovid and Fryer 1987). Its greatest advantage is in the simulation of the performance of large and complex systems. The accuracy of the simulation is highly dependent on the program input assumptions, which include the following:

(i)   Circuit design and component choice

(ii) Component used

(iii) Weather conditions used

(iv) Interpretation of building drawings and specifications

(v)   Operational characteristics

(vi) User experience

Furthermore, through such a tool, professionals can incorporate new technologies and innovations to their designs, thus enhancing energy savings.

10.6.2　Computation Fluid Dynamics

The origin of this field is credited to the advancements in the field of computing technologies which have reduced the problem-solving time to a few hours (depending on the computer configuration). There are many commercial computation fluid dynamics (CFD) packages available, but the solution technique is similar in all. Initially, there is a domain (model) creation and visualization to mark a problem. This is followed by the choice of meshing and setting the boundary conditions, such as air inlets, outlets, and exhaust. When parameters have been defined, the project is imported to the solver. In the solver, the solution is found by solving the equations of fluid motion, for the boundary conditions over the domain, till steady state is achieved. The k–ε turbulence model is commonly used to account for steady-state turbulence and advanced numerical schemes to solve the problems efficiently. Several types of k–ε turbulence models are available for use with the solver, including a re-normalized group (RNG) model and so on. Finally, the solution is loaded into a visualization program, where streamlines, vectors, scalars, and a variety of other features can be graphed to observe the results (Chen and Jiang 1992; Chen 1997a, 1997b; Bunn 1995; Howard and Winterkorn 1993).

Equations of Motion

The basic equations of motion are known as Navier–Stokes equations. These are a set of equations that relate the conservation of mass, momentum, and energy. Initially, we have five equations (the continuity equation, the conservation of momentum equation in each of the three dimensions, and the energy equation) and seven unknowns [pressure P, density ρ, temperature T, x-component velocity u_1, y-component velocity u_2, z-component velocity u_3, and internal energy (expressed in terms of the enthalpy H]. The continuity equation for an open Newtonian system with possible sources or sinks S_{mp} for mass at a point within the system can be written as follows:

$$\frac{\partial p}{\partial t} = \frac{\partial(\rho u_i)}{\partial x_i} = S_{mp}$$

where the subscript "i" identifies the directions through the faces of each sub-volume of the system. Similarly, the three separable equations for the conservation of momentum can each be written as follows:

$$\frac{\partial(\rho u_i)}{\partial t} + \frac{\partial(\rho u_i u_j)}{\partial x_i} = \frac{\partial \tau_{ij}}{\partial x_i} - \frac{\partial P}{\partial x_i} + \rho B_i + S_{u_i p}$$

where B represents the body forces on the sub-volume (for example, gravity), S is again the source or sink term, and τ_{ij} is a collection of cross product terms of velocity differential components identified as the viscous stress tensor. Finally, the conservation of energy equation can be written for a fixed sub-volume given as follows:

$$\frac{\partial \rho H}{\partial t} + \frac{\partial \rho u_i H}{\partial x_i} = \frac{\partial}{\partial x_i}\left[\frac{k}{C_p}\frac{\partial H}{x_i}\right] + \frac{\partial P}{\partial t} + u_i\frac{\partial P}{\partial x_i} + \varphi + Q_p + S_{HP}$$

where φ is another collection of cross product terms, identified as the Stokes molecular dissipation function, Q is the rate of energy added as heat to the sub-volume, and S is the source or sink term for enthalpy. The additional terms K (the thermal conductivity) and C_p (the specific heat), which are the measured data for a specific problem, are necessitated by our use of H instead of the internal energy. Such a system of equations is not solvable without additional assumptions. We can assume a relationship between fluid density and other thermodynamic variables. In our case, we use the perfect gas law, $T = P/\rho R$, where R is the gas constant, which is reasonable for air at the surface of the earth. In addition, we can assume that we are able to express the conservation of turbulent kinetic energy k and the dissipation rate ε of turbulent kinetic energy with similar transport equations (usually expressed in a more complex form than would be useful to reproduce here) that relate back to the other equations through the stress tensor τ and the molecular dissipation function φ, and do not include any more unknowns because we are able to assume the values for k and ε that are constant across the system. With the five initial equations, the assumption of the perfect gas law, and $k - \varepsilon$ assumption, we now have a system of equations that can be solved. Any number of scalar transport equations, as well as chemical reaction equations that couple to this set of equations, can also be used in the solution.

The tricky problems for CFD involve moving air and obstacles to flow which require the flow to have zero velocity at the object wall. This results in a gradient of velocity across the flow field, even with no changes in the air direction or velocity. Similarly, the real atmosphere under gravitational force yields a pressure gradient that must be considered even at a small scale of urban buildings. The perfect gas law assumption is then valid only locally, which limits the size of the solution space to microscale. The most useful steady-state solutions are in dynamic equilibrium, averaged over both space and time. The model does not provide information about the range of fluctuations about this equilibrium solution. Reynolds averaging is applied to equations to derive the steady-state solutions. With Reynolds averaging, the instantaneous value of a variable is assumed to be the sum of the mean value and a perturbation from that value. For example, if $u(t) = \bar{u} + u'(t)$, then the steady-state solution is the expected value of this quantity, or $\bar{u}(t) = $ mean $[\bar{u} + u'(t)]$. Replacing each velocity variable with the Reynolds-averaged variable results in the Reynolds-averaged Navier–Stokes equations, which can be solved for the steady-state flow. The vector cross products generated when the mean value is taken of the velocity perturbations are known as Reynolds stresses. These cross product terms allow information, for example, about the location of a wall, to be disseminated throughout the model space. Turbulence modelling is concerned with accounting for these terms and the viscous dissipation term. The $k - \varepsilon$ equations assume isotropic eddy viscosity, which simplifies the viscous dissipation term. The method of parameterized Reynolds stresses and viscous dissipation by turbulent kinetic energy transport is a popular one that is used rigorously, mostly owing to the comparatively lower computational burden of the method. The two turbulent kinetic energy and dissipation equations replace a complex numerical representation of the flow, which would be difficult to model. Furthermore, isotropic eddy viscosity is a good assumption in lower velocity atmospheric boundary layer flow. Though this method is not perfect and inaccurate in certain situations, it has proven to be sufficiently appropriate in representing the steady-state flow of situations where the length scales of velocity and obstacles are much greater than the length scales of turbulence.

10.7 APPLICATIONS OF CFD FOR BUILDING DESIGN

As discussed, CFD, owing to its many capabilities, can be used for many applications, but in the building design, it can be used for applications such as site planning, natural ventilation study, HVAC system design, and pollution dispersion and control. In site planning, CFD can help optimize building sites by predicting the distributions of air velocity, temperature, moisture, turbulence intensity, and contaminant concentration around buildings. It can also improve outdoor pedestrian comfort and increase energy efficiency of buildings by deploying passive HVAC strategies, such as using natural ventilation for summer and wind break for winter. Figure 10.1(a) and (b) presents an example of using CFD in which the movement of air inside the room (depicted by velocity vectors) coming from the air conditioner is shown. Walls are 0.3 m thick and subjected to varying temperature conditions, so as to imitate a room of a house which is not exposed from any side and give a variation for better understanding of simulation. The wall representation and the temperature set on that wall are given in Table 10.1.

Table 10.1   Wall representation and the temperature set on walls

Sl. No.	Wall	Placement	Temperature (°C)
01	01	Right of AC	56.85
02	02	AC mounted Wall	40
03	03	Opposite to Wall 02	50
04	04	Opposite to Wall 01	46

The velocity streamlines are also effected by the placement of the outlet duct. The variation of the temperature on walls and floor can also be observed. The simulation is done for a thermally comfortable condition inside the room; that is, the internal temperature is maintained at 26°C.

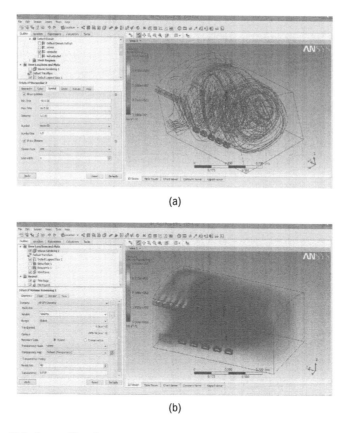

(a)

(b)

Fig. 10.1   *Velocity profiles shown as (a) velocity streamline and (b) velocity rendering*

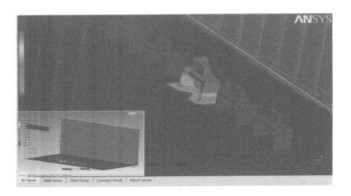

Fig. 10.2 *Convection of heat from the computer*

Figure 10.2 shows the distribution of heat inside the room. The explanation to this is given by considering a test row of 05 desktop computers. One of them is put in "ON" condition, and thus it becomes the source of heat generation inside the room. The assumed heat generated is 137°C. Heat generation rate is 70 W/m^3 and the heat flux is 89 W/m^2. The convection of the heat inside the air-conditioned room is shown with the help of a plane in Figure 10.3.

As mentioned earlier, the ambient conditions of exposure to buildings are considered. So, the ambient temperature is taken to be 27°C. The effect of the air conditioning and the heat generation by one computer inside the room on the temperature of walls and floor is shown in Figure 10.4 (a). For the ease of visualization, an aerial view of four walls and floor is shown. The temperature variation on the four walls and the floor along with the heating load can be seen in Figure 10.4(b).

The temperature is 16.5°C on floor and it varies from 27°C to 29°C on walls for a constant ambient temperature [Figure 10.4(b)].

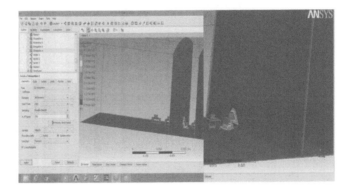

Fig. 10.3 *Plane showing the convection of heat from the heat source*

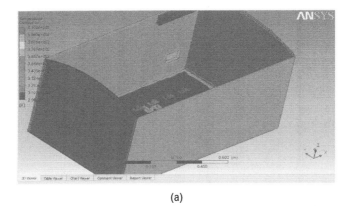

(a)

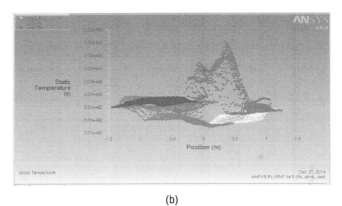

(b)

Fig. 10.4 (a) *Effect of air conditioning and heat generation on the temperature of walls and floor and* (b) *graphical representation of temperature contours on walls*

A wide application of CFD has shown its ability in the dispersion modelling of the indoor and outdoor contaminants. It is useful in the prognosis of extreme conditions, such as toxic scenarios, and can be easily used to assess the impact of the dispersion of a particular contaminant under parameters such as wind speed or air temperature.

Figure 10.5 (a) and (b) shows the case of interaction between two nearby buildings, considering a microenvironment, with respect to the smoke distribution originating from one building. Figure 10.5 (a) represents the condition with smoke and Figure 10.5(b) represents the same condition without smoke. Here, the wind speed is taken as 3 m/s and the smoke escaping from the chimney is 2 m/s.

Figure 10.6 depicts the same case for the wind speed of 4.4 m/s and the smoke escape velocity of 5 m/s. The streams of air along with the streams of smoke are shown in the figure. Such problems can be beneficial in designing a building with respect to its orientation, placement of windows

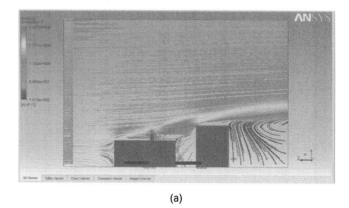

(a)

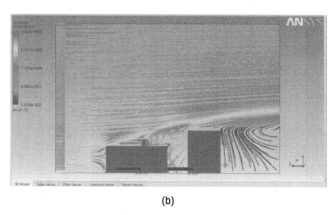

(b)

Fig. 10.5 *(a) Velocity streamlines of wind with smoke and (b) Velocity streamlines of wind without smoke*

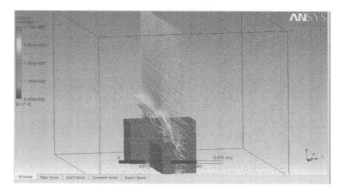

Fig. 10.6 *Velocity streamlines of airflow and smoke*

and ventilation systems, installation of HVAC systems, placement of chimneys, and so on.

Hence, there are two cases. In one case, the wind speed is greater than the smoke velocity, and in the other case the wind speed is lesser than the smoke velocity. Figure 10.6 shows this process with respect to the root

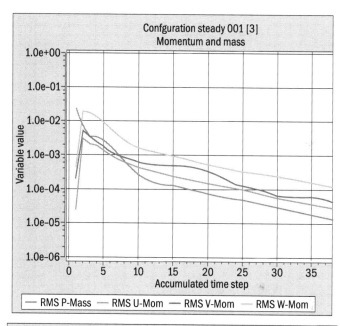

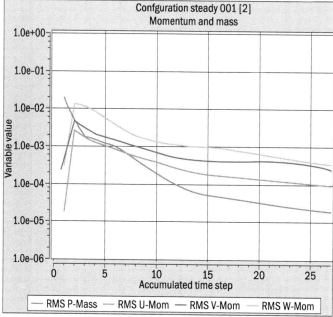

Fig. 10.7 (a) *Graphical representation of momentum and mass model*

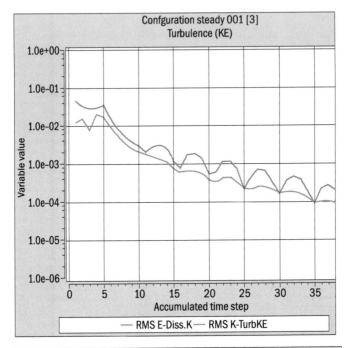

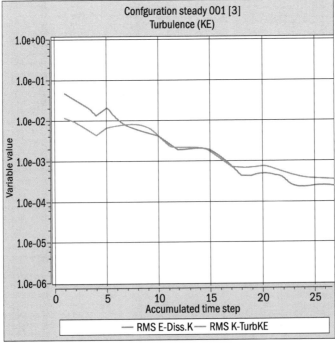

Fig. 10.7(b)   *Graphical representation of turbulence model*

mean square (RMS) values obtained from different turbulence models of the software.

Figure 10.7(a) shows the RMS for the mass and momentum (in x, y, and z directions) distribution for the smoke and wind interaction. The solution was deemed converged when the values were below 1×10^{-4} level for the first case, then the values were at the 1×10^{-3} level, and finally the flow reached a steady value. Convergence was generally achieved between 30–40 iterations and 20–30 iterations, which was dependent on the superficial wind and smoke velocity. Figure 10.7(b) shows the RMS for turbulence quantities.

When the wind speed is more than the velocity of smoke mixing with the wind, the addition of the smoke takes some time (that is, fractionally more time) to blend, and the momentum values suggest that it is evenly distributed afterwards. However, in the second case, in which that the velocity of smoke is greater than the wind velocity, there is more smoke being added to the mix (wind and smoke), and so the momentum values rise

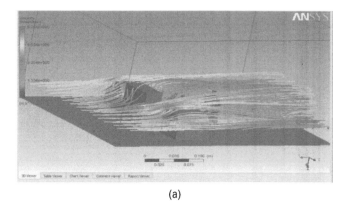

(a)

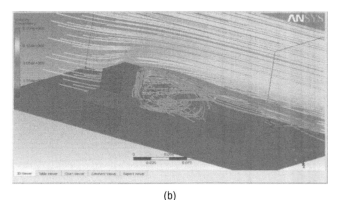

(b)

Fig. 10.8 *Representation of the wind flow over a spread of buildings*

compared to the values when the condition is reversed, but the momentum is still linear after the mixing has taken place.

Also, in the study of the microenvironment, the effect of the layout on the society can be examined. Figure 10.8 (a) and (b) shows how the spread of built structures within the locality hinders the flow of wind, causing it to, in turn, affect the built structures.

Figure 10.8 (a) and (b) shows the effect of flowing wind over the plan of the locality. It can be seen in Figure 10.9 that the mass of the wind is not linearly dispersing due to the formation of the whirlpool of wind as a result of the obstruction of buildings. Also, the convergence of the momentum is at the level between 1×10^{-2} and 1×10^{-4}.

10.8   CONCLUSION

The integration of CFD with other building modelling programs is the need of the hour. Even though it took a long time to subscribe to CFD as a tool for building simulation, it is being extensively applied to all spheres of building design. Many complex problems can be modelled through CFD, with limited knowledge of fluid mechanics and building science. Hence, the demand of the day is to develop a framework for using the package for building design as the nature of problem is complex. New directions for building study are being promoted to compute as many problems as possible, and if the simulation cannot be carried out, then the requisite alteration can be incorporated for the prediction of building performance. Also, as the understanding and implementation of the CFD accounts for the knowledge of the fluid flow, to make it more user-friendly, the simplification of the package is required. Improvement in algorithms towards the generation of the simulated results shall account for faster results, thus reducing the computing costs. Most building phenomena can be simulated by current CFD programs, such as the case of modelling explained, and the results are obtained for the following cases:

1. For the heat transfer in an air-conditioned room towards the walls of the room.
2. For the dispersion of smoke towards the neighbouring building.
3. For investigating the locality for effects of wind flow and its dispersion.

The advances in the building continue to impose new challenges to CFD, for example, modelling of new window designs and advanced heating/cooling systems. The problem may become more complicated when there is the involvement of mass transfer, phase changes, and multi-phase interactions, such as air condensation and combustion. Special methods and models need to be developed to accommodate such changes.

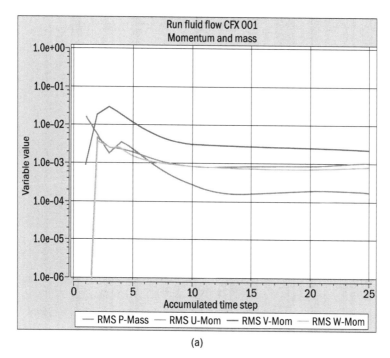

(a)

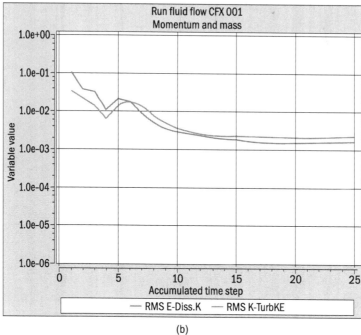

(b)

Fig. 10.9 *Graphical representation of (a) momentum and mass model and (b) turbulence model for the wind flow over the spread of buildings*

REFERENCES

Ahmad, Q. and S. Szokolay. 1993. Thermal design tools in Australia: A comparative study of TEMPER, CHEETAH, ARCHIPAK and QUICK. In *Proceedings of Building Simulation 99*, IBPSA Conference, Adelaide, Australia, August, pp. 351–357

Aizlewood, M. E. and P. J. Little fair. 1996. Daylight prediction methods: a survey of their use. *CIBSE National Lighting Conference Papers*, Bath, UK, pp. 126–140

Al-Homoud, M. S. 2001. Computer-aided building energy analysis techniques. *Building and Environment* 36: 421–433

Amistadi, H. 1993. Energy analysis software review. *Engineered Systems* 10: 34–45

Amistadi, H. 1995. CAD-building load software review. *Engineered Systems* 12(6): 50–67

ASHRAE (American Society of Heating, Refrigerating and Air-Conditioning Engineers). 1997. *ASHRAE Handbook: Fundamentals*. Atlanta, GA: American Society of Heating, Refrigerating and Air-Conditioning Engineers

Building Design Tool Council. 1984. *Evaluation Procedure for Building Energy Performance Prediction Tools*, Vol. 1. Washington, DC: ACEC Research and Management Foundation

Building Design Tool Council. 1985. *Design Tool Evaluation Reports for ASEAM, CALPAS3, CIRA, and SERI-RES*. Washington, DC: ACEC Research and Management Foundation

Bunn, R. (Ed.). 1995. Computer survey. Building Services. *CIBSE Journal* 11: 61

Chen, Q. 1997a. Computational fluid dynamics for HVAC: Successes and failures. *ASHRAE Transactions* 103(1): 178–187

Chen, Q. 1997b. Controlling urban climate: Using a computational method to study and improve indoor environments. *Journal of Urban Technology* 4(2): 69–83

Chen, Q. and Z. Jiang. 1992. Significant questions in predicting room air motion. *ASHRAE Transactions* 98(1): 929–938

Clark. J. A. and A. D. Irving (eds). 1988. Special issue on building energy simulation. *Energy and Buildings* 10(3)

Corson, G. C. 1990. *A Comparative Evaluation of Commercial Building Energy Simulation Software*. Gale G. Corson Engineering

Crawley, D. (Ed.). 1996. *Building Energy Tools Directory*. Washington, DC: U.S. Department of Energy, Office of Building Technology, State and Community Programs

Crawley, D. B., J. W. Hand, M. Kummert, and B. T. Griffith. 2008. Contrasting the capabilities of building energy performance simulation programs. *Building and Environment* 43: 661–673

de Boer, J. and E. Hans. 1999. Survey simple design tools. IEA SHC Task 21, Subtask C4 simple design tools. Stuttgart: Fraunhofer-Institut fur Bauphysik

Degelman, L. O. and A. Guillermo. 1986. *A Bibliography of Available Computer Programs in the Area of Heating, Ventilating, Air Conditioning, and Refrigeration*. Atlanta: ASHRAE

Fireovid, J. A. and L. R. Fryer. 1987. *ASEAM Users' Manual*, 2nd Edn. Washington, DC: ACEC Research and Management Foundation

Haglund, B. and K. Rathmann. 1996. Thermal mass in passive solar and energy-conserving buildings. Vital Signs Curriculum Materials Project, Center for Environmental Design, University of California, Berkley

Haltrecht, D., R. Zmeureanu, and I. Beausoleil-Morrison.1999. Defining the methodology for the next-generation HOT2000 simulator. In *Proceedings of Building Simulation 1999*, Kyoto, Japan, Vol. 1, pp. 61–68

Handler, B. A. 1970. *Systems Approach to Architecture*. New York: American Elsevier Publishing Co.

Hong, T., S. K. Chou, and T. Y. Bong. 2000. Building simulation: An overview of developments and information sources. *Building and Environment* 35: 347–361

Howard, R., E. Winterkorn, and I. Cooper. 1993. Building environmental and energy design survey. *Building Research Establishment Report*. IHS BRE Press

Jacobs, P. and H. Henderson. 2002. State-of-the-art review of whole building, building envelope, and HVAC component and system simulation and design tools. *Final report ARTI-21CR/30010-01*. Arlington, VA: Air Conditioning and Refrigeration Technology Institute

Jorgensen, O. 1983. Analysis model survey. Passive and Hybrid Solar Low Energy Buildings. Report No. 43. T.VIII.B.1.1983, IEA SHC Task 8. Lyngby: Thermal Insulation Laboratory, Technical University of Denmark

Kenny, P. and J. O. Lewis (eds). 1995. Tools and techniques for the design and evaluation of energy efficient buildings. European Commission DG XVII Thermie Action No. B 184. Dublin: Energy Research Group, University College Dublin

Khemani, M. 1997. Energy audit software directory. Ottawa: Natural Resources Canada

Lawrie, L., W. Klock, and D. Leverenz. 1984. Evaluation of Microcomputer Energy Analysis Programs, *Technical Report E-193*, US Army Construction Engineering Research Laboratory, Champaign

Lighting Design and Application. 1996. *IESNA Lighting Design Software Survey*, pp. 39–47. New York: Illuminating Engineering Society of North America

Lomas, K., H. Eppel, C. Martin, and D. Bloomfield. 1994. Empirical validation of thermal building simulation programs using test room data, Vol. 1. *Final report*. Leicester: De Montfort University

Matsuo, Y. 1985. Survey of simulation technology in Japan and Asia. In *Proceedings of Building Energy Simulation 1985,* Seattle, pp. 23–30

McAlpine, J. D. and M. Ruby. 2004. Using CFD to study air quality in urban microenvironments. *Environmental Sciences and Environmental Computing* 2: 1–31

Rittelmann, P. R. and S. F. Ahmed. 1985. Design tool survey. IEA SHC Task 8 passive and hybrid solar low energy buildings, Subtask C design methods. Washington, DC: U.S. Department of Energy

Shaw, M. (ed.). 1995. Computer modelling as a design tool for predicting building performance: Part 1. *Building Services Engineering Research and Technology* 16(4): B41–B54

Shaw, M. (ed.). 1996. Computer modelling as a design tool for predicting building performance: Part 2. *Building Services Engineering Research and Technology* 17 (2)

Solar Energy Laboratory. 1988. *TRNSYS Users' Manual*. Madison, MI: Solar Energy Laboratory, University of Wisconsin Madison

Sowell, E. F. and D. C. Hittle. 1995. Evolution of building energy simulation methodology. *ASHRAE Transactions* 101(1): 850–855

Underwood, C. 1997. A comparative assessment of two HVAC plant modelling programs. In *Proceedings of Building Simulation 1997*, Prague, Czech Republic, Vol. 1, pp. 385–392

Waltz, J. P. 2000. *Computerized Building Energy Simulation Handbook*. Lilburn, GA: The Fairmont Press, Inc.

Willman, A. J. 1985. Development of an evaluation procedure for building energy design tools. In *Proceedings of Building Energy Simulation*, Seattle, WA, pp. 302–307

Wiltshire, J. and A. Wright. 1987. The Evaluation of the Simulation Models ESP, HTB 2 and SERI-RES for the UK Passive Solar Program, *Report for Energy Technology Support Unit*. London: Department of Energy

Wiltshire, J. and A. Wright. 1989. The documentation and evaluation of building simulation models. Building Environmental Performance Analysis Club

Zmeureanu, R. 1998. Defining the methodology for the next-generation HOT 2000 simulator, Task 3 report. Ottawa: Natural Resources Canada

11

Cyanobacterial Biomass – A Tool for Sustainable Management of Environment

Vaishali Gupta[*] **and Jaishree Dubey**

Department of Botany, Dr Hari Singh Gour Central University, Sagar, Madhya Pradesh 470003

[*]E-mail: vaishaligupta28@gmail.com

11.1   INTRODUCTION

Sustainable environment requires a balance between the natural resources available and their uses. As the natural resources have been exploited for a long time, they cannot be replenished easily. However, the biological and natural resources need to be replenished in order to improve and cope up with the present environmental problems. In the last few decades, cyanobacteria have gained attention because of their potential application in environmental management. Studies have reported that oil components and surfactants, herbicides, and other organic compounds are oxidized by cyanobacteria. Carbon dioxide (CO_2) sequestration by cyanobacteria reduces greenhouse gases (GHGs) and mitigates global warming. As an alternative future energy source, cyanobacteria can be used because apart from being ecofriendly in nature, they are efficient, renewable, and CO_2 is not produced during their production and utilization. The most common limiting nutrient for plant productivity is nitrogen, which has led to the increase usage of fertilizer. The demand for chemical fertilizers has increased and it is important that their consumption be reduced. Environmental problems, such as GHG effect, depletion of ozone layer, and water acidification are increasing rapidly due to excessive use of chemical fertilizers. Currently, 100 million tonnes of nitrogen (N_2) per annum is applied as fertilizers for agricultural production worldwide. Of this 50% is consumed in the production of three major crops: wheat, maize, and rice (Heffer 2010). Worldwide, the demand for nitrogen fertilizer has increased from 108.2 million tonnes in 2011 to 109.9 million tonnes in 2012, at a 1.6% growth rate. It is expected to be around 116 million tonnes in 2016 at the annual growth rate of 1.3%. In 2007/08, India consumed 225.7 million tonnes of fertilizers and this reached 277.40 million tonnes in 2011/12 (IFFCO 2012).

Cyanobacteria should be the prospective environmental sustainability agent. They are one of the important factors in the ecosystem and a critical component of the earth's biosphere. Originating 3.5 billion years ago, these oxygen producing photosynthetic gram negative bacteria fall under Cyanophyta (myxophyta), a diverse taxon of prokaryotes. In the Archaean and Proterozoic eras (2.7 billion years ago), it was cyanobacteria that brought oxygen revolution through their photosynthetic activities. They are ubiquitous, inhabit a wide range of diverse and extreme habitats, and their numbers are also increasing with new studies; however, the total number of species in different taxonomic groups remains uncertain. At present, 2698 species of cyanobacteria have been formerly described out of 2000 to 8000 estimated species. As per Gompertz model (Nabout, Rocha, Melox, et al. 2013; Gompertz 1825), it is estimated that 6280 species exist. Their wide adaptability in diverse habitats makes them excellent materials to be explored and exploited by ecologists, biotechnologists, microbiologists, biochemists, and physiologists. Due to their bioremediation potential, cyanobacteria have been used for wastewater treatment. They immoblize heavy metals in their sheaths from effluent. Cyanobacteria, as a biofertilizer, are safe alternatives to chemical fertilizers. They are studied for their nitrogen fixation mechanism, morphology, photosynthetic activity, and certain other aspects of their structures. The key enzyme nitrogenase fixes atmospheric nitrogen (Stewart, Rowell, Kerbly, et al. 1987). Nitrogen fixing cyanobacteria are unique as they are able to fix both carbon and nitrogen symbiotically or asymbiotically. This dual ability of fixing nitrogen and assimilating carbon makes them a highly productive and efficient biological system (Singh, Dhar, Pabbi, et al. 2000). Besides, they also produce a wide variety of chemically unique secondary metabolites, which have apparently complex role in defending against aquatic invertebrates and their larvae.

11.1.2 Distribution

Cyanobacteria are cosmopolitan as they exhibit worldwide distribution including in extreme environment. They can adapt in extremes of environment and are widely spread both in aquatic and in terrestrial habitats. Aquatic cyanobacteria, such as *Dermocarpa* and *Trichodesmium*, mostly occur in fresh water. Cyanobacteria, such as *Oscillatoria*, *Nostoc*, and *Scytonema* inhabit damp surface and can thrive in hot springs, salt marshes, wet rocks, damp soil, and polar regions including Antarctic rocks. A few of the cyanophycean members exhibit cyclic growth, and their buoyancy results in appearance and forms plankton bloom. They are also found in almost every endolithic ecosystem (Rios, Martin, Sancho, *et al.* 2007). A few members of cyanphyceae grow symbiotically in association with other plants, and fungi, such as *Nostoc,* is found within the thalli of *Anthoceros* and *Anabaena cycadeae* is reported in coralloid root of cycas.

11.1.3   Structure

Cyanobacteria are among the simplest photosynthetic organisms living today. In some species, the thallus is unicellular (*Chroococcus*), colonial (*Gleocapsa*), and filamentous (*Oscillatoria*).

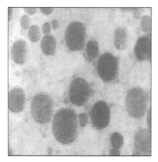
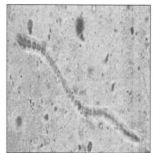

Unicellular form   Filamentous form

The colonies may be filamentous or non-filamentous. Each colony is generally enclosed in a gelatinous sheath. Filaments are either branched or unbranched, and sometimes it may be pseudobranched, such as in *Scytonema.*

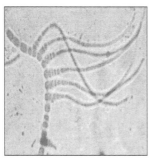
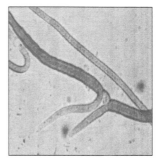

True branching   False branching

11.1.4   Nitrogen Fixation Sites

Heterocyst is the site of nitrogen fixation in cyanobacteria (Wolk, Ernst, and Elhai 2004). It is a specialized cell that develops from an ordinary vegetative cell, particularly the recently divided. One of the daughter cells, known as preheterocyst, develops into a heterocyst and the other daughter cell develops into a vegetative cell. Rarely do both the daughter cells develop into a pair of heterocyst. In rice fields, heterocysts are considered ecologically and agriculturally important because of their role in nitrogen fixation. Figure 11.1 shows the diagrammatic representation of the ultra structure of a heterocyst in L.S.

11.1.5   Mass Cultivation of Cyanobacteria

Cyanobacteria act as a potentially useful tool for bioremediation and soil improvement in the management of the ecosystem. As a result, cyanobacterial biomass should be mass cultivated. For this strain selection is very important. The selected strain should fulfil the following four minimum criteria:

1. Rapid growth rate over a wide range of temperature.
2. Nitrogen fixation ability in pH 6.5–8.5.
3. No adverse effects of agriculture chemicals and pesticides.
4. Able to survive with specific carrier.

Regional strain should be selected for mass cultivation. The Department of Biotechnology (DBT), Government of India has established several regional culture collection laboratories for studying, producing, and accessing starter inoculums. A mixture of five or six different types of acclimatized cyanobacterial strains are used for starter inoculums that

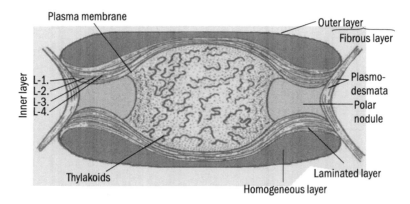

Fig. 11.1 *Diagrammatic representation of the ultra structure of a heterocyst in L.S.*

are cultured in the laboratories. Repetitive culture techniques are used in broth and agar media.

Mass production

Cyanobacterial species are mass cultured in large scale for various uses including biofuel, medicine, food, and biofertilizer production. Mass cultivation of cyanobacteria is done by the following methods:

(i) **Cemented tank method:** Permanent, measured cemented tanks are prepared to mass produce cyanobacteria. In large trays, soil is mixed well with superphosphate and insecticide, and filled with water. The pH is adjusted at neutral by adding lime. When the soil settles, cyanobacteria are spread over the tray. The unit is well aerated and exposed to sunlight. Water is added at regular intervals to make up its loss. Within a few days, cyanobacteria start growing and form a thick mat. Now these cyanobacterial mats are left to dry into flakes that are then collected and stored for further use. For further propagation, a few flakes are left as inoculums and the same process is repeated for large-scale production (Patterson 1996). See Figure 11.2.

(ii) **Pit method:** This technique follows the method similar to cemented tank method. However, it is used in small-scale production and is extremely easy and not expensive. The difference between

Fig. 11.2 *Cement tank for mass cultivation of cyanobacteria*

pit method and cemented tank method is that instead of troughs or tanks, shallow pits are dug in the ground and laminated with a thick polyethylene sheet to hold the water.

(iii) **Field method:** A bigger version of the cemented tank method, the field method is used to produce bulk quantities of cyanobacterial biomass material. In this 40–50 m^2 fields are selected. The soil has to be puddled well to facilitate water logging. The fields are filled with a thin layer of water above the soil. Insecticides and superphosphate are mixed well in the soil. Within 8–10 days, cyanobacterial growth reaches optimal level. The cyanobacterial mats are left to dry and after a few days the cyanobacterial flakes are collected. The process of cyanobacterial propagation and harvest can be repeated many times without further cyanobacterial inoculation. Figure 11.3 shows the polyhouse for mass cultivation of cyanobacteria.

(iv) **Nursery-cum-cyanobacterial method:** This method is useful for farmers who want to produce biofertilizers and rice seedlings in nurseries. The nursery is made adjoined with cyanobacterial production field as the latter is efficient in generating biofertilizer. This concurrent nursery method uses the whole area for transplantation so that the land neither gets wasted nor logged up, especially for cyanobacterial culture (Meeting 1988). As early as 1977, the Chinese Agricultural Research Institute, in Nanjing, used a mixture of *Anabaena* sp. and *Nostoc* sp. as inoculants at 750 kg per hectare to produce, within 10–15 days, very efficient cyanobacterial biofertilizer material amounting to 7.5 tonnes per hectare in normal cool weather and 15 tonnes per hectare when the temperature rose above 3°C (FAO 1977).

Fig. 11.3 *Polyhouse for mass cultivation of cyanobacteria*

Fig. 11.4 *Mass production by biofermentor*

(v) **Polythene-lined pit method:** The polythene-lined pit method is most beneficial for small and marginal farmers who wish to prepare algal biofertilizer. In this method, small pits are prepared in a field and lined with thick polythene sheets.

(vi) **Indoor production:** A biofermentor is used for indoor production of cyanobacteria (Figure 11.4). This method is used only for pure axenic culture. This technique is expensive as the apparatus used is very sophisticated.

11.2 APPROACHES TO BIOFERTILIZER APPLICATION

A cyanobacteria starter requires a carrier, which can be soil that is peat, coal, clay, inorganic soil, plant waste material composts, farmyard manure, soyabean meal, straw husk, wheat bran, press mud, perlite, calcium sulphate, or alginate beads. As per requirement, carrier materials are used for introducing inoculums in fields. When inoculums are used in large amounts, they exhibit visualizing effects on fields. Inoculum is mixed with different carriers to form various starter materials. Ready to use material is supplied to market with assured quality. The following is the list of general starter materials:

(i) **Soil-based algal inoculums:** 10–15 days of old blue green algae (BGA) culture is used as inoculums. Once a thick algal mat is formed, it is dried. After it completely dries, the algal mat forms flakes. The flakes are dried in sun before they are packed in sealed polybags. The reason of using soil is its easy availability. However, alternative carriers are also used because of the problem of soil-based contamination.

(ii) **Fuller's earth:** Due to good water holding capacity, microbial inert, good adhesion, and buffering capacity, Fuller's earth can be used as a carrier. Wet biomass of BGA is mixed with equal amount of Fuller's earth. Algal–clay mixture is sundried and can be introduced in the field. Approximately, 10^4 CFU/g of inoculums' shelf life is 2 years.

(iii) **Straw-based inoculums:** Soya bean meal straw, husk, and wheat bran are also used as carrier for cyanobacteria inoculums. These starters can be introduced in the field.

11.3   CYANOBACTERIA: A BIOFERTILIZER

Plants require many macro- and micronutrients, such as nitrogen, phosphorus, and potassium to maintain their life processes. However, soil nutrients get depleted with time and are not recycled when plants absorb these nutrients. The nutrients can be reintroduced to the soil either by the natural process of plant decomposition or by adding fertilizers. Soil fertility increases when chemical and natural substances are added. One of the important soil nutrients is nitrogen. Although 78% of the atmosphere is nitrogen, however plants are not able to absorb it directly. To meet the plants' requirement of nitrogen, fertilizers are added to soil. The demand for nitrogen between 2012 and 2016 is expected to increase to 6 million tonnes of which 60% would be from Asia, 19% from America, 13% from Europe, 7% from Africa, and 1% from Oceania. Among the Asian countries, the bulk of the demand for nitrogen is expected to come from India (30%), followed by China (7%), Pakistan (6%), Indonesia (5%), Bangladesh (3%), Vietnam (2%), and Malaysia (1%). As per estimates, more than 18 kg N_2/ha year was added to the soils by cyanobacteria (Watanabe and Cholitkul 1979). Chemical fertilizers extensively enhance the growth of plants. However, it decreases the quality of soil and is hazardous too. Excessive usage of chemical fertilizers may contaminate the soil and waterways. Due to increasing demand and adverse effects of chemical fertilizers, they are being replaced by biofertilizers.

Biofertilizers are microbial inoculants consisting of living organisms such as algae, bacteria, cyanobacteria, and fungi alone or combination which may help in crop productivity. They provide nutrients, especially by fixing the atmospheric nitrogen for the plants and maintain the soil structure.

11.4   NITROGEN FIXATION

The radio tracer technique can be used to study nitrogen fixation in BGA. Nitrogen fixation occurs in the heterocyst. Near about 50 species are known as active nitrogen fixers. Cyanobacteria form an important group of soil organisms that are of great agricultural importance because of their ability to synthesize organic substances and fix free atmospheric nitrogen. Nitrogenase enzyme is present in heterocyst and has a key role in nitrogen fixation. As nitrogenase enzyme is inactivated by oxygen, hence heterocyst must create a microanaerobic environment. In an anaerobic environment, both heterocyst and vegetative cell can fix nitrogen.

Nitrogen reduction from nitrogen to ammonia in the heterocyst

$$N \equiv N \implies HN = NH \implies H_2N - NH_2 \implies H_3N + NH_3$$
(Di-nitrogen)   (Di-imide)   (Hydrazine)   (Ammonia)

Di-nitrogen is inert because of the triple bond, thus plants are not able to absorb nitrogen directly. Chemical energy is required for enzymatic reduction of di-nitrogen into ammonia by nitrogenase enzyme. A nitrogenase enzyme complex is composed of 2 Fe–protein dimers and 1 MoFe–protein tetramer domain. Each component consists of a metallocluster that facilitates electron flow to the active centre within the complex, which reduces nitrogen. Adenosine triphosphate (ATP) supplies energy to transfer electron from Fe–protein to MoFe–protein component. Each electron transfers its reduction potential to MoFe–protein to breakdown the chemical bond of 1 di-nitrogen. These three cycles convert 1 molecule of nitrogen to ammonia. The nitrogenase enzyme binds each atom of nitrogen to 3 hydrogen atoms to form ammonia, which is further turned into glutamate and is glutamine in the presence of a glutamine synthetase enzyme. Glutamine transfers into the nearest vegetative cell, where glutamine changes into a glutamate amino acid. Through transamination, glutamate and glutamine change into amino acid. A suitable substrate is required for enzyme activity as it binds all the active sites of the enzyme and a product is produced (Figure 11.5).

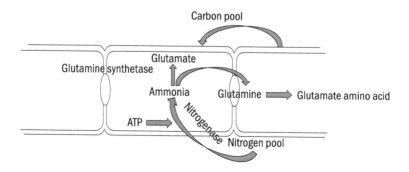

Fig. 11.5 *Transfer of nitrogenous compound from heterocyst to vegetative cells*

The following reactions reduce di-nitrogen into ammonia:

Reaction 1: Nitrogen fixation (Lowe and Thorneley 1984; Bryce, Hill, Smith, *et al.* 1993)

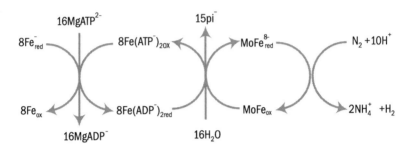

Reaction 2: Production of superoxide (O_2^-) (Thorneley and Ashby 1989)

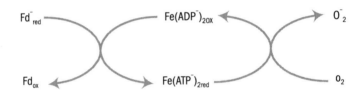

Reaction 3: Autoprotective oxygen consumption

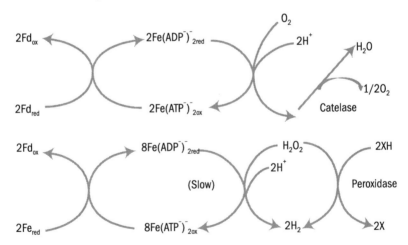

Note *Fd – Ferredoxin, $Fe(ADP^-)_2$ – Fe-protein (Mg $ADP^-)_2$ complex, $Fe(ATP^{2-})_2$ – Fe-protein ($MgATP^{2-}$) complex, X(XH) – Oxidized (reduced) form of peroxidase substrate*

The BGA heterocystous forms, such as *Nostoc, Anabaena, Tolypothrix,* and *Aulosira* are used in rice fields. Mycorrhiza, especially VAM fungi (vesicular arbuscular mycorhhiza), are widely used as biofertilizers to replenish phosphorus and potassium in soil. Apart from enhancing soil quality and crop yield, they are ecofriendly too. Nowadays, biofertilizers are being rapidly used for giving best results.

Majority of aerobic diazotrophic cyanobacteria associated with rice fields are multicellular heterocystous forms, although aerobic diazotrophy has been demonstrated in a few unicellular and non-heterocystous cyanobacteria belonging to *Aphanothece, Gleocapsa, Gleothece, Plectonema,* and *Trichodesmium* genera (Singh 1973; Singh and Tiwari 1988). Non-heterocystous cyanobacteria fix nitrogen only when grown under subatmospheric oxygen levels. A few species, such as *Gleocapsa, Gleothece,* and *Synechococcus,* are capable of aerobic nitrogen fixation also (Gallon and Chaplin 1988; Mitsui, Kumazawa, Takahashi, *et al.* 1986). But the best documented nitrogen-fixing heterocystous forms commonly found in rice fields are *Anabaena, Nostoc, Cylindrospermum, Gleotrichia, Aulosira,* and *Scytonema.* Free atmospheric nitrogen are fixed by the symbiotic association of nitrogen fixing cyanobacteria *Anabaena azollae* and free floating water fern *Azolla.*

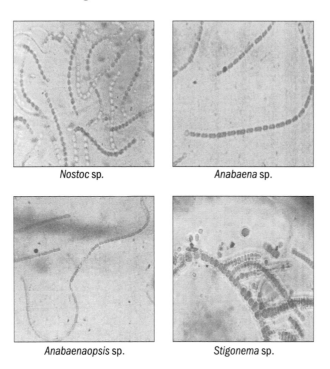

Nostoc sp.   Anabaena sp.

Anabaenaopsis sp.   Stigonema sp.

Cyanobacteria have fundamental affinity with photosynthetic cells of higher plants. They are vulnerable to herbicides that inhibit oxygenic photosynthesis. These cyanobacteria, native to rice fields, lack resistance to herbicides except 2, 4-D, which enhances heterocyst frequency and nitrogen-fixing growth of *Nostoc linckia* (Tiwari, Pandey, and Mishra 1981).

11.5   CYANOBACTERIA: A NUTRIENT SOURCE

Soil is the main and only source of nutrient for forest plants and crops. Microorganisms in soil promote root development and are ecofriendly. Cyanobacteria play an important role in accumulation of organic matter in the soil. Organic matter acts as a storehouse of nutrients, such as nitrogen and phosphorus. They release CO_2, increase the water holding capability, and buffer the soil pH. Cyanobacterial microflora in soil helps in fixing atmospheric nitrogen. Under field conditions, 5%–32% increase in the soil organic carbon has been observed as a result of cyanobacterial inoculations (Singh and Bisoyi 1989). After death and decay, BGA produce soil minerals which act as a reservoir of elements in soil available for higher plants. They are rich in proteins, contain beta-carotene, riboflavin, and thiamine, and are considered to be one of the richest sources of vitamin B_{12} (Abed, Dobretsov, and Sudesh 2008).

11.6   CYANOBACTERIA: A PESTICIDE TOLERANT

Pesticides are used in agriculture to safeguard plants from pathogens. However, these pesticides make soil inhospitable to sustain flora and fauna. They are also responsible for bioaccumulation and biomagnification. Thus, pesticides pose a threat for environmental sustainability. Prolonged application of pesticides has an adverse impact on soil health and its ability to sustain productivity (Kookana, Baskaran, Naidu, *et al.* 1998). Many cyanobacteria exhibit a wide range of pesticide tolerance (Ahmad and Venkatraman 1973; Kaushik and Venkatraman 1983; Pabbi and Vaishya 1992). Cyanobacterial microflora degrade the pesticides present in the soil and make the soil agrofriendly. Many gaseous, solid, and liquid recalcitrant pollutants, including those of natural and xenobiotic origins, namely. carbon dioxide, nitrogen, phosphorus, phenolics, pesticides, antibiotics, lignin, and detergents are detoxified or metabolized by cyanobacteria (Subramanian and Uma 1999). Algalization checks weeds proliferation by blocking supply of nutrient and light. It is well recognized that algalization is essential for increasing crop production and maintaining soil fertility. Economically feasible methods have been developed to produce quality inoculums. Cheap carrier materials, such as cyanostraw increase the yield of rice. Cyanobacteria serve as an environmentally safe biopesticide that can replace toxic organic chemical pesticides.

At present, algalization is followed in India and in other parts of the world.

11.7 CYANOBACTERIA: A SALINITY TOLERANT

Salinity is a wide spread problem around the world. In India about 70 lakh hectares are adversely affected by salinity, which leads to an appreciable loss of land area for effective crop production (Singh, Dhar, Pabbi, et al. 2000). As the problem of salinity further increases, the option of reclaiming salt affected soils through organic inputs including cyanobacteria may be a better alternative. Cyanobacteria synthesize and excrete biopolymer as a response to stress (Jha, Venkataraman, and Kaushik 1987). They can also bind with Na+ and K+ ions and reduce soil salinity (Subhashini and Kaushik 1981). Some studies have shown negative relationship with magnesium and sodium. A negative relation with sodium indicates the possibility of using cyanobacteria as an ameliorating agent for salt affected area. These organisms act as a sodium scavenger by way of metabolizing the sodium present in the soil. Salinity induces polysaccharide production. These microbial polysaccharides bind the soil particles and are considered to be the most important natural products in the formation and stabilization of the ecosystem (Kaushik, Krishnamurti, and Venkatraman 1981). They also increase the air and water movement in the soil, and improve root penetration and increase the uptake of nutrients by the plants. Single application of BGA can ameliorate saline soils in impounded fields. Cyanobacteria promote environmental health by oxygen evolution and enhance other physico-chemical parameters.

11.8 CYANOBACTERIA: A BIOREMEDIATION TOOL

Cyanobacteria have the potential of mobilization and immobilization of heavy metals from the environmental and biogeochemical cycles. This feature has an application in bioremediation too. Bioremediation is a technique that uses living organisms to reduce or neutralize environmental hazardous pollutants into less toxic or non-toxic substances. Cyanobacteria are used for bioremediation due to their non-toxic, higher biodegradability, better environmental compatibility, effective, and low-cost material option. Cyanobacteria oxidize oil components and other complex organic compounds, such as surfactants and herbicides (Yan, Jiang, Wu, et al.1998). Species, such as *Oscillatoria acutissima, Phormidium mucicola, Nostoc punctiforme, Chroococcus disperses*, and *Anabaena aequalis* can remove heavy metals from industrial effluent. Cyanobacteria used as a single cell protein, *Spirulina platensis* has shown to contain detectable level of mercury and lead when grown under contaminated conditions. They secrete extracellular polymeric substances (EPS) that are also used for bioremediation. Extracellular polymeric substances producing cyanobacteria also remove heavy metals from aquatic environment. Bioremediation is a good alternative to conventional clean-up technology. This technique has been adopted worldwide, including Europe and the USA.

11.9   CYANOBACTERIA: A CO_2 SEQUESTRATION AGENT

Anthropogenic activities, such as burning of fossil fuels, have increased the concentration of CO_2 in the atmosphere. This increase in atmospheric CO_2 from about 280 to more than 380 parts per million (PPM) over the last 250 years is leading to global warming (Sundquist, Burruss, Faulkner, *et al.* 2008). To reduce the effect of global warming and climate change, and to mitigate GHGs in the atmospheric and marine accumulation, CO_2 can be stored for a long term.

Carbon sequestration is a natural and deliberate process of removing CO_2 from environment or detracted from emission and from sources and can be stored in ocean and terrestrial environments. Cyanobacteria are the vast group of photoautotrophic microorganisms and use water as a reducing agent and increase the transfer rate of CO_2 from the atmosphere through photosynthesis (Pedorani, Lamenti, Prosperi, *et al.* 2005; Brown and Zeiler 1993; Benemann 1997). CO_2 sequestration by cyanobacteria is gaining attention in alleviating the impact of increasing CO_2 in the atmosphere. In addition to CO_2 fixation, cyanobacterial biomass generate renewable energy such as bioethanol and biodiesel. Carbon sequestration is largely affected by the characteristics of cyanobacteria, their tolerance to stress, and composition of flue gas. However, other physico-chemical parameters, such as pH and light also limit the process. Photobioreactors have been designed to estimate CO_2 sequestration process. Marine culture of *Synechocystis* sp. removal rate of CO_2 is 4.44 g CO_2/litre/day in a photobioreactor (IEA 1998), while *Chlorella–Synechocystis*–based systems fix approximately 50 g CO_2/m^2/day (Otuski 2001). *Spirulina* and *Scenedesmus* sp. are some of the species that have high biomass productivity, CO_2 fixation ability, and tolerance to relatively higher levels of stress (Kanhaiya and Das 2013).

11.10   NUTRIENT MANAGEMENT SYSTEM

To improve and maintain soil health and productivity, the Government of India is promoting Integrated Nutrient Management by justified application of chemical fertilizers, including secondary and micronutrients, in concurrence with organic manures and biofertilizers. The objectives of nutrient management systems are to enhance the quality of soil, sustain it, optimize plant production and yield, and, last but not the least, conserve resources. The government runs integrated nutrient management programmes. In these programmes, cyanobacteria biofertilizer and green manure are introduced in the soil along with chemical fertilizers which improve the soil quality and productivity. Green technologies are environmentaly safe and sustainable, and is based on use of indigenous native technologies, and environment friendly supplements.

REFERENCES

Abed, R. M. M., S. Dobretsov, and K. Sudesh. 2008. Applications of cyanobacteria in biotechnology. *Journal of Applied Microbiology* 106(1): 1–12

Ahmad, M. H. and G. S. Venkatraman. 1973. Tolerance of *Aulosira fertilissima* to pesticides. *Current Science* 42: 108

Benemann, J. 1997. CO_2 mitigation with microalgae system. *Energy Conversion and Management* 38(1): S475–S479

Brown, L. M. and K. G. Zeiler. 1993. Aquatic biomass and carbon dioxide trapping. *Energy Conversion and Management* 34: 1005–13

Bryce, J. H., S. A. Hill, C. J. Smith, D. J. Murphy, R. J. Smith, J. R. Gallon, G. Hendry, M. D. Watson, N. J. Robinson, A. H. Shirsat, J. A. Gatehouse, J. A. Bryant, J. A. Cuming, and R. Walden. 1993. *Plant Biochemistry and Molecular Biology*, P. J. Lea and R. C. Leehood (eds). Chichester, New York: Wiley

FAO. 1977. China: recycling of organic wastes in agriculture. *FAO Soils Bull.*, 40- Rome.

Gallon, J. R. and A. E. Chaplin. 1988. Recent studies on nitrogen fixation by non-heterocystous cyanobacteria. In *Nitrogen Fixation: Hundred Years After*, H. Bothe, F. J. de BruYn and W. E. Newton (eds), pp. 183–188. Stuttgart and NewYork: Fischer

Gompertz, B. 1825. On the nature of the function expressive of the law of human mortality, and on a new mode of determining the value of life contingencies. *Philosophical Transactions of the Royal Society of London* 115: 513–585

Heffer, P. 2010. Assessment of fertilizer use by crop at the global level. *International Fertilizer Association* (IFA), Paris, France.

IEA. 1998. *Carbon Dioxide Capture from Power Stations*. Paris: International Energy Agency

IFFCO (Indian Farmers Fertiliser Cooperative). 2012. *Indian Fertilizer Scenario 2012*. New Delhi: Ministry of Chemicals and Fertilizers

Jha, M. N., G. S. Venkataraman, and B. D. Kaushik. 1987. Response of *Westiellopsis prolifica* and *Anabaena* sp. to salt stress. *MIRCEN J.* 3: 307–317

Kanhaiya, K. and D. Das. 2013. CO_2 sequestration and hydrogen production using cyanobacteria and green algae. In *Natural and Artificial Photosynthesis: Solar Power as an Energy Source*, edited by R. Razeghifard, pp. 173–215. New York: Wiley

Kaushik, B. D. and G. S. Venkatraman. 1983. Response of cyanobacterial nitrogen fixation of insecticides. *Current Science* 52: 321–323

Kaushik, B. D., G. S. R. Krishnamurti, and G. S. Venkatraman. 1981. Influence of blue-green algae on saline alkali soils. *Science and Culture* 47: 169–170

Kookana, R. S., S. Baskaran, and R. Naidu. 1998. Pesticide fate and behavior in Australian soils in relation to contamination and management of soil and water: a review. *Austr. J. Soil Res.* 36: 715–764

Lowe, D. J. and R. N. F. Thrneley. 1984. The mechanism of *Klebisiella pneumonia* nitrogenase action Pre-steady-state kinetics of an enzyme-bound intermediate in N_2 reduction and of NH_3 formation. *Biochemical Journal* 224(3): 895–901

Meeting, B. 1988. Micro-algae in agriculture. In *Micro-algal Biotechnology*, edited by M. A. Borowitzka and L. J. Borowitzka, pp. 288–304. Cambridge: Cambridge University Press

Mitsui, A., S. Kumazawa, A. Takahashi, H. Ikemoto, S. Cao and T. Arai. 1986. Strategy by which nitrogen-fixing unicellular cyanobacteria grow photoautotrophically. *Nature* 323: 720–722

Nabout, João, Barbbara da Silva Rocha, Fernanda Melo Carneiro, and Célia Sant'Anna. 2013. How many species of Cyanobacteria are there? Using a discovery curve to predict the species number. *Biodiversity and Conservation* 22: 2907–2918

Otuski, T. 2001. A study for the biological CO_2 fixation and utilization system. *The Science of the Total Environment* 277(1–3): 21–25

Pabbi, S. and A. K. Vaishya. 1992. Effect of insecticides on cyanobacterial growth and nitrogen fixation. In the *Proceeding of the 1992 National Symposium on Cyanobacterial Nitrogen Fixation*, edited by B. D. Kaushik. Indian Agriculture Research Institute, New Delhi, pp. 389–493

Patterson, G. M. L. 1996. Biotechnolgical applications of cyanobacteria, *J. Sci. Ind. Res.* 55: 669–684

Pedorani, P. M., G. Lamenti, G. Prosperi, L. Ritorto, G. Scolla, F. Capuano, and M. Valdiserri. 2005. Enitecnologie R&D Project on Microalgae Biofixation of CO_2: Outdoor Comparative Tests of Biomass Productivity Using Flue Gas CO_2 from a NGCC Power Plant. *Proceedings of 7th International Conference on Greenhouse Gas Control Technologies,* Contributed Papers and Panel

Discussion, edited by M. Wilson, T. Morris, J. Gale and K. Thambimuthu, Elsevier Sciences, Oxford, Vol. 2, pp. 1037–42

Ríos, A. D. L., G. Martin, G. L. Sancho, and A. Carmen. 2007. Ultrastructural and genetic characteristics of endolithic cyanobacterial biofilms colonizing Antarctic granite rocks. *FEMS Microbiology Ecology* 59(2): 386–395

Singh, P. K. 1973. Nitrogen fixation by the unicellular blue-green alga Aphanothece. *Archives of Microbiology* 92: 59–62

Singh, L. J. and D. N. Tiwari. 1988. Effect of selected rice-field herbicides on photosynthesis, respiration and nitrogen assimilatory enzyme systems of paddy soil diazotrophic cyanobacteria. *Pesticide Biochemistry and Physiology* 31: 120–128

Singh, P. K., D. W. Dhar, S. Pabbi, R. Prasanna, and A. Arora (eds). 2000. Current areas of research in cyanobacteria with special reference to their use as biofertilizers. In *Biofertilizers, Blue Green Algae and Azolla*, pp. 55–63. New Delhi: Indian Agriculture Research Institute

Singh, P. K., D. W. Dhar, S. Pabbi, R. Prasanna, and A. Arora (eds). 2000. Reclamation of salt affected soil through cyanobacteria. In *Biofertlizers Blue Green Algae and Azolla*, pp. 107–121. New Delhi: Indian Agriculture Research Institute

Singh, P. K. and R. K. Bisoyi. 1989. Blue–green algae in rice fields. *Phykos* 28:18–195

Stewart, W. D. P., P. Rowell, N. W. Kerbly, R. H. Reed, and G. C. Machray. 1987. N_2 fixing cyanobacteria and their potential applications. *Philosophical Transactions of the Royal Society of London* 317: 245–258

Subhashini, D. and B. D. Kaushik. 1981. Amelioration of sodic soils with blue green algae. *Aust. J. Soli Res.* 19: 361–367

Subramanian, G. and L. Uma. 1999. The role of cyanobacteria in environmental management. *Bulletin de l' Institut Oceanographique - Marine Cyanobacteria* 19: 599–606

Sundquist, E., R. Burruss, S. Faulkner, R. Gleason, J. Harden, Y. Kharaka, L. Tieszen, and M. Waldrop. 2008. Carbon sequestration to mitigate climate change: U.S. Geological Survey Fact Sheet, 2008-3097. Details available at http://pubs.er.usgs.gov/publication/fs20083097

Thorneley, R. N. F. and G. A. Ashby. 1989. Oxidation of nitrogenase iron protein by dioxygen without inactivation cloud contribute to high respiration rates of *Azotobacter* species and facilitate nitrogen fixation in aerobic environments. *Biochemical Journal* 261(1): 18–187

Tiwari, D. N., A. K. Pandey, and A. K. Mishra. 1981. Action of 2, 4-dichlorophenoxyacetic acid and rifampicin on heterocyst differentiation in the blue-green alga, *Nostoc linckia. Journal of Bioscience* 3(1): 33–39

Watanabe, I. and W. Cholitkul. 1979. Field studies on nitrogen fixation in paddy soils. In *Nitrogen and Rice*. Los Banos, Philippines: IRRI

Wolk, C. P., A. Ernst, and J. Elhai. 2004. Heterocyst metabolism and development. *The Molecular Biology of Cyanobacteria* 769–823

Yan, G. A., J. W. Jiang, G. Wu, and X. Yan. 1998. Disappearance of linear alkylbenzene sulfonate from different cultures with *Anabaena* sp. HB 1017. *Bull Environ Contam Toxicol* 60: 329–334

12

Low-cost Production of Algal Biofuel from Wastewater and Technological Limitations

Neha Gupta and D. P. Singh[*]

Department of Environmental Science, Babasaheb Bhimrao Ambedkar University, Lucknow, Uttar Pradesh 226025
[*]*E-mail: dpsingh_lko@yahoo.com*

12.1 INTRODUCTION

Microalgae groups consist of a large group of photoautotrophic organisms that have extraordinary potential for cultivation as energy crops. The small unicellular forms of algae are known as microalgae. They exhibit simple reproductive structures and are widely distributed in waterbodies and wastewater streams. They can be easily cultivated under different environmental conditions and are capable of producing a wide array of products, such as fatty acid, pigment, sugars, and other functional bioactive compounds. Macroalgae, on the contrary, are larger in size and difficult to cultivate. The main viable option for the application of macroalgae for production of renewable bioenergy is through anaerobic digestion of biomass. However, algal carbon dioxide (CO_2) fixation contributes to approximately 40%–50% of the oxygen (O_2) in the atmosphere, which is an essential requirement to support life on this planet. Algae are considered to be one of the oldest living photoautotrophs that grow about 100 times faster than other terrestrial plants. Another advantage associated with the algal system is that they can easily double their biomass within a short span of time (Tredici 2010). It has been observed that solar energy harvesting by algal biomass is also about 10–50 times greater than that by other terrestrial plants (Khan, Rashmi, Hussian, et al. 2009). A number of workers, industries, and government institutions are exploring the possibility of a cost-effective technology to produce biofuel from algal biomass (Gonzales, Canizares, and Baena 1997). The most popular designs known so far include open ponds, open raceways, and closed photobioreactors. Meiers (1955) proposed the use of algal biomass for biofuel production (biomass to methane) in a concept note, which was later demonstrated by researchers under laboratory conditions (Golueke and Oswald 1955; FAO 2009b). However, a more comprehensive study on the use of algal biomass as a feedstock for biofuel production was carried out between 1978 and 1996 by Aquatic Species Program (ASP) on behalf of the U.S. National Renewable Laboratory (Sheehan, Dunahay, Benemann, *et al.* 1998). The main agenda of the program was to produce biodiesel from algae containing high lipid, grown by utilizing pond water and waste CO_2 from coal-fired power plants. The main advantage of this program was the development of a viable renewable fuel technology by utilizing wastewater and CO_2. Microalgae are now considered an ideal feedstock for biofuel generation due to their high growth rate, high lipid content, and ability to grow throughout the year using less water and in small areas (Stephens, Ross, King, *et al.* 2010). Many species of algae have been explored for their potential as feedstock for biodiesel production. More importantly, cultivation of algal biomass is carried out in non-agricultural wasteland. Thus, it helps in avoiding competition with agricultural production. At present, algal biofuel is economically uncompetitive with the conventional fossil fuels because of its high-production cost. If we use wastewater as

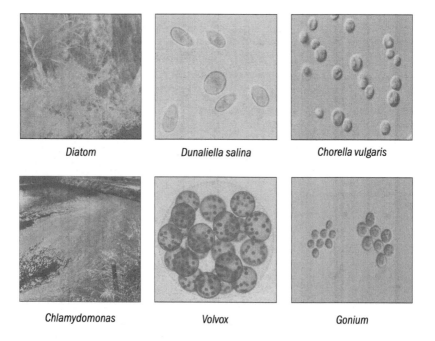

Diatom	Dunaliella salina	Chorella vulgaris
Chlamydomonas	Volvox	Gonium

Fig. 12.1 *Different species of algae found in the environment*

a medium for growing algae, it may reduce the input cost of nutrients and other minerals which are otherwise added to the growth medium for production of algal biomass. The potential benefits of using algal feedstock for biofuel production can be linked to high productivity, and non-agricultural resource grown on non-productive wasteland. A wide variety of industrial wastewater can be utilized for algal growth, which also helps in the mitigation of greenhouse gas (GHG), that is, CO_2 present in the atmosphere. Thus, biofuel produced from algal biomass can be considered a value-added eco-friendly by-product. See Figure 12.1.

12.2 CLASSIFICATION OF ALGAE

Algae possess diverse types of pigments, reserve food, cilia, and so on. According to their morphological and physiological characteristics, they have been divided into several classes (Table 12.1). Fritsch (1935) classified the whole group of algae into 11 classes based on the type of pigments, reserve food materials, mode of reproduction, and so on.

12.3 FACTORS AFFECTING GROWTH OF MICROALGAE

There are various environmental parameters which influence the growth and cultivation of microalgae. The fundamental requirement for algal growth is photosynthesis (Wogan, Ak, Webber, *et al.* 2008) which is

Table 12.1 Classification of algae

Class	Colour	Example
Chlorophyceae	Green algae	*Desmotetra, Volvox*
Phaeophyceae	Brown algae	*Ectocarpus, Sargassum*
Xanthophyceae	Yellow-green algae	*Vaucheria*
Bacillariophyceae	Diatoms	*Pinnularia*
Chrysophyceae	Golden algae	
Cryptophyceae	–	*Chroomonas*
Dinophyceae	–	*Dinoflagellate, Ceratium*
Rhodophyceae	Red algae	*Batrachospermum, Polysiphonia*
Myxophyceae	Blue-green algae	*Oscillatoria, Nostoc*
Euglenineae Chloromonadineae	–	*Euglena*

Source Details available at http://www.plantscience4u.com/2014/04/fritsch-classification-of-algae.html#.VsLBt_l97lV

basically a light-driven reaction that involves splitting of water molecules and fixing of atmospheric CO_2 into biomass with the help of light energy (da Rosa 2005). The algal cell absorbs energy in the form of photons, which convert CO_2 and H_2O into sugar and O_2. The algal cells are pretreated and carbohydrate, lipid, and proteins are extracted.

12.3.1   Nitrogen/phosphorus and Other Nutrients

The essential elements required for the growth of microalgae include nitrogen and phosphorus. Some of the microalgae can fix atmospheric nitrogen gas for their own use by nitrogen fixation process. Optimal phosphorus concentration dissolved in water is conducive for the growth of microalgae. The nutrient requirement for algal growth varies depending upon the type of species and growth cycle.

12.3.2   Light

Light is an essential requirement for the growth of microalgae. Microalgae use light energy to drive photosynthetic reaction. Pigments in microalgae help in absorption of all the photons, but the excess photons cause inhibition of light-driven reactions in photosynthesis, which is termed as photoinhibition.

12.3.3   Temperature

Temperature determines the activity and reaction rates of intracellular enzymes, which in turn will affect algal photosynthesis, intensity of respiration, growth, and distribution (Tan, Kong, Kong, *et al.* 2009).

12.3.4   pH and Salinity

The pH value will also affect the growth rate of microalgae. Microalgae capture CO_2 present in the atmosphere easily when the growing condition is alkaline, and there is higher production of biomass (Zang, Huang, Wu, *et al.* 2011; Melack 1981).

Microalgae have its own optimal growth salinity, and salinity higher or lower than this will be harmful to algal growing rate. For example, in the case of low-salinity conditions, the addition of NaCl and $NaSO_4$ will be helpful for the growth of algae, but if the salinity is more than 6 g/L, the growth rate of algae will be inhibited (Liu, Sun, Zhu, *et al.* 2006).

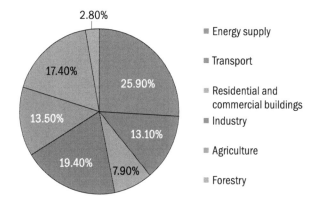

2.80%

17.40%

25.90%

13.50%

13.10%

19.40%

7.90%

- Energy supply
- Transport
- Residential and commercial buildings
- Industry
- Agriculture
- Forestry

Fig. 12.2 *Sources of CO$_2$ emission*

12.3.5 Mixing

Mixing is a problem in the growth of microalgae. It will prevent the effective absorption of light by microalgae, and, thus, affect biomass production. Proper mixing of culture must enhance the growth of microalgae.

12.4 SINK FOR CARBON DIOXIDE

The CO$_2$ present in the atmosphere is a colourless and odourless GHG which absorbs infrared radiation emanating from earth's surface and other sources. Apart from natural sources, CO$_2$ is also produced from burning of fossil fuels, oil, natural gas, and coal. The burning of fuels is the major source of CO$_2$ emission globally. Another reason for the increase in the level of atmospheric CO$_2$ is deforestation (Figure 12.2).

Algal sequestration of CO$_2$ and its conversion to biomass is a useful technology for mitigating the effect of CO$_2$ concentration in the atmosphere (Sheehan, Dunahay, Benemann, *et al.* 1998). The quantity of CO$_2$ fixed per unit biomass varies, depending upon the algal species. In general, it is about 1.8 tonnes CO$_2$ locked in 1 tonne algal biomass (Chisti 2007). Thus, the entire carbon requirement of algae for the production of biomass is directly obtained from atmosphere. The cultivation sites for algae are considered to be the most effective sites for CO$_2$ sequestration. Many countries are now signatories to the Kyoto Protocol related to carbon emission. The existing carbon credit and world trade are now dependent upon the efficiency of CO$_2$ sequestration system, where algae as a resource can be potentially exploited for monetizing the carbon credits as a rich source of biofuel. As the electric power plants account for up to one-thirds of the total CO$_2$ emissions in developed countries (Kadam 1997), the power plant flue gas can also be used as a source of CO_2 for cultivation of microalgae (Brown 1996; Hauck 1996). Besides, the nitrogen source used by algae is dilute nitrogen oxide. The quantity of flue gas needed per hectare will depend on the algal species to be cultivated and the day length and light intensity, which need to be optimized for a specific application.

12.5   WASTEWATER AS A RESOURCE FOR ALGAL GROWTH

Wastewater is discharged from homes, industries, and other commercial activities. This wastewater is categorized based on its source of origin. There are different kinds of wastewaters, such as commercial, industrial, and municipal. There are various other types of industrial wastewaters based on the nature of industrial contaminants, each sector producing its own mixture of pollutants. See Table 12.2.

Industrial effluents are usually a rich source of nutrients. Many organic and inorganic materials (Lara, Rodríguez-Malaver, Rojas, *et al.* 2003), if they are unrecovered, contribute to water pollution (González, Canizares, and Baena 2010). Microalgae growing in wastewater can be a tool to remove various elements, nutrients (such as nitrogen, phosphorous, silica, and calcium), and different materials. The microalgae biomass is formed by the accumulation of toxic compounds and heavy metals. In India, many industries, such as sugar factories, dairies, meat processing units, pulp and paper mills, and distilleries, release about 65%–70% of organic pollutants in waterbodies. There are many parameters for the commercial, industrial, and municipal wastewaters, such as hardness, alkanity, pH, total solids, total phosphorus, total nitrogen, biochemical oxygen demand, and chemical oxygen demand. The characteristics of agro-food industrial wastewater are given in Table 12.3.

Table 12.2   Specific contaminants from various types of industries

Industries	Types of contaminants	
	Inorganic	Organic
Pulp and paper	Solids, COD, chlorinated	BOD, organic compounds
Iron and steel	Metals, COD	BOD, oil, acids, phenols
Textile and leather	Solids, sulphate, chromium	BOD
Mining	SS, metal, acids, salts	
Petrochemical and refineries	COD, chromium	BOD, mineral oil
Chemicals	COD, heavy metals, SS, cyanide	Organic chemicals
Electronics	COD	Organic chemical

Note BOD–biochemical oxidation demand; COD–chemical oxidation demand; SS–suspended solids
Source Details available at http://www.eolss.net/sample-chapters/c09/e4-11-02-02.pdf

Table 12.3   Characteristics of agro-food industrial wastewater

Industries	TS (mg/L)	TP (mg/L)	TN (mg/L)	BOD (mg/L)	COD (mg/L)
Food industry	–	3	50	600–4,000	1,000–8,000
Palm oil mill	40	–	750	25	50
Sugar beet	6100	2.7	10	–	6600
Dairy	1100–1600	–	–	800–1,000	1,400–2,500
Corn milling	650	125	174	3,000	4,850
Baker's yeast	660	3	275	–	6,100
Winery	150–200	40–60	310–410	–	18,000–21,000
Cheese dairy	1,600–3,900	60–100	400–700	–	23,000–40,000
Olive mill	75,500	–	460	–	130,100

Note BOD–biochemical oxygen demand; COD–chemical oxygen demand; TN–total nitrogen; TP–total phosphorus; TS–total solids
Source Rajagopal, Saady, Torrijos, *et al.* 2013

Water is vital for cultivation of microalgae as these photosynthetic aquatic organisms derive their nutrients from water (wastewater treatment) and result into accumulation of biomass. However, the level of nutrients varies among different types of water, depending upon space and time. Therefore, nutrient-rich wastewater can be a good resource for the production of algal biomass without additional cost. Simultaneously, algal biomass can be a useful tool to remediate wastewater. Various methods can be employed in using microalgae to treat wastewater, such as aiding the oxidation of organic matter present in wastewater or the absorption of nutrients such as nitrogen, phosphorous, and other minerals from wastewater, thereby reducing the pollution load of the wastewater. The algal treatment of wastewater results into improved water quality. Microalgae are fed by different nutrients present in different environments (for example, silica, iron, and magnesium), sea water nutrients (for example, sulphate, chloride, calcium, sodium, magnesium, and potassium), and other nutrients. Wastewater nutrients are cheap nutrient resources available without additional cost for the growth of algal biomass (Pintoa, Guarieiroa, Rezendea, *et al.* 2005). Thus, the production of microalgae biomass using wastewater offers several advantages, such as the following:

- Improved water quality by algal treatment of wastewater
- Low energy requirement as solar energy is the source of energy for algal growth
- Reduced sludge formation
- Efficient growth of algal biomass

A large volume of wastewater is discharged from various industries, including agro-food industry and industry using livestock raw materials, and these wastewaters are disposed in aquatic ecosystems throughout the world. The quantum of wastewater is increasing day by day as a result of enhanced demand and production of food. Particularly, the agro-industrial effluents are characterized by a high level of organic loading, nitrogen and phosphorus contents, and a variable pH (Drogui, Asselin, Brar, *et al.* 2008). Both the flow rate and the characteristics of these wastewaters are industry specific and can vary significantly throughout the year because of the seasonal nature of the raw material processing. The uncontrolled disposal of such effluents in natural waterbodies often results in surface and groundwater contamination and other environmental problems, such as eutrophication and ecosystem imbalance. The main advantages of using microalgae in a variety of industrial applications include the following:

- They have a higher solar energy conversion efficiency as compared to most of the terrestrial plants.
- They grow faster and can be easily harvested in batch or continuous culture conditions throughout the year.
- Algal production facilities do not require additional space as the cultivation sites are located on non-productive, non-agricultural land.
- They can use minerals and wastewater resources faster than that in the conventional agriculture.
- They are capable of using waste CO_2, and, thus, have the potential to reduce the quantum of GHG from the atmosphere.
- They can produce a large amount of biomass, which provide various value-added products and biofuels.

Microalgae-based treatment of wastewater can overcome the oxygen limitation faced by microbial population as microalgae release oxygen via photosynthesis and concomitantly remove nutrients through assimilation and convert them to algal biomass in a very simple and economical manner. Besides, in the presence of sunlight, microalgae consume the CO_2 released during the bacterial mineralization of the organic matter and, in turn, release O_2 required by other microorganisms for mineralization and NH_4^+ oxidation (Oswald 2003). The microalgae-based wastewater treatment technology generates a large amount of residual microalgae biomass which constitute a valuable feedstock for renewable energy production. However, there is still a need for the development of a cost-effective and environment-friendly technology for treatment of industrial effluents. Culturing microalgae in wastewater offers a more sustainable and inexpensive alternative method for the treatment of wastewater (Posadas, Bochon, Coca, *et al.* 2014).

Microalgal biomass can be generated by utilizing nitrogen and phosphorus compounds present in wastewater. In comparison to physico-chemical removal of nutrients from wastewater, assimilation of nutrients by microalgae is less expensive and ecologically safer. Development of an algal biorefinery could be potentially integrated with several other fuel conversion technologies to produce biofuel, green gasoline, aviation fuel, ethanol, and methane, including valuable co-products such as oils, protein, pigments, and carbohydrates. However, the integration of bioremediation technology with biofuel production requires the complete separation of suspended algal biomass from wastewater. It is observed that separation and harvesting of the suspended microalgae is still not cost-effective (Molina Grima, Belarbi, Acién Fernández, et al. 2003). Species such as *Chlorella vulgaris* has shown promising results owing to its higher lipid content (42%). Thus, setting up of wastewater high-rate algal ponds (HRAPs) close to an industrial area would makes the system more economical and eco-friendly for biofuel production (Park and Craggs 2010). This kind of an integrated approach represents a unique model for an efficient energy management system to produce bio-oil and other value-added products such as glycerol and proteins from microalgae. It is reported that the cost of damage caused by nitrogen pollution from wastewater alone across Europe is about 70–320 billion Euros per year (Sutton, Oenema, Erisman, et al. 2011). Since the resources of phosphorus are limited and the production of phosphorus is approaching its highest rate (Cordell, Drangert, and White 2009), the recycling of resources is vital. The environmental sustainability is achieved by wastewater treatment assembling with nutrient recycling (Reid, Bower, and Lloyd-Jones 2008). The amount of water generated on per tonne of paper produced has reduced between $10-50\,m^3$ over a period of time (Pizzichini, Russo, and Di Meo 2005; Buyukkamaci and Koken 2010). The pulp and paper industry wastewater is also a rich source of carbon, but it has a limited amount of N_2 and P (Thompson, Swain, Kay, et al. 2001; Slade, Ellis, vanden Huevel, et al. 2004). However, microalgae have an efficient mechanism for the removal of COD, BOD, colour, suspended solids, and organic compounds from wastewater (Tarlan, Dilek, and Yetis 2002). The wastewater generated by dairy industry is rich in N_2 and P nutrients (Kothari, Pathak, Kumar, et al. 2012). A huge amount of wastewater sludge is produced by dairy industry (Ramasamy, Gajalakshmi, Sanjeevi, et al. 2004).

12.6 MICROALGAE CULTIVATION METHODS

There are various cultivation systems available for production of biomass feedstock from microalgae, and they are based on environmental conditions.

Table 12.4 Advantages and disadvantages of open pond and closed pond systems

Cultivation technique	Advantages	Disadvantages
Open pond system	• It is easy to construct • It is easy to operate • It is cheap • Low production cost for algal biomass	• Poor light utilization through cell • Poor evaporative losses • Requires a large area • Poor diffusion of CO_2 to the atmosphere
Closed pond system	• Process conditions can be controlled • It requires a small area • It is easier to harvest • Better control of CO_2 transfer • The temperature is more uniform • Large surface-to-volume ratio	• It is difficult to construct • It is very costly • It is difficult to operate • Sterilization is tough

Source (Modified from Al Darmaki, Govindrajan, Talebi, *et al.* 2012)

These systems include open pond system, closed pond system, wastewater, and marine environment. See Table 12.4.

12.6.1   Open Pond System

Algae can be grown in the open pond system, which can be categorized as lagoons, lakes, ponds, and artificial ponds. However, in this system all the environmental conditions cannot be regulated. There are several benefits of this system. It is easy to construct and operate. However, the open pond system also has some drawbacks, such as loss of water by evaporation, easily contaminated by pollutants, poor light utilization by algal cell, limited diffusion of gases, and the need for a large area. In addition, microalgae can be grown in the open pond system during a definite period of time, when all the conditions are favourable. See Figure 12.3.

Fig. 12.3 *Open pond system*
Source *Details available at www.intechopen.com*

12.6.2   Closed Pond System

Algae are cultivated by using photobioreactor and other reactors. The major advantage of using a closed pond system is that all the required conditions, such as land area, light utilization, percentage of CO_2 and so on, can be controlled in such systems. The high cost of construction and difficulty in operation are the major drawbacks of the closed pond system.

Photobioreactors

Microalgae growing in closed pond system can produce high biomass as it can be grown under controlled conditions. It is necessary to supply CO_2 and nutrients in these systems. Density, a suitable light condition, pH, water and mixing regime, standard temperature and pressure should also be maintained in such systems (Sheehan, Dunahay, Benemann, *et al.* 1998). Microalgae cultivation in photobioreactor makes harvesting easier compared to the open pond system. The photobioreactor can provide protection from outside pollutants to the growing biomass of microalgae. Therefore, microalgae can be grown in a photobioreactor anywhere unlike the open pond system. The limitations of the photobioreactor are that it has very high capital cost and it is difficult to sterilize the reactor. The types of photobioreactors used for the industrial work are flat-plate photobioreactor and tubular photobioreactor (Figure 12.4).

Flat-plate photobioreactor: The flat-plate photobioreactor is used for the cultivation of algae (Gonzales, Canizares, and Baena 1997). The beneficial features of the flat-plate photobioreactor are its suitability for open air application, good light path, better productivity of biomass,

Tubular PBR   Flatplate PBR

Fig. 12.4 *Photobioreactors*
Source *Details available at http://www.et.byu.edu/~wanderto/homealgaeproject/Photobioreactor.html*

immobilization of algae, cost-effectiveness, and reduced deposition of oxygen.

Tubular photobioreactor: The most widely used photobioreactor is the tubular photobioreactor (TBP) owing to its good features, such as availability of a very large surface area for illumination, open air application, high biomass productivity, and economical cost (Borowitzka 2007).

The open pond system is relatively inexpensive compared to the close pond system. Many designs of photobioreactors have been developed that satisfactorily fullfil the conditions required for generating biomass for biofuel production. Higher value products such as pharmaceuticals and biofuels are utilized by photobioreactors economically and efficiently.

12.7   HARVESTING TECHNOLOGIES FOR ALGAL BIOMASS

Harvesting of algal biomass accounts for about 20%–30% of the total cost of biomass generation (Molina Grima, Belarbi, Acién Fernández, *et al.* 2003). The cost of biomass production is gradually reducing; the ratio of total algal biomass to water requirement is also low. Thus, it is necessary to have an efficient and low-cost harvesting technology. For this reason, the selection of algal species for mass culturing is vital. Each harvesting method is described in the following sections. See Figure 12.5.

12.7.1   Settling and Sedimentation

Settling of biomass is considered to be a technically more simple method of harvesting algae from water streams. In this method, settling ponds are filled with fully grown algal culture and the water is drained out at the end of the day, leaving behind a concentrated mass of algae at the bottom. This can be further utilized for extraction of lipids and production of biofuel (Benemann and Oswald 1996). Thus, about 80%–85% (and up to 95%) of the algal biomass is concentrated at the bottom of the pond (Sazdanoff 2006). The density of cells collected at the bottom of the pond always depends on the algal species used for growth and the size of the settling pond.

12.7.2   Filtration

Filtration of algal biomass is another technique used for the separation of algae from water pond. The filtration of algal biomass is generally carried out through commonly used cellulosic membranes, aided by vacuum suction pump. Filtration is considered as a cheaper, easier, and more advantageous concentrating device that can be applied to even low-density cell suspension. However, this cell concentrating device has a limited use as it cannot be operated at a large scale and there is clogging of filters when suction is applied.

12.7.3 Centrifugation

The process of centrifugation involves the separation of algal biomass from the broth medium with the use of large volume of centrifuge. The concentrated form of algal cell is used under high centrifugal force for a few minutes. This technique is relatively expensive and consumes higher electric power. However, industrial and commercial establishments have now started using centrifuges.

12.7.4 Flotation

Flotation is a method in which water and algal biomass are aerated and the resulting froth is removed from the water. The flotation procedure used for harvesting algal biomass from dilute cell suspensions is considered to be highly efficient. The froth flotation method is found to be a commercially viable technology applicable to large volumes of algal cell suspension.

12.7.5 Flocculation

Flocculation technique is employed for the separation of algal biomass from aqueous medium by using certain chemicals, which allows lump formation by algal biomass. The chemical flocculants induce the formation of colloidal suspension of particles in liquid phase and result into the formation of aggregates. The most commonly used flocculants for harvesting algal biomass are alum and ferric chloride.

The lack of a cost-effective and efficient technology for harvesting of algal biomass is one of the major barriers for the expansion of microalgae-based biofuel technology. For this reason, the main focus of the present researches is towards the development of a reliable and low-cost harvesting methodology.

12.8 LIPID SYNTHESIS IN MICROALGAE

Little research in the past on algal biofuel, and fluctuations in oil prices have generated renewed interest of scientists on the technology of algal biodiesel for its high yield and production of hydrogen (H_2), methane, hydrocarbons, and ethanol from algal biomass. Once the dried biomass is obtained, for the production of biomass, two steps are followed: (i) extraction

Fig. 12.5 *Schematic representation of algae growth and harvesting process*

of lipids from biomass and (ii) transesterification reaction (Urréjola, Rocio Pérez, *et al.* 2012). There is an established and successful protocol for these processes (Bligh and Dyer 1959). The current technology for algal biodiesel production involves the grinding or sonication process for lysis of algal cells and the use of organic solvent for extraction of lipids from biomass. This method is commonly used for harvesting of biochemical products in laboratories. In the case of industrial production on large scale, batch reactors are used for harvesting in which extraction takes place in large vats. Then the transesterification reaction occurs for harvested lipids in which a catalyst and alcohol are used. Normally, for a standard reaction, the ratio of organic solvent to methanol is 1:1, the molar ratio for oil to methanol is 6:1, and the molar ratio for NaOH to oil is 1:100 (Molina Grima, Belarbi, Acién Fernández, *et al.* 2003). The continuous flow system is reported to be used in the tranesterification process for large-scale harvesting for commercial production; however, details are not available on the mode of operation of these systems. The tubular reactor is filled with a solid catalyst which remains in the reactor; oil is poured on the reactor for transesterification. The biodiesel is generated at a higher rate using such a tubular reactor design because the tube does not need to be emptied and refilled. As the design of this tubular reactor is not like a vat and smaller in size, it is easier to transport. As the separation of product and catalyst is not required for following the reaction in this system, the cost is considerably reduced when this system is used.

12.9   CONVERSION TECHNOLOGIES FOR ALGAL BIOFUEL

A successful biofuel conversion process mainly depends on the selection of an efficient algal strain, characteristics of growth media, and harvesting

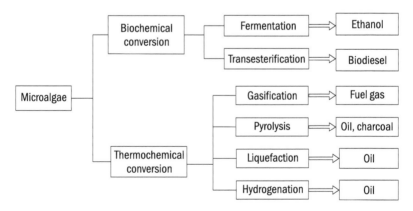

Fig. 12.6 *Energy conversion process from microalgae*

procedure. The other determining factors are usually the nutrient status of water and the fatty acid content of biomass (Amin 2009). The biomass harvesting and conditioning steps of the process are the key factors contributing to both efficiency and energy requirement of the process. See Figure 12.6.

12.9.1   Thermochemical Conversion

Thermochemical conversion of the feedstock into fuel is one of the established processes used by industries. This process involves the steps discussed in the following sections.

Gasification

Gasification can be described as a thermochemical process which involves the conversion of carbonaceous materials (hydrocarbon) into gaseous form by partial oxidation in the presence of air, oxygen, and steam at high temperatures, typically in the range 800–900°C (Figure 12.7). Minowa and Sawayama (1999) have reported an efficient energy production system that allows the low-temperature catalytic gasification of the microalgae combined with nitrogen cycling. Elliot and Sealock (1999) developed a process for low-temperature catalytic gasification of algal biomass with high moisture content. In this process, algal biomass (without drying) is directly gasified to methane-rich fuel in the presence of high moisture content.

Liquefaction

Microalgae biomass with high moisture content is initially collected by centrifugation and is considered to be a good raw material for the liquefaction conversion process. In several studies conducted by Patil, Tran, and Giselrød (2009), FAO (2009a), and Murakami, Yokoyama, Ogi, *et al.* (1990), it has been shown that the direct hydrothermal liquefaction under

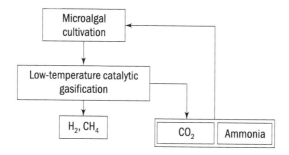

Fig. 12.7 *Flow diagram of an algal system for fuel production by gasification*

subcritical water conditions is a suitable technology for the conversion of wet biomass to liquid fuel.

Pyrolysis

Pyrolysis involves the conversion of biomass to biofuel and charcoal by heating the biomass to around 500°C in the absence of air (McKendry 2003) or, alternatively, by heating the biomass in the presence of a catalyst. The major advantage of fast pyrolysis is that it can directly generate a liquid fuel (Bridgwater and Peacocke 2000). In the case of flash pyrolysis, an efficiency of up to 80% can be achieved during the conversion of biomass to biofuel.

Hydrogenation

Hydrogenation is a chemical reduction of a substrate that results into the addition of H_2 to unsaturated organic compounds. The study performed by FAO (2009a) demonstrated a process in which hydrogenation of algal biomass was carried out in the presence of a catalyst and a solvent in an autoclave under high temperature and pressure conditions. Using this process, algal biomass can be converted to liquid hydrocarbons at temperatures between 400°C and 430°C and operating pressures about 7–14 MPa in the presence of a cobalt molybdate catalyst (Amin 2009). The maximum oil yield obtained by this method was about 46.7% of the algal feedstock.

12.9.2   Biochemical Conversion

Biochemical conversion process involves the use of enzymes derived from different microorganisms for breakdown of biomass. The two processes generally employed for biofuel production from algal biomass are fermentation and transesterification.

Fermentation

Fermentation is a process generally used for the production of ethanol from sugar crops and starch crops. Production of ethanol by using microalgal

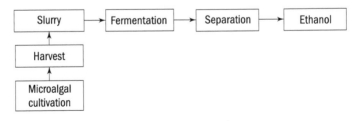

Fig. 12.8 *Fermentation process of algae*

biomass as the raw material involves the following procedure: first, the microalgae biomass is collected and then cells are broken with the aid of mechanical equipment. Following the cell breakage, *Saccharomyces cerevisiae* (yeast) is added to the biomass to start the fermentation process; the resulting product of fermentation, ethanol, is drained into a tank, followed by feeding to distillation unit for further processing. See Figure 12.8.

Transesterification

Amin (2009) defined transesterification process in terms of exchange of the alkoxy group of an ester compound by another alcohol group. In another study by Agarwal (2007), the reactions were catalysed by an acid or a base using a homogeneous or heterogeneous catalytic process. Finally, the conversion of triglycerides or oil into biodiesel as an alternative fuel for diesel engines was about 98% (Noureddini, Harkey, and Medikonduru 1998) or more (Anderson, Masterson, and McDonald 2003).

12.10   LIMITATIONS WITH BIOMASS/BIOFUEL PRODUCTION

12.10.1   Limitation of Biomass Growth

When algae are grown under controlled conditions for biomass production, processes are optimized to overcome the limitations of algal growth. Recently, a great deal of attention has been focused on maximizing the production of algal biomass for biofuel by manipulating the algal physiology and overcoming other technological limitations. Currently, more than 100,000 strains of microalgae are known the world over. It is a great challenge to identify suitable strains of microalgae for biofuel production. Several research programmes have been launched by different institutions to find microalgae strains that have high oil content and rapid growth rate (Patil, Tran, and Giselrød 2009). Some strains capable of producing large amounts of high-grade lipids are found to be highly susceptible to environmental conditions and exhibit growth limitations owing to poor nutritional conditions. Nevertheless, recent researches have emphasized on genetic modification of algae as a favoured option for biofuel production (Borowitzka 1988; Roessler 1990). There is ample scope for mutations and genetic engineering in the photosynthetic apparatus at various levels with an aim to increase the efficiency of light harvesting and the rate of solar energy conversion per unit biomass to enhance biofuel production. More specifically, there is enough scope to increase the wavelength range used by the photosynthetic pigments and photochemical reaction centres. Usually the energy transducing thylakoids in oxygenic photosynthesis involve the coupling of excitation energy, coming from two photons moving one electron from water to $NADP^+$, with the movement of three protons from the stroma (cytosol) into the lumen of the thylakoid (Sacksteder, Kanazawa, and Jacoby, *et al.* 2000; Falkowski and Raven 2007; Raven, Beardall, and Giordano 2014). The bioenergetics of adenosine triphosphate (ATP) synthesis and ATP synthetase enzyme can also be improved. The fluctuating light effect could be utilized and photosynthetic efficiency can be changed by replacing part or all the current machinery for photooxidation and fixation of inorganic carbon. There is another possibility of decreasing the wasteful carbon loss in dark respiration or carbon loss in the form of dissolved organic carbon.

If the electron (or photon) transfer efficiency of an algal strain can be improved, then the overall efficiency of biofuel production can be increased (Raven and Peter 2015). The efficiency of biofuel production can also be enhanced by means of artificial irradiance such as LEDs (Matthijs, Balke, van Hes, *et al.* 1996). Generally, the transfer of excitation energy is energetically an uphill process; the approximate limit for such energy transfer in photosystem II is 730 nm, and it is 750 nm for photosystem I. Not all of the energy in photons can be utilized at shorter wavelengths, as in the case of photosynthesis, the high-energy photons can be fully used in the photochemistry at energy contents corresponding to 680 nm and 700 nm (Falkowski and Raven 2007; Williams and Laurens 2010; Blankenship, Tiede, Barber, *et al.* 2011). This unavoidable loss of energy in every photon, at each wavelength shorter than 700 nm, can be estimated. There is possibility of having mutants with a high photosynthetic efficiency which converts light energy (electromagnetic radiation in a specific wavelength range) to the added value of biofuel production. If we assume that there are pigments capable of absorbing the same fraction of photons at each wavelength, it can be found that the calculated loss of energy in the range 350–700 nm is about 22% of the total energy contained within the photons when compared to the use of photons at 680 nm and 700 nm (Raven, Beardall, and Larkum 2013). Some analyses of energy use do not explicitly acknowledge this factor in the energy losses associated with the use of absorbed photons in photochemistry, and, therefore, we overestimate the potential photon harvest (Weyer, Bush, Darzins, *et al.* 2010). To optimize the use of such incident irradiances, mixed cultures of algae with specific preferences for absorption of different wavelengths such as in *Acaryochloris marina* (Scott, Davey, Dennis, *et al.* 2010). In this way, chlorophyll *(a)* containing alga would not only absorb the majority of irradiance but also allow the second algal strain to capture the unused wavelengths such as far-red region. The most widely studied algal strain that absorbs far-red region of spectrum is the Ulvophycean marine green alga *Ostreobium*, which typically lives inside massive corals where shading by photobionts (*Symbiodinium* spp.) permits irradiances (in the wavelength range 340–680 nm) at the outer surface of the *Ostreobium* zone that may be as low as 0.19%–0.15% of those incident on corals (Halldal 1968, Shibata and Haxo 1969, Magnusson, Fine, and Kühl 2007). The action spectrum of photosynthesis in *Ostreobium* shows a major shoulder at 720 nm (Halldal 1968; Wilhelm and Jakob 2006), as does the action spectrum fluorescence excitation (Fork and Larkum 1989), and growth can also occur under monochromatic 700 nm irradiation (Wilhelm and Jakob 2006).

12.10.2 Limitation of Biofuel Production

For biofuel applications, the light absorption properties of industrial-scale cultures with a large optical thickness clearly present an important challenge (Scott, Davey, Dennis, *et al.* 2010; Weyer, Bush, Darzins, *et al.* 2010; Williams and Laurens 2010, Beardall and Raven 2013). High photosynthetically active irradiance (PAR) will only pass through a few centimetres of dense algae, where the shading effect limits the photon use efficiency, while the surface cells suffer from photoinhibition (further reducing efficiency); although weaker light is used more efficiently, it penetrates less far. It is suggested that high-intensity sunlight is photoinhibitory near the surface, whereas it is limiting in the deeper layer of the culture (Bar-Even, Noor, and Milo 2012).

In most algae tested, xanthophyll cycles play a crucial role in non-photochemical quenching, although all algae show energy-dependent quenching (qE), which works on a more rapid timescale (Goss and Jakob 2010; Depauw, Rogato, d'Alcalá, *et al.* 2012). Major means of non-photochemical quenching in cyanobacteria and Rhodophyta is generally qE (high-energy state), as these organisms lack conventional xanthophyll cycles. The water solubility of phycobilins, which are the major antenna pigment in PSII for these organisms, poses problems for the usual xanthophyll cycle related to non-photochemical quenching in the pigment bed. The xanthophyll cycle independent of the qE component of non-photochemical quenching occurs in a number of algae (Goss and Jakob 2010), including some members of Cryptophyta whose antenna pigments include chlorophyll a–c pigment–protein complexes in the thylakoid membrane, as well as phycobiliproteins in the thylakoid lumen (Kana, Kotabova, and Sobotka, *et al.* 2012).

Besides, it has been observed that nitrogen source plays a crucial role in the overall carbon fixation process. The quantity of individual proteins required, based on each enzyme's specific reaction and rate saturating substrate level, for CO_2 assimilation and stoichiometric relation of carbon and nitrogen is expected to reflect the total protein cost of the pathway related to a given rate of CO_2 fixation. The protein breakdown and re-synthesis in microalgae has an energy cost equivalent to one-thirds of the energy transformation occurring in respiration during the dark phase. This protein turnover and associated energy costs can be determined for cells cultured under constant temperature and with continuous illumination based on environmental conditions. The cell cycle and other endogenous cycles, and repair of photodamage to PSII, do not seem to account for the entire observed protein turnover (Falkowski and Raven 2007).

12.11 CONCLUSION

The screening of algal strain is important to valuate the biofuel outlook of algae. The native species of algae nourished in the wastewater by nutrients such as nitrogen and phosphorous could be advantageous than commercially available strains of algae. In this chapter, an effort has been made to examine the technological advantage of the microalgae growth in wastewater, which offers the twin benefits of growing biomass for bioremediation of wastewater and production of biofuel. An integrated approach to bioremediation of wastewater and reduction in the cost of biofuel production is a novel idea that needs to be further researched and studied.

REFERENCES

Agarwal, A. K. 2007. Biofuel (alcohol and biodiesel) applications as fuel for internal combustion engine. *Progress in Energy and Combustion Science* 33:233–271

Al Darmaki, Ahmed , L Govindrajan, Sahar Talebi, Sara Al-Rajhi, Tahir Al-Barwani, and Zainab Al-Bulashi. 2012. Cultivation and characterization of microalgae for wastewater treatment. *World Congress on Engineering* I: 2078–0966

Alrie, J. 2010. Cyclic electron flow around photosystem I in unicellular green algae. *Photosynthesis Research* 106: 47–56

Amin, S. 2009. Review on biofuel oil and gas production processes from microalgae. *Energy Conversion and Management* 50: 1834–1840

Anderson, D., D. Masterson, B. McDonald, and L. Sullivan. 2003. Industrial biodiesel plant design and engineering: practical experience. *International Palm Oil Conference (PIPOC)*, Putrajaya. Details available at <http:// www. crowniron.com/userImages/Biodiesel.pdf>, last accessed on 7 August 2009

Bar-Even, A., E. Noor, and R. Milo. 2012. A survey of carbon fixation path- ways through a quantitative lens. *J Exp Bot* 63: 2325–2342

Bar-Even, A., E. Noor, N. E. Lewis, and R. Milo. 2010. Design and analysis of synthetic carbon fixation pathways. *Proceedings of the National Academy of Sciences of the United States of America* 107: 8888–8894

Beardall, J. and J. A. Raven. 2013. Limits to phototrophic growth in dense cultures: CO_2 supply and light. In *Algae for Biofuels and Energy*, M. A. Borowitzka and NR Moheimani (eds), pp. 91–97. Dordrecht: Springer

Becker, E. W. 1994. *Microalgae: Biotechnology and Microbiology*, J. Baddiley, *et al.* (ed.), p. 178. Cambridge, NY: Cambridge University Press

Behzadi, S. and M. M. Farid. 2007. Review: Examining the use of different feedstock for the production of biodiesel. *Asia-Pacific Journal of Chemical Engineering* 2: 480–486

Benemann, J. 2009. Microalgae biofuels: a brief introduction. Details available at http://advancedbiofuelsusa.info/wp-content/uploads/2009/03/microalgae-biofuels-an-introduction-july23-2009-benemann.pdf

Benemann, J. 2003. Biofixation of CO_2 and greenhouse abatement with microalgae—technology roadmap. *Final Report for the U.S. Department of Energy*, National Energy Technology Laboratory

Benemann, J. and W. Oswald. 1996. Systems and economic analysis of microalgae ponds for conversion of CO_2 to biomass. *Final report for the U.S. Department of Energy,* Pittsburgh Energy Technology Center

Blankenship, R. E., D. M. Tiede, J. Barber, G. W. Brudvig, G. Fleming, M. Ghirardi, M. R. Gunner, W. Junge, D. M. Kramer, A. Melis, T. A. Moore, C. C. Moser, D. G. Nocera, A. J. Nozik, D. R. Ort, W. W. Parson, R. C. Prince, and R. T. Sayre. 2011. Comparing photosynthetic and photovoltaic efficiencies and recognising the potential for improvement. *Science* 332: 805–809

Bligh, E. G. and W. J. Dyer. 1959. A rapid method of total lipid extraction and purification. *Canadian Journal of Biochemistry and Physiology* 37: 911–917

Borowitzka, M. A. 1988. Fats, oils and hydrocarbons. In *Micro-algal Biotechnology*, M. A. Borowitzka and L. J. Borowitzka (eds), pp. 257–287. Cambridge: Cambridge University Press

Borowitzka, M. A. 1999. Commercial production of microalgae: ponds, tanks, tubes and fermenters. *Journal of Biotechnology* 70: 313–321

Borowitzka, M. A. 2007. Algal biotechnology products and processes matching science and economics. *Journal of Applied Phycology* 4: 267–279

Bridgwater, A. and G. Peacocke. 2000. Fast pyrolysis processes for biomass. *Renewable and Sustainable Energy Reviews* 4: 1–73

Brown, P. M. 1996. OLDLIST: A database of maximum tree ages. In *Tree Rings, Environment, and Humanity*, J. S. Dean, D. M. Meko and T. W. Swetnams (eds), pp. 727–731. Radiocarbon 1996. Department of Geosciences, The University of Arizona, Tucson

Buyukkamaci, N. and E. Koken 2010. Economic evaluation of alternative wastewater treatment plant options for pulp and paper industry. *Sci. Total Environ.* 408: 6070–6078

Chisti, Y. 2007. Biodiesel from microalgae. *Biotechnology Advances* 25: 294–206

Cordell, D., J. O. Drangert, and S. White. 2009. The story of phosphorus: global food security and food for thought. *Global Environmental Change* 19: 292–305

da Rosa, A. 2005. *Fundamentals of Renewable Energy Processes*, p. 515. Amsterdam: Elsevier

Depauw, F.A., A. Rogato, A. R. d'Alcalá, A. Falcatiore. 2012. Exploring the molecular basis of responses to light in marine diatoms. *J Exp Bot* 63: 1575–1581

Drogui, P., M. Asselin, S. K. Brar, H. Benmoussa, and J. F. Blais. 2008. Electrochemical removal of pollutants from agro-industry wastewaters. *Separation and Purification Technology* 61: 301–310

Elliot, D. C. and L. J. Sealock. 1999. Chemical processing in high-pressure aqueous environments: low temperature catalytic gasification. *Transactions of the Institution of Chemical Engineers* 74: 563–566

FAO (Food and Agriculture Organization of the United Nations). 2009a. Oil production. FAO Corp. Doc Repository. Details available at www.fao.org/docrep/w7241e/ w7241e0h.htm

FAO (Food and Agriculture Organization of the United Nations). 2009b. Algae-based biofuels: a review of challenges and opportunities for developing countries. *Environment and Natural Resources Working Paper* No. 33, FAO, Rome

Falkowski, P. G. and J. A. Raven. 2007. *Aquatic Photosynthesis*, 2nd edn. Princeton: Princeton University Press

Fork, D. C. and A. W. D. Larkum. 1989. Light harvesting in the green alga *Ostreobium* sp., a coral symbiont adapted to extreme shade. *Marine Biology* 103: 381–385

Golueke, C. G. and W. J. Oswald. 1955. Biological conversion of light energy to the chemical energy of methane. *Applied Microbiology* 7: 219–245

Gonzales, L. E., R. O. Canizares, and S. Baena. 2010. Efficiency of ammonia and phosphorus removal from a Colombian agroindustrial wastewater by the microalgae. *Chlorella vulgaris* and *Scenedesmus dimorphus*. *Bioresource Technology* 60: 259–262

Goss, R. and T. Jakob. 2010. Regulation and function of xanthophyll cycle-dependent photoprotection in algae. *Photosynthesis Research* 106: 103–122

Hankamer, B. 2010. An economic and technical evaluation of microalgal biofuels. *Nature Biotechnology* 28: 126–128

Halldal, P. 1968. Photosynthetic capacities and photosynthetic action spectra of endozoic algae of the massive coral *Favia*. *Biological Bulletin* 134: 411–424

Hauck, M. 1996. Die Flechten Niedersachsens. Bestand, Ökologie, Gefährdung und Naturschutz. *Naturschutz und Landschaftspflege in Niedersachsen* 36: 1–210

Huang, G., F. Chen, D. Wei, X. Zhang, and G. Chen. 2010. Biodiesel production by microalgal biotechnology. *Applied Energy* 87: 38–46

Jazrawi, C., P. Biller, Y. He, A. Montoya, A. B. Ross, T. Maschmeyer, B. S. Haynes. 2015. Two-stage hydrothermal liquefaction of a high-protein microalga. *Algal Research* 8: 15–22

Kadam, K. L. 1997. Microalgae production from power plant flue gas: environmental implications on a life cycle basis. NREL/TP-510-29417. Colorado: National Renewable Energy Laboratory

Kana, R., E. Kotabova, R. Sobotka, and O. Prášil. 2012. Non-photochemical quenching in cryptophyte alga *Rhodomonas salina* is located in the chlorophyll a/c antenna. *PLoS ONE 7: e29700*

Khan, S. A., A. Rashmi, M. Z. Hussian, S. Prasad, and U. C. Banerjee. 2009. Prospects of biodiesel production from microalgae in India. *Renewable and Sustainable Energy Reviews* 13: 2361–2372

Kothari, R., V. V. Pathak, V. Kumar, and D. P. Singh. 2012. Experimental study for growth potential of unicellular alga *Chlorella pyrenoidosa* on dairy waste water: An integrated approach for treatment and biofuel production. *Bioresource Technology* 116: 466–470

Lam, M. K. and K. Lee. 2012. Microalgae biofuels: A critical review of issues, problems and the way forward. *Biotechnology Advances* 30(3): 673–690

Lara, M. A., J. Rodríguez-Malaver, O. J. Rojas, O. Hoimquist, A. M. González, J. Bullón, N. Peñaloza, and E. Araujo. 2003. Black liquor lignin biodegradation by *Trametes elegans. International Biodeterioration and Biodegradation* 52: 167–173

Laurens, L. M., S. Van Wychen, J. P. McAllister, S. Arrowsmith, T. A. Dempster, J. McGowen, and P. T. Pienkos. 2014. Strain, biochemistry, and cultivation-dependent measurement variability of algal biomass composition. *Analytical Biochemistry* 452: 86–95

Levin G. V., J. R. Clendenning, A. Gibor, and F. D. Bogar. 1962. Harvesting of algae by froth flotation. *Applied Microbiology* 10(2): 169–175

Liu, C., H. Sun, L. Zhu, *et al.* 2006. Effects of salinity formed with two inorganic salts on freshwater algae growth. *Acta Scientiae Circumstantiae* 26(1): 157–161

Magnusson SH, Fine M, Kühl M. 2007. Light microclimate of endolithic phototrophs in the scleractinian corals Montipora montasteriata and Porites cylindrica. *Mar. Ecol. Prog. Ser.* 332:119–128

Mata, T. M., A. A. Martins, and N. S. Caetano. 2010. Microalgae for biodiesel production and other applications: A review. *Renewable and Sustainable Energy Reviews* 14: 217–223

Matthijs, H. C. P., H. Balke, U. M. van Hes, B. M. A. Kroon, L. R. Mur and R. A. Binot. 1996. Application of light-emitting diodes in bioreactors: flashing light effects and energy economy in algal culture (*Chlorella pyrenoidosa*). *Biotechnology and Bioengineering* 50: 98–107

McKendry, P. 2003. Energy production from biomass (part 2): Conversion technologies. *Bioresource Technology* 83: 47–54

Meher, L. C., D. Vidya Sagar, and S. N. Naik. 2006. Technical aspects of biodiesel production by transesterification–A review. *Renewable and Sustainable Energy Reviews* 10(3): 248–268

Meiers, R. L. 1955. Biological cycles in the transformation of solar energy into useful fuels. In *Solar Energy Research*, F. Daniels and T. A. Duffie (eds), pp. 179–189. Madison, WI: University of Wisconsin Press

Melack, M. J. 1981. Photosynthetic activity of phytoplankton in tropical African soda lakes. *Hydrobiology* 81: 71–85

Miao, X. L. and Q. Y. Wu. 2004. High yield bio-oil production from fast pyrolysis by metabolic controlling of *Chlorella protothecoides*. *Journal of Biotechnology* 110: 85–93

Minowa, T. and S. Sawayama. 1999. A novel microalgal system for energy production with nitrogen cycling. *Fuel* 78: 1213–1215

Molina G., E., E. H. Belarbi, F. G. Acién Fernández, A. Robles Medina, and Y. Chisti. 2003. Recovery of microalgal biomass and metabolites: Process options and economics. *Biotechnology Advances* 20: 491–515

Murakami, M., S. Yokoyama, T. Ogi, and K. Koguchi. 1990. Direct liquefaction of activated sludge from aerobic treatment of effluents from the corn starch industry. *Biomass* 23: 215–228

Noureddini, H., D. Harkey, and V. Medikonduru. 1998. A continuous process for the conversion of vegetable oils into methyl esters of fatty acids. *Journal of the American Oil Chemists' Society* 75(12): 1775–1783

National Science Foundation (NSF). 2009. Algae: A New Way to Make Biodiesel. Details available at <http://www.nsf.gov/discoveries/disc_summ.jsp?org=NSF&cntn_id=114934>

Olsson, G. 2012. *Water and Energy—Threats and Opportunities,* pp. 68–69. London: IWA Publishing

Oswald, W. J. 2003. My sixty years in applied algology. *Journal of Applied Phycology* 15: 99–106

Park, J. B. K. and R. J. Craggs. 2010. Wastewater treatment and algal production in high rate algal ponds with carbon dioxide addition. *Water Science and Technology* 61: 633–639

Patil, V., K.-Q. Tran, and H. R Giselrød. 2009. Towards sustainable production of biofuels from microalgae. *International Journal of Molecular Sciences* 9: 1188–1195

Pedroni, P., J. Davison, H. Beckert, P. Bergman, J. Benemann. 2001. A proposal to establish an international network on biofixation of CO_2 and greenhouse gas abatement with microalgae. *Journal of Energy and Environmental Research* 1(1): 136–150

Pintoa, A. C., L. L. N. Guarieiroa, G. M. J. C. Rezendea, N. M. Ribeiroa, E. A. Torresb, and W. A. Lopesc, *et al.* 2005. Biodiesel: An overview. *Journal of Brazilian Chemical Society* 15(5B): 1313–1330

Pittman, J. K., A. P. Dean, and O. Osundeko. 2011. The potential of sustainable algal biofuel production using wastewater resources. *Bioresource Technology* 102: 17–25

Pizzichini, M., C. Russo, and C. Di Meo. 2005. Purification of pulp and paper wastewater, with membrane technology, for water reuse in a closed loop. *Desalination* 178: 351–359

Posadas, E., S. Bochon, M. Coca, M. C. García-González, P. A. García-Encina, and R. Muñoz. 2014. Microalgae-based agro-industrial wastewater treatment: A preliminary screening of biodegradability. *Journal of Applied Phycology* 26: 2335–2345

Prescott, G. W. 1959. *How to Know the Fresh Water Algae*, Vol. 1. Michigan: Cranbrook Press.

Rajagopal, R., N. M. C. Saady, M. Torrijos, J. V. Thanikal, and Y. T. Hung. 2013. Sustainable agro-food industrial wastewater treatment using high rate anaerobic process. *Water* 5(1): 292–311

Ramasamy, E. V., S. Gajalakshmi, R. Sanjeevi, M. N. Jithesh, and S. A. Abbasi. 2004. Feasibility studies on the treatment of dairy wastewaters with upflow anaerobic sludge blanket reactors. *Bioresource Technology* 93: 209–2012

Raven, J. A., J. Beardall, A. W. D Larkum, and P. Sanchez-Baracaldo. 2013. Interactions of photosynthesis with genome size and function. *Philosophical Transactions of the Royal Society B* 368: 20120264

Raven, J. A., J. Beardall, and M. Giordano. 2014. Energy costs of carbon dioxide concentrating mechanisms in aquatic organisms. *Photosynthesis Research* 121(2–3): 111–124

Raven, J. A. and J. Peter. 2015. Enhanced biofuel production using optimality, pathway modification and waste minimization. *Journal of Applied Phycology* 27: 1–31

Reid, N. M., T. H. Bower, and G. Lloyd-Jones. 2008. Bacterial community composition of a wastewater treatment system reliant on N_2 fixation. *Applied Microbiology and Biotechnology* 79: 285–292

Roessler, P. G. 1990. Environmental control of glycerolipid metabolism in microalgae: Commercial implications and future research directions. *Journal of Phycology* 26: 393–399

Rosenberg, J. N., G. A. Oyler, L. Wilkinson, and M. J. Betenbaugh. A green light for engineered algae: Redirecting metabolism to fuel a biotechnology revolution. *Current Opinion in Biotechnology* 19(5): 430–436

Ruiz-Martinez, A., N. Martin Garcia, I. Romero, A. Seco, and J. Ferrer. 2012. Microalgae cultivation in wastewater: Nutrient removal from anaerobic membrane bioreactor effluent. *Bioresource Technology* 126: 247–253

Sacksteder, C. A., A. Kanazawa, M. E. Jacoby, and D. M. Kramer. 2000. The proton to electron stoichiometry of steady-state photosynthesis in living plants: A proton-pumping Q cycle is continuously engaged. *Proceedings of the National Academy of Sciences of the United States of America* 97: 14283–14288

Slade, A. H., R. J. Ellis, M. vanden Heuvel, and T. R. Stuthridge. 2004. Nutrient minimisation in the pulp and paper industry: an overview. *Water Sci. Technol.* 50: 111–122

Santiago, Urréjola, Rocio Maceiras, Leticia Pérez, Angeles Cancela, and Angel Sánchez. 2012. Analysis of microalgae oil transesterification for biodiesel production. *Chemical Engineering Transactions* 12: 1153–1158

Sazdanoff, N. 2006. Modeling and simulation of the algae to biodiesel fuel cycle. Ohio: Department of Mechanical Engineering, The Ohio State University

Scott, S. A., M. P. Davey, J. S. Dennis, I. Horst, C. J. Howe, D. J. Lea-Smith, and A. G. Smith. 2010. Biodiesel from algae: Challenges and prospects. *Current Opinion in Biotechnology* 21: 277–288

Sheehan, J., T. Dunahay, J. Benemann, and P. Roessler. 1998. A look back at the U.S. Department of Energy's Aquatic Species Program—Biodiesel from algae (NREL/TP-580-24190). Golden, CO: National Renewable Energy Laboratory (NREL), U.S. Department of Energy

Shibata, K. and F. T. Haxo. 1969. Light transmission and spectral distribution through epi-and endozoic algal layers in the brain coral. *Favia. Biol. Bull.* 136: 461–468

Slamovits, C. H., N. Okamoto, L. Burri, E. R. James, and P. J. Keeling. 2011. A bacterial proteorhodopsin proton pump in marine eukaryotes. *Nature Communications* 2: 183

Spoehr, H. A. and H. W. Milner. 1949. The chemical composition of *Chlorella*: Effect of environmental conditions. *Plant Physiology* 24: 120–149

Song, D., J. Fu, and D. Shi. 2008. Exploitation of oil-bearing microalgae for biodiesel. *Chinese Journal of Biotechnology* 24: 341–348

Stephens, E., I. L. Ross, Z. King, J. H. Mussgnug, O. Kruse, C. Posten, M. A. Borowitzka, and M. R. Tredici. 2010. Photobiology of microalgae mass cultures: Understanding the tools for the next green revolution. *Biofuels* 1: 143–162

Steuer, R., H. Knoop, and R. Machné. 2012. Modeling cyanobacteria: From metabolism to integrative models of phototrophic growth. *Journal of Experimental Botany* 63: 2259–2274

Sutton, M. A., O. Oenema, W. Erisman, A. Leip, H. van Grinsven, and W. Winiwarter. 2011. Too much of a good thing. *Nature* 472: 159–161

Tan, X., F. Kong, Y. Yu, *et al.* 2009. Effects of enhanced temperature on algae recruitment and phytoplankton community succession. *China Environmental Science* 29(6): 578–582

Tarlan, E., F. B. Dilek, and U. Yetis. 2002. Effectiveness of algae in the treatment of a wood-based pulp and paper industry wastewater. *Bioresource Technology* 84: 1–5

Thompson, G., J. Swain, M. Kay, and C. F. Forster. 2001. The treatment of pulp and paper mill effluent: A review. *Bioresource Technology* 77: 275–286

Tiwari, A. and P. Pospisil 2009. Superoxide oxidase and reductase activity of cytochrome b559 in photosystem II. *Biochimica et Biophysica Acta* 1787: 985–994

Tredici, M. R. 2010. Photobiology of microalgae mass cultures: Understanding the tools for the next green revolution. *Biofuels* 1: 143–162

Trentacoste, E. M., R. P. Shrestha, S. R. Smith, C. Glé, A. C. Hartmann, M. Hlidebrand, W. H. Gerwick. 2013. Metabolic engineering of lipid catabolism increases microalgal lipid accumulation without compromising growth. *Proceedings of the National Academy of Sciences of the United States of America* 110: 19748–19753

Weyer, K. M., R. D. Bush, A. Darzins, and B. D. Willson. 2010. Theoretical maximum algal oil production. *Bioenergy Research* 3: 204–213

Williams, P. J. L. B. and L. M. L. Laurens. 2010. Microalgae as biodiesel and biomass feedstocks: review and analysis of the biochemistry, energetics and economics. *Energy Environ. Sci.* 3:553–590

Wilhelm, C. and T. Jakob. 2006. Uphill energy transfer from long-wavelength absorbing chlorophylls to PSII in *Ostreobium* sp. is functional in carbon assimilation. *Photosynthesis Research* 87: 323–329

Wogan, D., D. M. S. AK, M. E. Webber, and E. Stauberg. 2008. Algae: Pond Powered Biofuels. ATI Clean Energy Incubator, pp. 1–23

Zang, C., S. Huang, M. Wu, *et al.* 2011. Comparison of relationships between pH, dissolved oxygen and chlorophyll a for aquaculture and non-aquaculture waters. *Water Air and Soil Pollution* 219(1–4): 157–174

13

Physical and Chemical Exergy Analysis and Assessment of Biogas as an Energy Source in Hybrid Cooling Machine

S. Anand[a,*], A. Gupta[a], Y. Anand[a], and S. K. Tyagi[b]

[a]*Faculty of Engineering, School of Energy Management, Shri Mata Vaishno Devi University, Katra, Jammu and Kashmir 182320*

[b]*Biomass Management Division, Sardar Swaran Singh National Institute of Renewable Energy (SSS-NIRE), Kapurthala, Punjab 144601*

*E-mail: anandsanjeev12@gmail.com

13.1   INTRODUCTION

Refrigeration/air-conditioning systems in the present time are the most important thermal systems which are required regularly and have also become the basic necessity. Industrial development has increased the demand of cooling systems. At present, issues related to the food preservation and their safe transportation have resulted in the demand of chillers and refrigerating units. The present lifestyle has also raised the demand of refrigeration units as people have started relying more on the preserved packaged foods. Various items such as medicines, dairy products, beverages, meat, and fish products are prone to bacterial attack if they are not properly stored. Bacterial growth, which can spoil food, needs to be inhibited by reducing both the moisture content and the temperature, and it can be achieved easily by installing refrigerating units. Different perishable products have different temperature ranges for storage, which are very closer to their freezing point. Different methods, such as dehydration, canning in the airtight containers, drying in room temperature, open sun drying, and so on, can be employed for preserving food items. Certain food products are in demand throughout the year and their demand can be met only through food preservation. The spoilage of such products can be prevented by maintaining the temperature and humidity of the place where they are preserved.

Refrigeration is a thermodynamic phenomenon in which heat from a low temperature source can be pumped to a high temperature sink, and for this purpose work is required to be done on the device. Such a device is called the refrigerating unit and the mode of operation of the cycle is called the refrigeration cycle. The working substance used in the cycle is called the refrigerant. The refrigerants can be both organic compounds (hydrocarbons or derivatives of alkanes) and inorganic substances (ammonia, carbon dioxide, and so on). Earlier, the halogenated derivatives of alkanes were used as refrigerants. However, as these refrigerants are found to be ozone depleting, they are not frequently considered for use or avoided because of their environmental threats. At present, technocrats and environmental protection agencies are emphasizing on the increased use of environmentally friendly refrigerants. The choice of refrigerants used in refrigeration cycles depends on the compatibility of the temperature range of the refrigerant with the particular application. The refrigeration cycles can be of various types; for example, vapour compression refrigeration cycle and vapour absorption refrigeration cycle. In the vapour compression refrigeration cycle, the refrigerant is compressed in the compressor, heat is rejected into the surroundings (condenser), expansion is carried out, and, finally, the heat is extracted from the space to be cooled. However, the vapour absorption refrigeration cycles do not require compressor work and can be operated with the use of low temperature heat from solar collectors, waste heat, biomass, or biogas. Such systems are easily operated in remote areas where electricity is not readily available. In the hybrid absorption refrigeration systems, a compressor is used between the generator and the condenser as this is an ideal position because of the availability of dry saturated vapours at the outlet of the generator. The compressor increases the pressure and temperature of the refrigerant before the entry into the condenser and the heat rejected in the condenser is utilized for process application. The hybrid refrigeration systems can also be channelled for sustainable future because they do not use both conventional energy and CFC-based refrigerants. Carbon emission into the environment can also be reduced with such systems.

Energy consumption for cooling and space air-conditioning in both domestic and industrial sectors owing to the increased level of thermal comfort and a variety of applications constitutes a huge share across the globe. The refrigeration and air-conditioning industry is facing two major problems, which include phasing out of halogenated fluorocarbon as proposed in the Montreal Protocol and minimizing the use of the conventional energy which leads to environmental degradation and global warming. The energy saving trends have also prompted the development of new and sustainable technologies. Absorption systems are becoming popular because they use not only non-ozone depleting environmentally friendly refrigerants but also cheap and readily available alternative sources of energy, subsequently reducing electricity consumption (Abdullah and Hien 2011; Anand, Gupta, and Tyagi 2014a). A recent energy saving trend has led to the development of a methodology called the second law, which helps in analysing thermal systems and their technologies (Lee and Sherif 2001). Several authors have worked on different types of refrigeration systems. The results obtained from the analysis of application of geothermal and solar energy in combined compression absorption refrigeration systems have revealed the potential of such energy sources (Ayala, Heard, and Holland 1998; Goktun 1999; Santayo-Gutierrez, Siqueiros, Heard, *et al.* 1999; Tarique and Siddiqui 1999; Kececiler, Acar, and Dogan 2000). Again, the generation of cooling at low temperature from the compression–absorption cascade refrigeration system powered by cogeneration is appreciable from the viewpoint of system design (Jose, Sieres, and Vazquez 2006). Another comparative study using different refrigerant–absorbent combinations for compression–absorption machine for simultaneous cooling and heating revealed that performance of the working fluid is influenced by the capacity of the system (Satapathy, Ram Gopal, and Arora 2007). The results obtained from the detailed study of compression–absorption refrigeration system having the capacity of 400 kW show that performance of the system is affected by the mass flow rate of weak solution (Pratihar, Kaushik, and Agarwal 2010). An analysis of cascaded absorption–compression refrigeration cycle using lithium bromide for absorption system and subcritical CO_2 for vapour compression cycle showed higher COP and energy saving when compared to a conventional vapour compression system (Garimella, Brown, and Nagavarapu 2011). Another study of a cascaded refrigeration system of 9 kW capacity using NH_3 and CO_2 as refrigerants revealed that the system can provide COP of 1.6 (Alberto Dopazo and Fernandez-Seara 2011). An experimental study of a 15-tonne cooling capacity system for preserving agricultural products concluded that operation of a system using hot water (95–98°C) and cooling water of temperature 26°C could maintain the cooling chamber at the temperature of 5°C (Borde and Jelinek 1987). The selection of a proper refrigerant–absorbent pair is also stressed in the study to meet the requirement of continuous refrigeration. The results obtained from biogas operated absorption refrigeration system reveal that generator is the worst component from irreversibility viewpoint. Also, in response to the actual case, the irreversibility in absorber is found to be more compared to that of a condenser (Anand, Gupta, and Tyagi 2014b). The analysis of NH_3–H_2O and $LiBr$–H_2O based hybrid heat pumps using solar energy/gas meant for refrigeration, power generation, and sea water purification demonstrated that hybrid systems have higher cooling efficiency and purified water output (Nguyen, Riffat, and Whitman 1996). The results obtained from the review of different renewable energy sources are found to be beneficial for the professionals involved in performance assessment, design, and analysis of such systems (Hepbasli 2008). Again, the results of an experimental study of hybrid air-conditioning system using solar biomass energy found COP to be around 0.11, which is comparatively better than other similar systems (Prasartkaew and Kumar 2013). This chapter deals with the first and second law analyses of biogas operated hybrid cooling machine. The analysis has been carried out to determine both the operational potential and the performance of such systems. The influence of various operating conditions on different performance indicators have been assessed and calculated. The work done by the compressor is also calculated in the analysis. The effect of the ambient temperature on the exergy loss (that is, Ψ^P_{loss} and Ψ^C_{loss}) and exergy efficiency are also discussed.

13.2   SYSTEM DESCRIPTION AND WORKING PRINCIPLE

The proposed system is an ammonia–water vapour hybrid cooling machine. The biogas fired boiler produces energetic hot water, which is an energy input to the generator of refrigeration machine. The system, shown in Figure 13.1 (Anand, Gupta, and Tyagi 2014c), consists of a biogas operated boiler, generator, compressor, condenser, an absorber, an evaporator, solution heat exchanger (SHE), throttling device, pressure reducing valves, flow controlling valves (V_1, V_2, and V_3), and a pump. The compressor is introduced between the generator and the condenser because dry saturated vapours are readily available at the outlet of the generator.

In the hybrid system, the solution containing a large concentration of refrigerant exits the absorber (8) and is pumped with the help of solution pump (9) into the SHE, where it is heated by the solution containing a small concentration of refrigerant (weak solution) that comes from the generator. The concentrated refrigerant solution moves into the generator (1), where the refrigerant is raised into vapours by the external heat supply in terms of hot energetic water from biogas-fired boiler and gets separated from the solution. The refrigerant vapours come out of the generator and are allowed to pass into the compressor (2–4) through valves V_1 and V_2 to raise the temperature and pressure of the working fluid, while the weak solution goes back to the absorber via SHE (3–10), where PR (pressure reducing) valve lowers its pressure (10–11). The compressed refrigerant is condensed in the condenser, releasing latent heat of condensation into the cooling water flowing through the condenser. The refrigerant

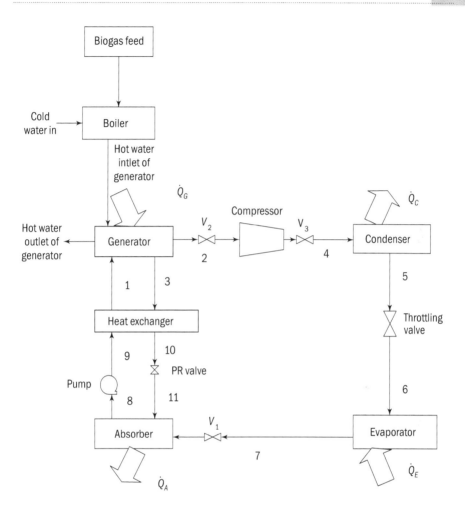

Fig. 13.1 *A line diagram of a biogas powered hybrid (VCA) cooling machine*

liquid produced in the condenser (5) has higher pressure and to reduce the pressure, the liquid is allowed to move through the throttling device (5–6). The saturated liquid ammonia formed at this low pressure moves into the evaporator and again gets converted into vapours by absorbing the heat from the chilled water circulated between the evaporator and the cooling space. The ammonia vapours generated enter the absorber (7) through valve V_1 and absorb the solution containing a large concentration of the refrigerant, subsequently releasing the heat of absorption into the circulating fluid moving through the absorber. The strong solution then comes out from the absorber either in saturated or in a slightly subcooled state. The cycle repeats and generates simultaneous heating and cooling effect.

13.3   THERMAL ANALYSIS OF BIOGAS BOILER

A simplified energy balance is done for biogas fired boiler. The biogas acts as an input energy to produce hot water which acts as an energy feed to the generator of the system without any compromise of temperature loss to the ambient. Therefore,

$$\dot{m}_{W,b} (h_{W,o} - h_{W,i})_G = \dot{Q}_G \tag{13.1}$$

Also,
$$\dot{m}_{W,b} (h_{W,o} - h_{W,i})_b = \frac{CVg}{\rho} \times \dot{m}_g \tag{13.2}$$

Comparing Equations (13.1) and (13.2), we get

$$\dot{m}_g = \frac{\dot{Q}_G}{CV_g} \times \rho \tag{13.3}$$

Also, the density of biogas is 1.214 kg/m^3 (about 60% CH_4 and 40% CO_2). The flow rate of the gas in terms of volume used for the generation of hot water of the required temperature can be calculated by the following equation:

$$\dot{V}_g = \frac{\dot{m}_g}{\rho} = \frac{\dot{Q}_G}{CV_g} \tag{13.4}$$

13.4   ENERGY AND EXERGY

The capacity to do work is called energy and it is the potential of the object to transfer its energy to other objects in different forms. The first law of thermodynamics deals with conservation of energy. The first law of thermodynamics gives no information about the direction in which a process can occur. Spontaneous process can proceed only in a definite direction. Also energy involved is conserved regardless of feasibility of the process. The first law of thermodynamics provides no information about the work done by any system in an isothermal process. It also gives no idea about the inability of the system to convert heat entirely into mechanical work. The energy can be of various types, such as kinetic energy, potential energy, gravitational energy, and chemical energy.

However, the second law of thermodynamics establishes the difference in the quality of different forms of energy and provides an insight into the spontaneity of some processes. It defines the fundamental quantity called entropy as a randomized energy state which is unavailable for direct conversion into work. It also states that all spontaneous processes, that is, physical and chemical, move in a direction to maximize entropy, which tends to convert energy in a less available form. The second law of thermodynamics also states that the entropy of a system is found to be maximum at thermodynamic equilibrium, thus indicating that no further increase in disorder is possible without changing the thermodynamic state of the system by some external forces (example, heat and work).

The second law of thermodynamics helps to determine the maximum possible work that can be obtained from a process. Therefore, exergy is a useful quantity that is used to analyse different thermal systems and processes. The exergy measures both the quality and the quantity of the available energy. Exergy has a characteristic that it is conserved only when all the processes occurring in a system and the environment are reversible. Exergy is destroyed whenever an irreversible process occurs. When energy loses its quality, exergy is destroyed. Exergy is part of the energy which is useful and, therefore, has an economic value and is worth managing carefully. Exergy, by definition, depends not only on the state of a system or flow, but also on the state of the environment.

Exergy is known to be of four types: physical, kinetic, potential, and chemical. The physical exergy is the maximum useful work obtained by passing unit mass of a substance from its respective generic state of pressure and temperature to the environmental pressure and temperature conditions through purely physical processes. Physical exergy can be determined with the enthalpy and entropy values both at the generic state and at the environmental temperature and pressure states. The maximum work that can be done when the considered system is brought into reaction with reference substances present in the environment is called chemical exergy. It depends on the temperature, pressure, and composition of a system.

13.5 THERMODYNAMIC ANALYSIS

The analysis of a hybrid system has been carried out by developing a computational model (Klein and Alvarado 2013) based on first and second laws of thermodynamics. The numerical model was developed to analyse and assess various performance indicators such as $COP_{cooling}$, $COP_{heating}$, second law efficiency, and irreversibility loss in different components. The analysis is carried out for 1 TR cooling capacity system using various operating conditions as given in Table 13.1 and the results obtained from the use of numerical approach are given in Table 13.2. The analysis is based on the following assumptions (Anand, Gupta, and Tyagi 2014c):

(i) Steady-state condition is applied and the heat loss through the conducting pipes is negligible.

(ii) Solution leaving the absorber and the generator is assumed to be saturated at their respective temperatures and concentrations.

Table 13.1 Fixed operating parameters and conditions used in the simulation

Operating parameters	Fixed values
Cooling capacity	1 TR
Heat source pressure	10 bar
Cooling unit pressure	0.9 bar
Heating unit temperature, T_G	348.15 – 398.15 K
Cooling unit temperature, T_E	271.15 K
Condensing unit temperature, T_C	308.15 K
Absorbing unit temperature, T_A	308.15 K
Mass flow rate of strong solution ($\dot{m}_1$)	0.125 kg/s
Mass flow rate of refrigerant ($\dot{m}_2$)	0.002808 kg/s
Mass flow rate of weak solution ($\dot{m}_3$)	0.1222 kg/s
Dead state temperature, T_0	298.15 K
Dead state pressure	1 bar
Feed water temperature to the boiler	30°C
CV of biogas	$(20 \times 10^3 \text{ kJ/m}^3)^a$

[a] Details available at http://mnre.gov.in/file-manager/UserFiles/case-study-Biogas-Generation_Purification_and_Bottling-Development-In-India.pdf, last accessed on December 2014

Table 13.2 Outcomes generated from the thermodynamic simulation ($T_E = -2°C$, $T_A = T_C = 35°C$, $T_G = 75°C$ and cooling capacity = 3.516 kJ/s)

State point	Tempera-ture (K)	x (%NH3)	Pressure (bar)	Entropy (kJ/(kg K))	Enthalpy (kJ/kg)	Ψ^P (kJ/s)	Ψ^C (kJ/s)	Ψ^t (kJ/s)
1.	323.1	0.35	10	0.6182	7.438	−21.53	−2.936	−24.466
2.	348.1	0.9816	10	4.816	1444	0.0382	1.205	1.2432
3.	348.1	0.4374	10	0.9228	100.3	−20.79	1.902	−18.88
4.	403.1	0.9816	10	5.183	1582	0.1182	1.285	1.4032
5.	308.1	0.7506	10	0.4438	−5.934	−0.3752	0.5177	0.1425
6.	237.9	0.9816	0.9	0.09581	−5.934	−0.08401	1.083	0.9989
7.	271.1	0.9816	0.9	5.288	1246	−0.9127	0.2543	−0.6584
8.	308.1	0.35	0.9	1.202	178.8	−21.35	−3.176	−24.526
9.	308.1	0.35	0.9	1.202	178.8	−21.35	−3.176	−24.526
10.	358.4	0.4374	10	1.416	275.6	−17.34	5.352	−11.988
11.	358.4	0.4374	10	1.416	275.6	−17.34	5.352	−11.988

Also, the refrigerant leaving the condenser and the evaporator is assumed to be saturated.

(iii) The pressure drop in valves is considered to be negligible.

(iv) The pump work is neglected in the analysis as it is considered to be working at isenthalpic conditions.

(v) The mass flow rate of the refrigerant at state points 2, 4, 5, 6, and 7, mass flow rate of the weak solution at state points 3, 10, and 11, and mass flow rate of the strong solution at state points 1, 8, and 9 are assumed to be the same.

(vi) The temperarture of the source water which is to be fed into the boiler is required to be heated with biogas and maintained at 30°C. It can be used further for the working of hybrid compression absorption refrigeration system without any temperature drop.

(vii) The temperature of water at the exit of the generator is also considered to be at 30°C, which again can be fed into the biogas boiler for closed cycle operation.

(viii) The reference enthalpy (h_o) and the entropy (s_o) are taken at the temperature and pressure of 25°C and 1 bar, respectively.

(ix) The efficiency of the compressor is assumed to be 0.7 and the effectiveness of solution heat exchanger is calculated to be 0.53.

(x) The analysis of the system involves heat rejection to the cooling water at the condenser and the absorber.

The following are the standardized equations of energy, species, and mass conservation:

Mass conservation

$$\sum \dot{m}_i - \sum \dot{m}_o = 0 \tag{13.5}$$

Species conservation

$$\sum (\dot{m}x)_i - \sum (\dot{m}x)_o = 0 \tag{13.6}$$

Energy conservation

$$\sum (\dot{m}h)_i - \sum (\dot{m}h)_o + \left[\sum \dot{Q}_i - \sum \dot{Q}_o \right] \pm \dot{W} = 0 \tag{13.7}$$

13.5.1 Energy Analysis

The thermodynamic analysis of hybrid machine involves the energy balance of different components on the basis of various assumed conditions as given in Table 13.1. The energy balance equations of different components are as follows:

Energy balance at generator

$$\dot{Q}_G = \dot{m}_2 h_2 + \dot{m}_3 h_3 - \dot{m}_1 h_1 \tag{13.8}$$

Energy balance at absorber

$$\dot{Q}_A = \dot{m}_2 h_7 + \dot{m}_3 h_{11} - \dot{m}_1 h_8 \qquad (13.9)$$

Energy balance at condenser

$$\dot{Q}_C = \dot{m}_2 (h_4 - h_5) \qquad (13.10)$$

Energy balance at evaporator

$$\dot{Q}_E = \dot{m}_2 (h_6 - h_7) \qquad (13.11)$$

Energy balance at solution heat exchanger

$$\dot{Q}_{SHE} = \dot{m}_3 (h_3 - h_{10}) = \dot{m}_1 (h_1 - h_9) \qquad (13.12)$$

Work done by compressor

$$\dot{W}_{COMP} = \frac{\dot{m}_2 (h_2 - h_4)}{\eta_C} \qquad (13.13)$$

The expansion valves and flow valves are considered to be isenthalpic in the analysis. The proposed hybrid refrigeration system can be a perfectly reversible system and the net cooling effect is the heat absorbed by the refrigerant in the evaporator to the sum of compressor work and heat load of the generator. Therefore, performance is given as follows:

$$COP_{cooling} = \frac{\dot{Q}_E}{(\dot{Q}_G + \dot{W}_{COMP})} \qquad (13.14)$$

The rejection of heat from the condenser and the absorber to the circulating fluid can also be utilized for heating applications and, therefore, the respective heating performance of the machine is given as

$$COP_{heating} = \frac{\dot{Q}_C + \dot{Q}_A}{(\dot{Q}_G + \dot{W}_{COMP})} \qquad (13.15)$$

13.5.2 Exergy Analysis

This is also called the second law analysis and can be used to evaluate the irreversible nature of different thermal processes. The major factors that lead to irreversibility in a thermal system are friction, heat transfer under temperature difference, and unrestricted expansion (ASHRAE 1997).

Kinetic exergy and potential exergy are assumed to be neglected in the analysis. The physical exergy can be determined with the enthalpy and entropy values both at the generic state and at the environmental temperature and pressure states. The irreversibility measures the process imperfection and, therefore, helps in determining the optimum operating conditions. However, the physical exergy can also be defined as the amount of work required to bring a stream of matter from an initial state (generic state) to a state that is in thermal and mechanical equilibrium with the environment (Bejan, Tsatsaronis, and Moran 1995), and it is given as follows:

$$\Psi^P = \dot{m}[(h - h_o) - T_o(s - s_o)] \tag{13.16}$$

Here, Ψ is the exergy of the fluid at temperature T. The terms h_o and s_o are the enthalpy and entropy, respectively, of the fluid at the environmental temperature T_o. The physical exergy balance equations of different components of the absorption system can be given as follows (Anand, Gupta, and Tyagi 2014c):

Physical exergy balance at the generator

$$\psi_G^P = \dot{m}_1[(h_1 - h_o) - T_o(s_1 - s_o)] - \dot{m}_2[(h_2 - h_o) - T_o(s_2 - s_o)] - \dot{m}_3$$

$$[(h_3 - h_o) - T_o(s_3 - s_o)] + \dot{Q}_G\left(1 - \frac{T_o}{T_G}\right) \tag{13.17}$$

Physical exergy balance at the condenser

$$\psi_C^P = \dot{m}_2[(h_4 - h_5) - T_o(s_4 - s_5)] - \dot{Q}_C\left(1 - \frac{T_o}{T_C}\right) \tag{13.18}$$

Physical exergy balance at the absorber

$$\psi_A^P = \dot{m}_2[(h_7 - h_o) - T_o(s_7 - s_o)] + \dot{m}_3[(h_{11} - h_o) - T_o(s_{11} - s_o)] - \dot{m}_1$$

$$[(h_8 - h_o) - T_o(s_8 - s_o)] - \dot{Q}_A\left(1 - \frac{T_o}{T_A}\right)$$

$$\tag{13.19}$$

Physical exergy balance at the evaporator

$$\psi_E^P = \dot{m}_2[(h_6 - h_7) - T_o(s_6 - s_7)] + \dot{Q}_E\left(1 - \frac{T_o}{T_E}\right) \tag{13.20}$$

Physical exergy balance at the compressor

$$\psi_{COMP}^P = \dot{m}_2[(h_4 - h_2) - T_o(s_4 - s_2)] \tag{13.21}$$

Physical exergy balance at the solution heat exchanger

$$\psi_{SHE}^P = \dot{m}_3[(h_3 - h_o) - T_o(s_3 - s_o)] - \dot{m}_3[(h_{10} - h_o) - T_o(s_{10} - s_o)]$$

$$\tag{13.22}$$

The maximum work that can be done when the considered system is brought into reaction with reference substances present in the environment is called the chemical exergy. It depends on the temperature, pressure, and composition of a system. The major difference between chemical exergy and the thermo-mechanical exergy in their calculation is that thermo-mechanical exergy does not consider the difference in the chemical composition of a system and environment. The generalized equation to calculate chemical exergy can be given as (Morosuk and Tsatsaronis 2008):

$$\Psi^C = \dot{m}\,[x \times \Psi^C_{ref} + (1-x)\,\Psi^C_{abs} + w^{rev}_x] \tag{13.23}$$

where $\quad w^{rev}_x = [h_o - xh_{ref,o} - (1-x)\,h_{abs,o}] - T_o\,[s_o - xs_{ref,o} - (1-x)s_{abs,o}]$

$$\tag{13.24}$$

Equation (13.24) represents the specific work done when mixing of pure refrigerant and pure absorbent takes place. $\Psi^C_{ref} = 5.790$ kJ/kg and $\Psi^C_{abs} = 0.810$ kJ/kg are the chemical exergy values of the refrigerant and the absorbent, respectively, at standard environmental conditions. $h_{ref,o}$, $h_{abs,o}$, $S_{ref,o}$, and $S_{abs,o}$ are the enthalpy and entropy values of refrigerant and absorbent, respectively. The chemical exergy balance equations of different components of the absorption system are as follows:

Chemical exergy balance at the generator

$$\Psi^C_G = \dot{m}_1\,\{[(x_1\,\Psi^C_{NH_3}) + (1-x_1)\,\Psi^C_{H_2O}] + [(h_1 - x_1 h_{NH_3,o}) - (1-x_1)h_{H_2O,o}]$$
$$-T_o\,[(s_1 - x_1 s_{NH_3,o}) - (1-x_1)s_{H_2O,o}]) - \dot{m}_2\,([(x_2\Psi^C_{NH_3}) + (1-x_2)\Psi^C_{H_2O}]$$
$$+[(h_2 - x_2 h_{NH_3,o}) - (1-x_2)h_{H_2O,o}] - T_o\,[(s_2 - x_2 s_{NH_3,o}) - (1-x_2)s_{H_2O,o}])$$
$$-\dot{m}_3([(x_3\,\Psi^C_{NH_3}) + (1-x_3)\Psi^C_{H_2O}] + [(h_3 - x_3 h_{NH_3,o}) - (1-x_3)h_{H_2O,o}]$$
$$-T_o\,[(s_3 - x_3 s_{NH_3,o}) - (1-x_3)s_{H_2O,o}]\}$$

$$\tag{13.25}$$

Chemical exergy balance at the absorber

$$\Psi^C_A = \dot{m}_3\,\{[(x_{11}\,\Psi^C_{NH_3}) + (1-x_{11})\Psi^C_{H_2O}] + [(h_{11} - x_{11}h_{NH_3,o}) - (1-x_{11})h_{H_2O,o}]$$
$$-T_o[(s_{11} - x_{11}s_{NH_3,o})] - (1-x_{11})s_{H_2O,o}]) + \dot{m}_2([(x_7\Psi^C_{NH_3}) + (1-x_7)\Psi^C_{H_2O}]$$
$$+[(h_7 - x_7 h_{NH_3,o}) - (1-x_7)h_{H_2O,o}] - T_o[(s_7 - x_7 s_{NH_3,o}) - (1-x_7)s_{H_2O,o}])$$
$$-\dot{m}_1([(x_8\,\Psi^C_{NH_3}) + (1-x_8)\Psi^C_{H_2O}] + [(h_8 - x_8 h_{NH_3,o}) - (1-x_8)h_{H_2O,o}]$$
$$-T_o[(s_8 - x_8 s_{NH_3,o}) - (1-x_8)s_{H_2O,o}]\}$$

$$\tag{13.26}$$

Chemical exergy balance at the condenser

$$\Psi_C^C = \dot{m}_2 \{[(x_4 \Psi_{NH_3}^C) + (1 - x_4)\Psi_{H_2O}^C] + [(h_4 - x_4 h_{NH_3,o}) - (1 - x_4)h_{H_2O,o}]$$
$$- T_o[(s_4 - x_4 s_{NH_3,o}) - (1 - x_4)s_{H_2O,o}]) - \dot{m}_2([(x_5 \Psi_{NH_3}^C) + (1 - x_5)\Psi_{H_2O}^{CH}]$$
$$+ [(h_5 - x_5 h_{NH_3,o}) - (1 - x_5)h_{H_2O,o}] - T_o[(s_5 - x_5 s_{NH_3,o}) - (1 - x_5)s_{H_2O,o}]\}$$

$$(13.27)$$

Chemical exergy balance at the evaporator

$$\Psi_E^C = \dot{m}_2 \{[(x_6 \Psi_{NH_3}^C) + (1 - x_6)\Psi_{H_2O}^C] + [(h_6 - x_6 h_{NH_3,o}) - (1 - x_6)h_{H_2O,o}]$$
$$- T_o[(s_6 - x_6 s_{NH_3,o}) - (1 - x_6)s_{H_2O,o}]) - \dot{m}_2([(x_7 \Psi_{NH_3}^C) + (1 - x_7)\Psi_{H_2O}^C]$$
$$+ [(h_7 - x_7 h_{NH_3,o}) - (1 - x_7)h_{H_2O,o}] - T_o[(s_7 - x_7 s_{NH_3,o}) - (1 - x_7)s_{H_2O,o}]\}$$

$$(13.28)$$

Chemical exergy balance at the compressor

$$\Psi_{COMP}^C = \dot{m}_2 \{[(x_4 \Psi_{NH_3}^C) + (1 - x_4)\Psi_{H_2O}^C] + [(h_4 - x_4 h_{NH_3,o}) - (1 - x_4)h_{H_2O,o}]$$
$$- T_o[(s_4 - x_4 s_{NH_3,o}) - (1 - x_4)s_{H_2O,o}]) - \dot{m}_2([(x_2 \Psi_{NH_3}^C)$$
$$+ (1 - x_2)\Psi_{H_2O}^C] + [(h_2 - x_2 h_{NH_3,o}) - (1 - x_2)h_{H_2O,o}]$$
$$- T_o[(s_2 - x_2 s_{NH_3,o}) - (1 - x_2)s_{H_2O,o}]\}$$

$$(13.29)$$

Chemical exergy balance at the solution heat exchanger

$$\Psi_{SHE}^C = \dot{m}_3 \{([(x_3 \Psi_{NH_3}^C) + (1 - x_3)\Psi_{H_2O}^C] + [(h_3 - x_3 h_{NH_3,o})$$
$$- (1 - x_3)h_{H_2O,o}] - T_o [(s_3 - x_3 s_{NH_3,o}) - (1 - x_3)s_{H_2O,o}]\}$$
$$- \dot{m}_3 \{[(x_{10} \Psi_{NH_3}^C) + (1 - x_{10})\Psi_{H_2O}^C] + [(h_{10} - x_{10} h_{NH_3,o})$$
$$- (1 - x_{10})h_{H_2O,o}] - T_o[(s_{10} - x_{10} s_{NH_3,o}) - (1 - x_{10})s_{H_2O,o}]\}$$

$$(13.30)$$

13.5.3 Exergy Efficiency

The performance of the system based on the second law can be measured in terms of exergetic efficiency as given in Equation (13.31) and is defined as the ratio of the net exergy produced by the evaporator (desired exergy output) and input exergy to the generator plus mechanical work of the solution pump and compressor.

$$\eta_\Psi = \frac{\dot{Q}_E \left| \left(1 - \dfrac{T_o}{T_E} \right) \right|}{\dot{Q}_G \left| \left(1 - \dfrac{T_o}{T_G} \right) \right| + \left| (\psi_{COMP}^P)_i - (\psi_{COMP}^P)_o \right|} \qquad (13.31)$$

Here, η_Ψ is the exergy efficiency and $(\Psi_{COMP}^P)_o$ and $(\Psi_{COMP}^P)_i$ are the output and input exergies (kW) at the compressor, respectively.

Also, the total exergy loss in a component is given as

$$\Psi^t = \Psi^P + \Psi^C \qquad (13.32)$$

Again, the non-dimensional exergy loss in a component is defined as the ratio of exergy loss in an individual component to the total exergy loss of the system (Aman, Ting, and Henshaw 2014), and it is given as

$$\text{Non-dimensional exergy loss } (\Psi_{loss,ND}^P \text{ or } \Psi_{loss,ND}^C) \qquad (13.33)$$

$$= \frac{\text{Exergy loss in an individual component}}{\text{Total exergy loss of a system}}$$

where

$$\text{Total physical exergy loss} = \Psi_{loss,G}^P + \Psi_{loss,E}^P + \Psi_{loss,C}^P + \Psi_{loss,A}^P$$

$$+ \Psi_{loss,SHE}^P + \Psi_{loss,COMP}^P \qquad (13.34)$$

$$\text{Total cemical exergy loss} = \Psi_{loss,G}^C + \Psi_{loss,E}^C + \Psi_{loss,C}^C + \Psi_{loss,A}^C$$

$$+ \Psi_{loss,SHE}^C + \Psi_{loss,COMP}^C \qquad (13.35)$$

13.6   RESULTS AND DISCUSSION

Using the concept of exergy and energy, a hybrid cooling machine has been assessed by a numerical method approach (Klein and Alvarado 2013). The computational methodology has been developed to study the various performance parameters such as $COP_{cooling}$, $COP_{heating}$, exergy efficiency, and irreversibility loss in different components. The various operating conditions used for calculating outcomes and their respective values are given in Tables 13.1 and 13.2. The input parameters taken in the analysis are $T_E = -2°C$, $T_G = 75°C$, and $T_C = T_A = 35°C$. The results obtained are also presented graphically in Figures 13.2–13.7.

The effect on the volume flow rate of the biogas and compressor work with generator temperatures is presented in Figure 13.2, and from the figure it becomes evident that the volume flow rate of biogas in the boiler increases with an increase in the generator temperature and attains the value 3.13 m³/h at the generator temperature of 125°C. The reason is that in order to attain the temperature of 125°C, more energy is required, which

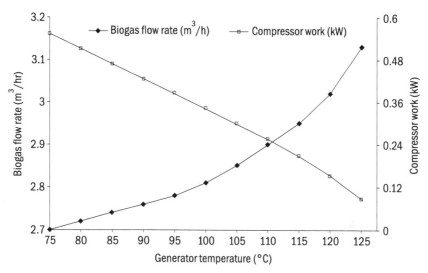

Fig. 13.2 *Biogas flow rate and compressor work with generator temperature ($T_E = -2°C$, $T_A = T_C = 35°C$, $\varepsilon_{SHE} = 0.53$)*

subsequently increases the biogas demand. However, the reduced pressure and temperature difference at the inlet and outlet of the compressor is the reason for decreasing compressor work with increasing generator temperature.

The effect on performance and second law efficiency with heat source temperature (generator) is presented in Figure 13.3, which clearly shows

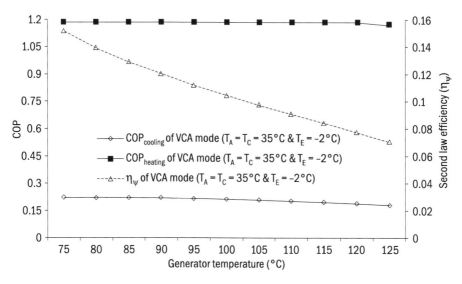

Fig. 13.3 *COP and second law efficiency of hybrid machine with generator temperature ($T_E = -2°C$, $T_A = T_C = 35°C$, $\varepsilon_{SHE} = 0.53$)*

that both the performance indicators show a descending behaviour. It is also observed that both $COP_{cooling}$ and $COP_{heating}$ decrease with a rising generator temperature and move up to 0.1781 and 1.174, respectively, at 125°C generator temperature. The second law efficiency also decreases with the rising generator temperature and moves up to 0.07 at 125°C. The reason for this behaviour is that the circulation ratio of the solution increases with an increase in the generator temperature, which consequently increases the generator heat load. This elevation in energy input brings down the compressor work to some extent, but the heat of condensation still increases and reduces the performance and second law efficiency at higher generator temperatures.

The effect on performance and second law efficiency with the evaporator temperature is shown in Figure 13.4, which clearly indicates that both $COP_{cooling}$ and $COP_{heating}$ ascend up to 0.222 and 1.189, respectively, at the evaporator temperature of 10°C. The behaviour of second law efficiency is also seen to be descending with the evaporator temperature (0.08 at 10°C) because the pressure ratio across the compressor descends with rising evaporator temperature, which subsequently lowers the compressor work. The reduction in compressor work also brings down the total energy input required and, therefore, improves the COP. The change in second law efficiency is because of the major alterations observed in the numerator part of Equation (13.31).

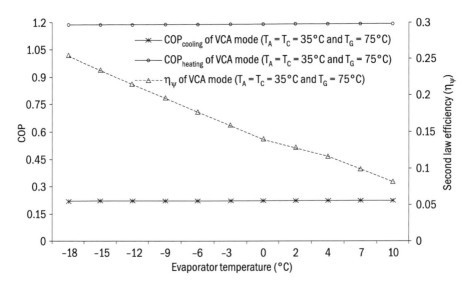

Fig. 13.4 *COP and second law efficiency of hybrid machine with evaporator temperature* $(T_G = 75°C, T_A = T_C = 35°C, \varepsilon_{SHE} = 0.53)$

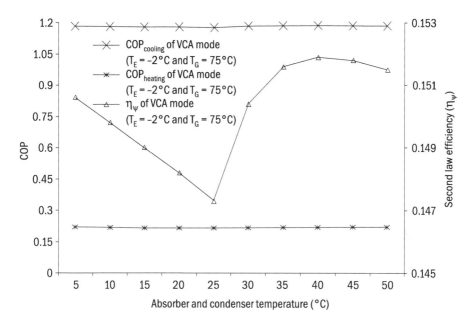

Fig. 13.5 *COP and second law efficiency hybrid machine with absorber and condenser temperature ($T_G = 75°C$, $T_E = -2°C$, $\varepsilon_{SHE} = 0.53$)*

Figure 13.5 presents the behaviour of performance and second law efficiency with the absorber and the condenser temperature, which clearly shows that both $COP_{cooling}$ and $COP_{heating}$ show a marginal increase up to 0.2204 and 1.186, respectively, at 50°C with increasing absorber and condenser temperature. The second law efficiency descends up to 25°C, and with further enhancement in the absorber and condenser temperature, it ascends up to 40°C and again descends further. The behaviour of second law efficiency is attributed to the reduction in solution circulation ratio because pressure ratio increases with rising condenser and absorber temperature, which subsequently increases the mass flow rate, thereby reducing the concentration of the strong solution, while the concentration of the weak solution remains constant.

Figure 13.6 shows the effect on physical (Ψ^P_{loss}) and chemical (Ψ^C_{loss}) irreversibility losses in different components along with second law efficiency of the system with ambient temperature ($T_G = 75°C$, $T_A = T_C = 35°C$, $T_E = -2°C$, and cooling capacity = 3.516 kJ/s), and it becomes evident from the figure that the maximum physical irreversibility is found in solution heat exchanger which subsequently goes on decreasing with an increase in the ambient temperature. The second worst component from the physical exergy viewpoint is the generator, followed by absorber, evaporator, condenser, and compressor. The physical irreversibility in the

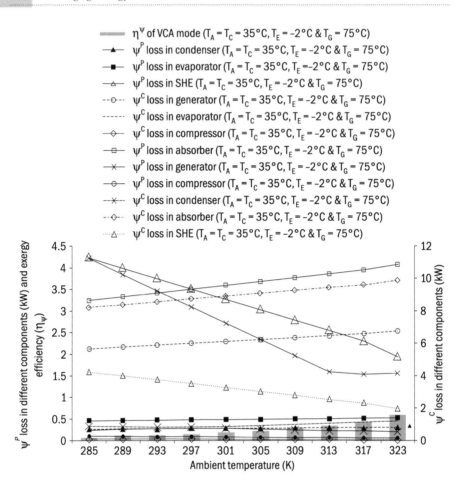

Fig. 13.6 *Physical and chemical irreversibility in different components of hybrid machine and second law efficiency with dead state (ambient) temperature (T_G = 75°C, T_E = –2°C, T_A = T_C = 35°C, ε_{SHE} = 0.53 and cooling capacity is 3.516 kJ/s)*

generator and compressor decreases with ambient temperature, while the physical irreversibility in absorber, evaporator, and condenser increases with ambient temperature. It also becomes evident from the figure that the maximum chemical irreversibility is found in the absorber, which ascends with an increase in dead state temperature. The second worst component from the viewpoint of chemical irreversibility is the generator, followed by solution heat exchanger, condenser, evaporator, and compressor. The chemical irreversibility in generator and evaporator ascends, while chemical irreversibility in solution heat exchanger, condenser, and compressor descends, with an increase in dead state temperature. This chemical irreversibility is because of a large temperature difference in heat exchange in the absorber, high concentration gradient mass transfer, and intermixing losses (Aman, Ting, and Henshaw 2014). In addition, ammonia leaving the generator is superheated, which requires higher temperature under the same pressure, leading to more losses in the generator and absorber. The superheated temperature of vapours from the generator requires more cooling for the condenser, which leads to irreversibility in the condenser (Aman, Ting, and Henshaw 2014). The second law efficiency also shows an ascending trend with ambient temperature because it is directly proportional to the ambient temperature as evident from Equation (13.31). The heat load of the components is directly affected by ambient temperature, thereby increasing the exergetic efficiency.

The percentage of non-dimensional physical ($\Psi^P_{loss,ND}$) and chemical exergy losses ($\Psi^C_{loss,ND}$) for different components of the systems are shown in Figure 13.7. It becomes evident that the maximum percentage of non-dimensional physical and chemical exergy losses is found to be in the absorber. The second worst component from the non-dimensional physical irreverrsibilty viewpoint is solution heat exchanger whereas from chemical irreversibility viewpoint is the generator. However, for absorber, condenser, and generator, the comparative percentage of non-dimensional chemical exergy losses is more, while in evaporator, solution heat exchanger, and compressor, the percentage of non-dimensional physical exergy losses is more.

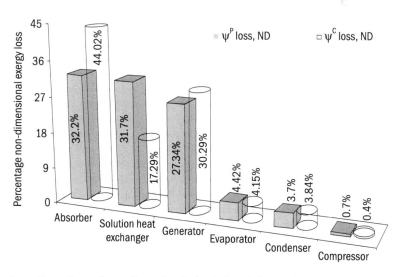

Fig. 13.7 *Percentage of non-dimensional physical and chemical exergy losses in different components of vapour compression absorption hybrid system ($T_G = 75°C$, $T_E = -2°C$, $T_A = T_C = 35°C$, $\varepsilon_{SHE} = 0.53$)*

13.7   CONCLUSION

A computational approach is applied using EES (Engineering Equation Solver) (Klein and Alvarado 2013) to assess the performance of biogas operated hybrid VCA (vapour compression absorption) cooling machine. The analysis is based on physical and chemical exergy evaluation and the conclusions drawn from the analysis are as follows:

- The $COP_{cooling}$, $COP_{heating}$, and exergy efficiency decrease with an increase in the generator temperature. The $COP_{cooling}$ and $COP_{heating}$ increase with an increase in the evaporator temperature, whereas second law efficiency decreases with an increase in the evaporator temperature.

- The $COP_{cooling}$ and $COP_{heating}$ show a marginal improvement with an increase in the absorber and condenser temperature levels, but the second law efficiency shows both increasing and decreasing trends.

- It is worth mentioning that the highest percentage of non-dimensional physical and chemical exergy losses is found in the absorber. The worst component which stands next to absorber from the viewpoint of non-dimensional physical exergy loss is the solution heat exchanger, but from chemical irreversibility viewpoint, it is the generator, followed by evaporator, condenser, and compressor. The increased irreversibility in the generator is because higher heat is required for vaporization of ammonia from refrigerant–absorbent solution compared to that of the pure refrigerant. However, intermixing of two different streams of solutions is the cause for the higher chemical exergy loss in absorber.

- The exergy losses found in the absorbing process and in the generator are large, and, therefore, efforts should be made to improve the absorber and generator. It is also established that proper matching of heat source and generator temperature, as well as evaporator temperature and space to be cooled, could lead to an improved efficiency of the system.

- The proposed hybrid machine is evaluated to be a good option in terms of performance and second law efficiency. However, certain modifications for the operation of components, such as generator and evaporator, can also improve the performance of the hybrid machine.

- The proposed system can also be channelled for sustainable future because they do not use the conventional energy and CFC-based refrigerants. Carbon emission into the environment can also be reduced with such systems.

REFERENCES

Abdullah, M. O. and T. C. Hien. 2011. Comparative analysis of performance and techno-economics for a H_2O–NH_3–H_2 absorption refrigerator driven by different energy sources. *Applied Energy* 87: 1535–1545

Alberto Dopazo, J. and Jose Fernandez-Seara. 2011. Experimental evaluation of a cascade refrigeration system prototype with CO_2 and NH_3 for freezing process applications. *International Journal of Refrigeration* 34: 257–267

Aman, J., D. S. K. Ting, and P. Henshaw. 2014. Residential solar air conditioning: Energy and exergy analyses of an ammonia–water absorption cooling system. *Applied Thermal Engineering* 62: 424–432

Anand, S., A. Gupta, and S. K. Tyagi. 2014a. Renewable energy powered evacuated tube collector refrigerator system. *Mitigation and Adaptation Strategies for Global Change* 19: 1077–1089

Anand, S., A. Gupta, and S. K. Tyagi. 2014b. Critical analysis of a biogas powered absorption system for climate change mitigation. *Clean Technologies and Environmental Policies* 16: 569–578

Anand, S., A. Gupta, and S. K. Tyagi. 2014c. Exergetic analysis and assessment of hybrid refrigeration system for dairy applications. *International Journal of Air Conditioning and Refrigeration* 22(4): 1–13

ASHRAE. 1997. *ASHRAE Handbook of Fundamentals.* Atlanta, GA: American Society of Heating, Refrigeration, and Air-Conditioning Engineers.

Ayala R., C. L. Heard, and F. A. Holland. 1998. Ammonia/lithium nitrate absorption/compression refrigeration cycle. Part II. Experimental. *Applied Thermal Engineering* 18: 661–670

Bejan, A., G. Tsatsaronis, and M. Moran. 1995. *Thermal Design and Optimization.* New York: Wiley

Borde, I. and M. Jelinek. 1987. Development of absorption refrigeration units for cold storage of agricultural products. *International Journal of Refrigeration* 10: 53–56

Garimella, S., A. M. Brown, and A. K. Nagavarapu. 2011. Waste heat driven absorption/vapor-compression cascade refrigeration system for megawatt scale, high-flux, low-temperature cooling. *International Journal of Refrigeration* 234: 1776–1785.

Goktun, S. 1999. Optimal performance of an irreversible, heat engine-driven, combined vapour compression and absorption refrigerator. *Applied Energy* 62: 67–79

Hepbasli, A. 2008. A key review on exergetic analysis and assessment of renewable energy resources for a sustainable future. *Renewable and Sustainable Energy Reviews* 12: 593–561

Jose, Fernandez-Seara, J. Sieres, and M. Vazquez. 2006. Compression–absorption cascade refrigeration system. *Applied Thermal Engineering* 26: 502–512

Kececiler, A., H. I. Acar, and A. Dogan. 2000. Thermodynamic analysis of the absorption refrigeration system with geothermal energy: An experimental study. *Energy Conversion and Management* 41: 37–48

Klein, S. A. and F. Alvarado. 2013. Engineering Equation Solver, Version 9.462. Middleton, WI: F-Chart Software

Lee, S. F. and S. A. Sherif. 2001. Second law analysis of various double-effect lithium bromide/water absorption chillers. *ASHRAE Transactions* 107(1): 664–676

Morosuk, T. and G. Tsatsaronis. 2008. A new approach to the exergy analysis of absorption refrigeration machines. *Energy* 33: 890–907

Nguyen, S., S. B. Riffat, and D. Whitman. 1996. Solar/gas-driven absorption heat-pump systems. *Applied Thermal Engineering* 16: 347–356

Prasartkaew, B. and S. Kumar. 2013. Experimental study on the performance of a solar biomass hybrid air-conditioning system. *Renewable Energy* 57: 86–93

Pratihar, A. K., S. C. Kaushik, and R. S. Agarwal. 2010. Simulation of an ammonia–water compression–absorption refrigeration system for water chilling application. *International Journal of Refrigeration* 33: 1386–1394

Santayo-Gutierrez, S., J. Siqueiros, C. L. Heard, E. Santoyo, and F. A. Holland. 1999. An experimental integrated absorption heat pump effluent purification system. Part I: Operating on water/lithium bromide solutions. *Applied Thermal Engineering* 19: 461–475

Satapathy, P. K., M. Ram Gopal, and R. C. Arora. 2007. A comparative study of R22–E181 and R134a–E181 working pairs for a compression–absorption system for simultaneous heating and cooling applications. *Journal of Food Engineering* 80: 939–946

Tarique, S. M. and M. A. Siddiqui. 1999. Performance and Economic study of the combined absorption/compression heat pump. *Energy Conversion and Management* 40: 575–591

14

Lignocellulosic Biomass to Bioenergy Production: Process and Techniques for Biomass Assessment

Preeti Vyas[a], Ashwani Kumar[a,*], and Rama Chandra Pradhan[b]

[a]*Metagenomics and Secretomics Research Laboratory, Department of Botany, Dr Harisingh Gour University, Sagar, Madhya Pradesh 470003*

[b]*Department of Food Process Engineering, National Institute of Technology, Rourkela, Odisha 769008*

[*]*E-mail: ashwaniiitd@hotmail.com*

14.1 INTRODUCTION

Excessive use of fossil fuels leads to detrimental impact on the environment and is recognized as an unsustainable practice all over the globe. This acknowledgment has given rise to a lot of interest in developing alternative energy options (Raheem, Azlina, Yap, *et al.* 2015). It is common knowledge that energy is a basic and fundamental requirement for economic and social upliftment as it plays an important role in improving livelihood (Jakhrani, Othman, Rigit, *et al.* 2012). Undesirable activities such as unchecked use of fossil fuels in various industries have enhanced the emission of greenhouse gases (GHGs) and released many harmful pollutants, resulting in global climate change (Cuellar-Bermudez, Garcia-Perez, Rittmann, *et al.* 2015). To tackle this problem, we need to develop renewable and sustainable sources of energy that can be accessible to the masses. To this end, utilization of biomass for biofuel production is a viable option (Raheem, Azlina, Yap, *et al.* 2015; Singh, Kumar, and Rai 2014). Biofuel generation from waste has also become a strategic research area as it holds the potential to improve the problems of energy security, air pollution, and carbon dioxide (CO_2) accumulation in the atmosphere (Raghunandan, Mchunu, Kumar, *et al.* 2014; Singh, Kumar, and Rai 2014). Biofuels generated from biomass can be in liquid, gas, and solid states. For the production of biofuels, diverse energy sources are utilized (Zhang

2011). The use of biorefining platforms for biofuel generation provides an opportunity to open new markets for the agricultural sector and increase employment, which in turn can contribute to the development of emerging economies. In this respect, second-generation bioethanol produced from lignocellulosic biomass is considered advantageous as it can avoid the fuel versus food competition (Dionisi, Anderson, Aulenta, et al. 2015). Biomass that comes from agriculture and food processing industries and wastewater treatment plants can be utilized for biofuel production. Chemically, plant biomass is mainly composed of cellulose (hexose sugar), hemicelluloses (pentose and hexose sugars), and lignin (phenolic compound). The degradation of biomass by chemical or enzymatic methods is challenging (Yang and Wyman 2008). The main challenge is the recalcitrant nature of biomass, which obstructs the conversion of biomass to sugars or other intermediates (Wyman 2007; Lynd, Laser, Bransby, et al. 2008). Application of biological approach which is mediated by microbes or their enzymes, chemical catalysis (example, gasification, pyrolysis, and aqueous phase reforming), or a combination of these two can be used for the production of biofuel (Serrano-Ruiz and Dumesic 2011; Wang, Huang, Sathitsuksanoh, et al. 2011; Zhang, Xiao, Jin, et al. 2013). Pretreatment of biomass is very important for reducing the recalcitrance of biomass constituents to other downstream biological and chemical processes. Pretreatment methods are broadly classified as chemical and physical, which include heat, acid, and chemicals, usually in aqueous-based reactions (Yang and Wyman 2008; Mosier, Wyman, Dale, et al. 2005; Kumar, Barrett, Delwiche, et al. 2009a). Thermochemical pretreatment unfolds the biomass for saccharification using a mixture of enzymes that releases sugars (Mosier, Wyman, Dale, et al. 2005; Wyman 2013). Although the enzymes used in the saccharification process must be selective and highly stable, their efficiency is subject to the efficacy of the pretreatment used. Even higher enzyme loadings have resulted in an increased yield of sugar (Zhang, Ding, Mielenz, et al. 2007; Kumar and Wyman 2013; Nguyen, Cai, Kumar, et al. 2015). Several previous articles have reviewed various aspects of biomass to bioethanol conversion. This chapter focuses on the process of biomass deconstruction, methods of biomass assessments, and future challenges.

14.2 PROCESS OF BIOMASS DECONSTRUCTION

Technically speaking, the commercial viability of lignocellulosic biomass depends on the initial conversion rate of cell wall polymers to sugar monomer, followed by further conversion into value-added products through fermentation (Prakasham, Nagaiah, Vinutha, et al. 2014). However, when it comes to the processes, the first process of biomass deconstruction is rate limiting due to the complex and xenobiotic nature

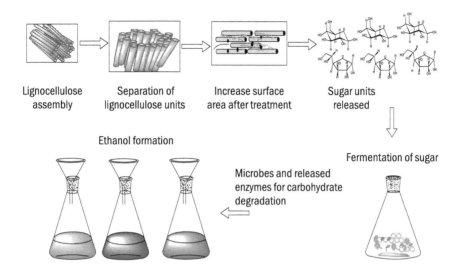

Lignocellulose assembly   Separation of lignocellulose units   Increase surface area after treatment   Sugar units released

Ethanol formation

Fermentation of sugar

Microbes and released enzymes for carbohydrate degradation

Fig. 14.1 *Process showing the production of ethanol from biomass*

of lignocellulose. Biomass is composed of cellulose, hemicellulose, and lignin, and among these components, lignin does not degrade easily in comparison to cellulose (Ravichandra, Yaswanth, Nikhila, *et al.* 2015). Conversion of lignocellulosic biomass into water-soluble sugars entails cleaving of ether bonds in hemicellulose (primarily xylan) and cellulose (glucan) chains, while minimizing further degradation of the resulting C5 and C6 sugars (primarily xylose and glucose) to insoluble degradation products (Luterbacher, Rand, Alonso, *et al.* 2014). Figure 14.1 shows the production process of ethanol from biomass.

14.2.1 Feedstock Pretreatment

Lignocellulose present in biomass is a heteropolymer compound which is essentially composed of cellulose (hexose), hemicellulose (pentose, hexose), and lignin (aromatic polymers). In lignocellulose, the lignin acts as a protecting layer of cellulose and hemicellulose against enzymatic hydrolysis. Therefore, pretreatment of biomass is the first step to breaking down lignin layers and increasing the accessibility of enzymes so that the cellulose and hemicellulose become more prone to degradation, which then results in the efficient conversion of biomass to ethanolic sugars. Previous researches have shown that pretreatment methods happen to be the most expensive processing step in the bioconversion of lignocellulosic biomass to ethanolic sugars and also have a negative effect on the environment (Mosier, Wyman, Dale, *et al.* 2005). In other words, biological pretreatments are a secure and eco-friendly way of degrading lignocellulosic biomass

(Binod, Sindhu, Singhania, *et al.* 2010). Out of the different biological pretreatments, enzymatic hydrolysis plays an important role in the production of ethanol from lignocellulosic materials by depolymerization of cellulose and hemicellulose into hexose and pentose sugars (Taherzadeh and Niklasson 2004). Researchers reported that the efficiency of enzymatic hydrolysis of lignocellulosic biomass can be increased by using cellulolytic and hemicellulolytic enzymes in combination rather than individually (Zhong, Lau, Balan, *et al.* 2009; Waghmare, Kadam, Saratale, *et al.* 2014). Some cultivated microbes (*Anaerocellum thermophilum* DSM 6725, an anaerobic bacteria) are able to degrade lignocellulosic biomass without any pretreatment (Yang, Kataeva, Hamilton-Brehm, *et al.* 2009). Table 14.1 lists the different types of pretreatments and principles involved in lignocelllulose breakdown.

14.2.2 Fermentation

Transformation of plant biomass into fermentable sugars is the key requirement of fermentation. After the generation of sugar monomers from lignocellulose degradation, fermentable bacteria are added into the medium for ethanol and CO_2 production. The process of fermentation takes place in three steps: (i) conversion of lignocellulosic biomass into sugars, (ii) bioethanol production by fermentation of these sugars, and (iii) separation and purification of ethanol. Microbes play an important role in this conversion, except in the case of lignin, which is non-fermentable. Although lignin is an extremely high-energy substance, its existence in the lignocellulosic biomass hinders the breakdown of biomass into fermentable sugars.

Process optimization using dilute and concentrated acids is essential for biomass pretreatment. High sugar recovery efficiency, which can be of the order of over 90% in the case of both hemicellulose and cellulose

Table 14.1 Types of pretreatments and principles involved in lignocellulose breakdown

Pretreatments	Principles
Uncatalysed steam explosion	Treatment with high-pressure steam and termination with rapid decompression
Liquid hot water	Treatment with water at high pressure and temperature
Dilute acid	Treatment with dilute acid at high temperature
AFEX (ammonia fibre explosion)	Treatment with aqueous ammonia at high temperature and pressure

sugars, is the primary advantage of the concentrated process. Ethanol-generating bacteria have a faster growth rate than fungi, owing to which they generate additional fermentative enzymes and utilize both pentose and hexose sugars. The capability to ferment cellulose directly to ethanol is a significant benefit of utilizing some bacterial strains (Singh, Bhatia, and Srivastava 2015). Researchers are looking into the environment for searching novel microbial isolates that are able to degrade biomass into ethanol.

Separate hydrolysis and fermentation

In separate hydrolysis and fermentation (SHF), cellulose is first hydrolyzed by different pretreatment methods. After the process of hydrolysis, the cellulose is transformed into bioethanol by a saccharification reaction during fermentation. The process of SHF could be conducted at its optimum conditions, which subsequently enhances the possibility of product recovery (Sharma and Arora 2010). The controlling parameters to be taken into consideration for the saccharification step are the availability of cellulose for conversion to sugar, reaction time, pH, temperature, optimal enzyme dose, and substrate input (Oberoi, Babbar, Sandhu, *et al.* 2012). However, it should be noted that the relative amount of arabinan and xylan in biomass hinders the bioconversion of cellulose into glucose. Previously, researchers used a non-conventional approach by applying biphasic media for saccharification, which resulted in higher glucose concentrations when hydrolysis ended (about 150 g/L achieved in aqueous phase) (Cantarella, Alfani, Cantarella, *et al.* 2001).

Simultaneous saccharification and fermentation

The simultaneous saccharification and fermentation (SSF) process is more effective than SHF in terms of a high-ethanol yield with low-energy consumption. In SSF, the cellulases for hydrolysis of biomass along with microbes for fermentation are inoculated together in the same process unit, which results in glucose formation and its immediate conversion by microbes into ethanol. Consequently, this process results in the neutralization of inhibitory action of sugars on cellulases. However, in this case, the problem of low concentration of the end product can be overcome by the use of concentrated media. Furthermore, the addition of a high concentration of enzymes and the use of surfactant increase the substrate conversion rate. It was reported that the addition of non-ionic surfactant (Tween-20 and Tween-80) to the steam-exploded wood in a batch of SSF by using *Saccharomyces cerevisiae* increases 8% ethanol yield and reduces 50% cellulases dosage (from 44 FPU/g to 22 FPU/g of cellulose). It also shortens the duration necessary for attaining the maximum ethanol productivity (Alkasrawi, Eriksson, Börjesson, *et al.* 2003). It has been observed that the surfactant prevents the non-useful adsorption of cellulases to lignin. In another study, an SHF experiment was performed that included optimal conditions for sugarcane leaves (Krishna, Prasanthi, Chowdary, *et al.* 1998). The researchers concluded that a temperature of 40°C and a pH of 5.1 were optimum for 3 days cultivation, and there was a yield of 31 g/L of

Table 14.2   Rapid methods for biomass analysis

Component of biomass	Methods
Cellulose	PyroMBMS LC FT-Raman FTIR NIR Raman Imaging TGA UV–VIS
Carbohydrates	PyroMBMS LC FT-Raman FTIR NIR TGA UV–VIS
Structural analysis of polysaccharides	Raman Imaging
Lignin	PyroMBMS LC FT-Raman FTIR NIR Raman Imaging TGA UV–VIS Fluorescence
Products (glucose and ethanol)	LC FT-Raman FTIR NIR UV–VIS

ethanol when the substrate loading was about 15%, but it was with a high enzyme dosage (100 FPU/g of cellulase), which is again less economical.

14.2.3 Fermentation of Pentose Sugars

Fermentation technology using specific microbes such as bacteria and yeast to produce various chemicals from hexoses is well known, but not much information is available about the ability of these organisms to ferment pentose sugars. The most promising yeast species identified so far are *Pichia stipitis, Candida shehatae, S. cerevisiae*, and *Pachysolen tannophilus. S. cerevisiae* is unable to ferment pentose sugar, mainly xylose. *S. cerevisiae* has a superior ability to utilize pentoses compared to *P. stipitis, C. shehatae*, and *P. tannophilus*. However, modification of the strain using genetic engineering can be a solution to this problem. Researchers transformed various genes related to pentose utilization into *S. cerevisiae, Zymomonas mobilis, Escherichia coli*, and so on. Use of pentose-fermenting yeasts and bacteria during the process to separately ferment pentose and hexose can be a useful approach to overcome this problem. Pentose fermentation is tedious compared to hexose fermentation, as the productivity of pentose-fermenting yeasts is less than that of hexose-fermenting yeasts. If the fermentation period is short and pentoses are left out in the medium, it also reduces the consumption rates of the lignocellulosic complex (Gong, Cao, Du, *et al.* 1999). For example, ethanol production using *S. cerevisiae* can yield 170 g/(L/h) in continuous systems along with cell recycling, while the productivity for *C. shehatae* with high cell concentrations is about 4.4 g/(L/h) (Olsson and Hahn-Hägerdal 1996; Shrivastava, Tekriwal, Kharkwal, *et al.* 2014).

14.3 TECHNIQUES FOR ASSESSMENT OF BIOMASS BREAKDOWN

It is necessary to assess lignocellulosic biomass and its structural modification during the process of bioconversion of sugars to ethanol. Although both destructive and non-destructive methods are available and applied frequently for analysis, the non-destructive methods are more rapid because they allow the measurement to be completed within milliseconds. A preconstructed computer model and available computer program for data analysis could automatically analyse the data and perform predictions (Silva and Northen 2015). Table 14.2 lists the rapid methods for biomass analysis. There are several methods available for analysing biomass and the specific choices depend upon the objectives of researchers. This section focuses on the different methods used for biomass assessments.

14.3.1   Liquid Chromatography-Mass Spectrometry

Liquid chromatography is applicable to a wide range of molecular polarities and is capable of separating thermally unstable or non-volatile molecules. The combination of exo-metabolomics and LC-MS (liquid chromatography-mass spectrometry) can provide critical information on the distributed metabolism occurring in microbial and other cellular communities (Silva and Northen 2015). The use of exo-metabolomics is a robust way of providing phenotypic information about the cell. This approach can be used for diverse applications, including gene annotation, biofuel development, bio-processing, and drug mechanisms of action. LC-MS is a highly sensitive instrument that provides the largest dynamic range for characterization of large numbers of metabolites, and their localization through MSI and targeted MSI, and also provides useful information on how microbes interact in complex environments.

14.3.2   Fourier Transform Infrared Analysis

In Fourier transform infrared spectroscopy (FTIR) process, more intense spectra are generated than those in Raman spectroscopy. FTIR has high-throughput capabilities and is non-destructive and inexpensive. The use of FTIR is also important for finding further information regarding chemical changes occurring during pretreatment and in the process of fermentation. The spectra generated by FTIR are also used to evaluate changes in the crystallinity of cellulose and study the influence of accompanying hemicelluloses (Elliston, Wood, Soucouri, *et al.* 2015).

14.3.3   High-Performance Liquid Chromatography

High-performance liquid chromatography (HPLC) is the most commonly used method for diverse analysis of pentose, hexose, and fermented sugar. It is accompanied by standard sugars, such as D-glucose, D-xylose, L-arabinose, and ethanol (Thomsen, Kádár, and Schmidt 2014). Concentration of carbohydrates and levels of organic acids are also analysed by HPLC (Elliston, Wood, Soucouri, *et al.* 2015).

14.3.4   X-ray Diffraction

A Scintag diffractometer is employed to obtain X-ray powder diffraction patterns of untreated and pretreated biomass. XRD (X-ray diffraction) is the only technique capable of yielding qualitative/quantitative data of compounds in a mixture and is also useful for the analysis of cellulose crystallinity. XRD also provides information about the modification of lignocellulosic biomass during the process of bioethanol production (Kumar, Mago, Balan, *et al.* 2009b).

14.3.5 Scanning Electron Microscopy

Scanning electron microscopy (SEM) is also a good technique for the analysis of structural modifications in biomass during the process of biomass degradation by surface images.

14.3.6 ATR–FTIR Spectroscopy

Attenuated total reflection-Fourier transform infrared spectroscopy (ATR–FTIR) is a great advantage in the analysis of structural modification in lignocellulosic biomass. This instrument provides information about the primary and secondary product formations (Singh, Simmons, Vogel, *et al.* 2009).

14.3.7 Raman Spectroscopy

With this technique, the sample preparation becomes very easy; in fact, there is sometimes little requirement for such preparation. In this technique, multiple excitation sources (UV, VIS, NIR) can be used and so analysis can be tailored to analyte. Raman spectroscopy is a non-destructive and high-throughput method, and it can measure solid, liquid, and gas. Analysis is also not hindered by the presence of water. This technique is used for the analysis of carbohydrate degradation and fermentation product formation (Li, Knierim, Manisseri, *et al.* 2010; Lupoi, Singh, Simmons, *et al.* 2014).

14.3.8 Atomic Force Microscopy

Atomic force microscopy (AFM) is used for measuring the surface topography and morphology of samples. Different methods are used to get the image. Silicon cantilevers with a typical resonant frequency of 240 kHz and spring constant of 11.8 N/m are used to acquire images in tapping mode at the room temperature under ambient conditions. The scanning rate is around 1.5 Hz (Deepa, Abraham, Cordeiro, *et al.* 2015).

14.3.9 Transmission Electron Microscopy

The size of the elementary particles is determined by transmission electron microscopy (TEM). No staining is used in this process (Deepa, Abraham, Cordeiro, *et al.* 2015).

14.3.10 Thermogravimetric Analysis

Thermogravimetric analysis (TGA) is faster than the standard techniques for biomass analysis. Through this technique, lignocellulosic biomass undergoes thermal degradation and is used to predict cellulose bioconversion (Deepa, Abraham, Cordeiro, *et al.* 2015).

14.3.11   Near Infrared Studies in Biomass

The NIR (near infrared) technique is useful for the analysis of chemical, physical, structural properties of wood, as well as other properties such as moisture, ash and char content (Lestander and Rhén 2005). It is also helpful in predicting the composition of lignocellulosic biomass during the degradation process (Hames, Thomas, Sluiter, *et al.* 2003). NIR can be used for the compositional analysis of corn stover and suggested a good prediction for glucan, xylan, lignin, protein, and ash. Similar studies also reported a good prediction capability for major components of corn stover (Wolfrum and Sluiter 2009). Lignin composition has been studied extensively by NIR and FTIR. However, NIRS (near-infrared spectroscopy) is unable to probe the information of trace elements, non-structural components, and those compounds with a concentration less than 1 g/L (or 1 g/kg) (Xu, Yu, Tesso, *et al.* 2013).

14.3.12   Differential Scanning Calorimetry

A Mettler–Toledo DSC 823 instrument is used for all measurements. Both dry and weighted samples can be used for compositional analysis (Ibbett, Gaddipati, Davies, *et al.* 2011).

14.3.13   Gel Permeation Chromatography Analysis of Cellulose and Hemicelluloses

The number average molecular weight (M_n) and weight average molecular weight (M_w) of cellulose are determined by gel permeation chromatography (GPC) (Foston and Ragauskas 2010; Evans, Bali, Foston, *et al.* 2015).

14.3.14   Determination of Cellulose Crystallinity with ^{13}C NMR

Nuclear magnetic resonance (NMR) plays a very important role in the structural analysis of biomass degradation products. Cellulose structure and crystallinity are determined by ^{13}C NMR (Evans, Bali, Foston, *et al.* 2015).

14.3.15   UV Spectrophotometry

This is the most commonly used instrument for biochemical analysis. The presence of pentose and hexose sugars before and after the fermentation of lignocellulosic biomass is analysed by UV spectrophotometry at 205 nm (Martín-Davison, Ballesteros, Manzanares, *et al.* 2015).

14.3.16   GC–MS

GC–MS is a susceptible and hasty technique for characterizing the chemical structure of lignin, which allows the analysis of a very small amount of the sample without prior manipulation or isolation (Gutiérrez, del Río, Rencoret, *et al.* 2006).

14.3.17   X-ray Photoelectron Spectrometry

X-ray photoelectron spectrometry (XPS) is a technique used for the analysis of the surface structure of biomass. This technique also helps in determining how much area is covered by lignin or cellulose. When monochromatic X-rays liberate electrons from core orbitals, only electrons from the surface region are emitted and analysed.

14.4   CURRENT CHALLENGES AND FUTURE PROSPECTS

The next generation of biofuel is dependent on the introduction of new techniques which can be directly used for the development of sustainable bioenergy. Research has to be oriented towards the use of genetic engineering. The application of genetic engineering helps in transgenic plant production and structural modification of cell wall. With the help of gene modification, we can produce microbes that can easily digest lignocellulosic biomass into pentose and hexose sugars and ferment these sugars into bioethanol. These techniques can be beneficial for the efficient conversion of lignocellulosic biomass into biofuel. A greater understanding of biomass traits will play a key role in the better utilization of biomass for biofuel production. We also need to develop techniques that can provide insight into the interaction of microbes and biomass fermentation, low energy use, low-cost treatments, eco-friendly product formation, low inhibitory product formation in the process of fermentation, and high sugar recovery after the fermentation process.

REFERENCES

Alkasrawi, M., T. Eriksson, J. Börjesson, A. Wingren, M. Galbe, F. Tjerneld, and G. Zacchi. 2003. The effect of Tween-20 on simultaneous saccharification and fermentation of softwood to ethanol. *Enzyme and Microbial Technology* 33(1): 71–78

Binod, P., R. Sindhu, R. R. Singhania, S. Vikram, L. Devi, S. Nagalakshmi, N. Kurien, R. K. Sukumaran, and A. Pandey. 2010. Bioethanol production from rice straw: An overview. *Bioresource Technology* 101(13): 4767–4774

Cantarella, M., F. Alfani, L. Cantarella, A. Gallifuoco, and A. Saporosi. 2001. Biosaccharification of cellulosic biomass in immiscible solvent–water mixtures. *Journal of Molecular Catalysis B: Enzymatic* 11(4): 867–875

Cuellar-Bermudez, S. P., J. S. Garcia-Perez, B. E. Rittmann, and R. Parra-Saldivar. 2015. Photosynthetic bioenergy utilizing CO_2: An approach on fuel gases utilization for third-generation biofuels. *Journal of Cleaner Production* 98: 53–65

Deepa, B., E. Abraham, N. Cordeiro, M. Mozetic, A. P. Mathew, K. Oksman, M. Faria, S. Thomas, and L. A. Pothan. 2015. Utilization of various lignocellulosic biomass for the production of nanocellulose: A comparative study. *Cellulose* 22(2): 1075–1090

Dionisi, D., J. A. Anderson, F. Aulenta, A. McCue, and G. Paton. 2015. The potential of microbial processes for lignocellulosic biomass conversion to ethanol: A review. *Journal of Chemical Technology and Biotechnology* 90(3): 366–383

Elliston, A., I. P. Wood, M. J. Soucouri, R. J. Tantale, J. Dicks, I. N. Roberts, and K. W. Waldron. 2015. Methodology for enabling high-throughput simultaneous saccharification and fermentation screening of yeast using solid biomass as a substrate. *Biotechnology for Biofuels* 8(1): 1–9

Evans, B. R., G. Bali, M. Foston, A. J. Ragauskas, H. M. O'Neill, R. Shah, J. McGaughey, D. Reeves, C. S. Rempe, and B. H. Davison. 2015. Production of deuterated switchgrass by hydroponic cultivation. *Planta* 242(1): 215–222

Foston, M. and A. J. Ragauskas. 2010. Changes in lignocellulosic supramolecular and ultrastructure during dilute acid pretreatment of *Populus* and switchgrass. *Biomass and Bioenergy* 34(12): 1885–1895

Gong, C. S., N. J. Cao, J. Du, and G. T. Tsao. 1999. Ethanol production from renewable resources. *Recent Progress in Bioconversion of Lignocellulosics*, pp. 207–241. Berlin: Springer-Verlag

Gutiérrez, A., J. C. del Río, J. Rencoret, D. Ibarra, and A. T. Martínez. 2006. Main lipophilic extractives in different paper pulp types can be removed using the laccase–mediator system. *Applied Microbiology and Biotechnology* 72(4): 845–851

Hames, B. R., Thomas, S. R., Sluiter, A. D., Roth, C. J. and Templeton, D. W., 2003. Rapid biomass analysis. In Biotechnology for Fuels and Chemicals, pp. 5-16. Humana Press

Ibbett, R., S. Gaddipati, S. Davies, S. Hill, and G. Tucker. 2011. The mechanisms of hydrothermal deconstruction of lignocellulose: New insights from thermal-analytical and complementary studies. *Bioresource Technology* 102(19): 9272–9278

Jakhrani, A. Q., A. K. Othman, A. Rigit, S. R. Samo, and S. A. Kamboh. 2012. Estimation of carbon footprints from diesel generator emissions. Green and Ubiquitous Technology (GUT) International Conference 2012

Krishna, S. H., K. Prasanthi, G. Chowdary, and C. Ayyanna. 1998. Simultaneous saccharification and fermentation of pretreated sugar cane leaves to ethanol. *Process Biochemistry* 33(8): 825–830

Kumar, P., D. M. Barrett, M. J. Delwiche, and P. Stroeve. 2009a. Methods for pretreatment of lignocellulosic biomass for efficient hydrolysis and biofuel production. *Industrial and Engineering Chemistry Research* 48(8): 3713–3729

Kumar, R., G. Mago, V. Balan, and C. E. Wyman. 2009b. Physical and chemical characterizations of corn stover and poplar solids resulting from leading pretreatment technologies. *Bioresource Technology* 100(17): 3948–3962

Kumar, R. and C. E. Wyman. 2013. Physical and chemical features of pretreated biomass that influence macro-/micro-accessibility and biological processing. *Aqueous Pretreatment of Plant Biomass for Biological and Chemical Conversion to Fuels and Chemicals*, pp. 281–310. Chichester: John Wiley & Sons Lestander, T. A. and C. Rhén. 2005. Multivariate NIR spectroscopy models for moisture, ash and calorific content in biofuels using bi-orthogonal partial least squares regression. *Analyst* 130(8): 1182–1189

Li, C., B. Knierim, C. Manisseri, R. Arora, H. V. Scheller, M. Auer, K. P. Vogel, B. A. Simmons, and S. Singh. 2010. Comparison of dilute acid and ionic liquid pretreatment of switchgrass: Biomass recalcitrance, delignification and enzymatic saccharification. *Bioresource Technology* 101(13): 4900–4906

Lupoi, J. S., S. Singh, B. A. Simmons, and R. J. Henry. 2014. Assessment of lignocellulosic biomass using analytical spectroscopy: An evolution to high-throughput techniques. *BioEnergy Research* 7(1): 1–23

Luterbacher, J. S., J. M. Rand, D. M. Alonso, J. Han, J. T. Youngquist, C. T. Maravelias, B. F. Pfleger, and J. A. Dumesic. 2014. Nonenzymatic sugar production from biomass using biomass-derived *y*-valerolactone. *Science* 343(6168): 277–280

Lynd, L. R., M. S. Laser, D. Bransby, B. E. Dale, B. Davison, R. Hamilton, M. Himmel, M. Keller, J. D. McMillan, and J. Sheehan. 2008. How biotech can transform biofuels. *Nature Biotechnology* 26(2): 169–172

Martín-Davison, J. S., M. Ballesteros, P. Manzanares, X. P. B. Sepúlveda, and A. Vergara-Fernández. 2015. Effects of temperature on steam explosion pretreatment of poplar hybrids with different lignin contents in bioethanol production. *International Journal of Green Energy* 12(8): 832–842

Mosier, N., C. Wyman, B. Dale, R. Elander, Y. Lee, M. Holtzapple, and M. Ladisch. 2005. Features of promising technologies for pretreatment of lignocellulosic biomass. *Bioresource Technology* 96(6): 673–686

Nguyen, T. Y., C. M. Cai, R. Kumar, and C. E. Wyman. 2015. Co-solvent pretreatment reduces costly enzyme requirements for high sugar and ethanol yields from lignocellulosic biomass. *ChemSusChem* 8(10): 1716–1725

Oberoi, H. S., N. Babbar, S. K. Sandhu, S. S. Dhaliwal, U. Kaur, B. Chadha, and V. K. Bhargav. 2012. Ethanol production from alkali-treated rice straw via simultaneous saccharification and fermentation using newly isolated thermotolerant *Pichia kudriavzevii* HOP-1. *Journal of Industrial Microbiology and Biotechnology* 39(4): 557–566.

Oberoi, H. S., P. V. Vadlani, L. Saida, S. Bansal, and J. D. Hughes. 2011. Ethanol production from banana peels using statistically optimized simultaneous saccharification and fermentation process. *Waste Management* 31(7): 1576–1584

Olsson, L. and B. Hahn-Hägerdal. 1996. Fermentation of lignocellulosic hydrolysates for ethanol production. *Enzyme and Microbial Technology* 18(5): 312–331

Prakasham, R. S., D. Nagaiah, K. S. Vinutha, A. Uma, T. Chiranjeevi, A. V. Umakanth, P. S. Rao, and N. Yan. 2014. Sorghum biomass: A novel renewable carbon source for industrial bioproducts. *Biofuels* 5(2): 159–174

Raghunandan, K., S. Mchunu, A. Kumar, K. S. Kumar, A. Govender, K. Permaul, and S. Singh. 2014. Biodegradation of glycerol using bacterial isolates from soil under aerobic conditions. *Journal of Environmental Science and Health, Part A* 49(1): 85–92

Raheem, A., W. W. Azlina, Y. T. Yap, M. K. Danquah, and R. Harun. 2015. Thermochemical conversion of microalgal biomass for biofuel production. *Renewable and Sustainable Energy Reviews* 49: 990–999

Ravichandra, K., V. Yaswanth, B. Nikhila, J. Ahmad, P. S. Rao, A. Uma, V. Ravindrababu, and R. Prakasham. Xylanase production by isolated fungal strain, *Aspergillus fumigatus* RSP-8 (MTCC 12039): Impact of agro-industrial material as substrate. *Sugar Tech.* 18: 29–38

Serrano-Ruiz, J. C. and J. A. Dumesic. 2011. Catalytic routes for the conversion of biomass into liquid hydrocarbon transportation fuels. *Energy and Environmental Science* 4(1): 83–99

Sharma, R. K. and D. S. Arora. 2010. Production of lignocellulolytic enzymes and enhancement of in vitro digestibility during solid state fermentation of wheat straw by *Phlebia floridensis*. *Bioresource Technology* 101(23): 9248–9253

Shrivastava, S., K. G. Tekriwal, A. C. Kharkwal, and A. Varma. 2014. Bio-ethanol production by simultaneous saccharification and fermentation using microbial consortium. *International Journal of Current Microbiology and Applied Sciences* 3: 505–511

Silva, L. P. and T. R. Northen. 2015. Exometabolomics and MSI: Deconstructing how cells interact to transform their small molecule environment. *Current Opinion in Biotechnology* 34: 209–216

Singh, N., A. Kumar, and S. Rai. 2014. Potential production of bioenergy from biomass in an Indian perspective. *Renewable and Sustainable Energy Reviews* 39: 65–78

Singh, R., A. Bhatia, and M. Srivastava. 2015. Biofuels as alternate fuel from biomass—The Indian scenario. *Energy Sustainability Through Green Energy*, pp. 287–313. Berlin: Springer

Singh, S., B. A. Simmons, and K. P. Vogel. 2009. Visualization of biomass solubilization and cellulose regeneration during ionic liquid pretreatment of switchgrass. *Biotechnology and Bioengineering* 104(1): 68–75.

Taherzadeh, M. J. and C. Niklasson. 2004. Ethanol from lignocellulosic materials: Pretreatment, acid and enzymatic hydrolyses, and fermentation. *ACS Symposium Series* 889: 49–68

Thomsen S. T., Z. Kádár, and J. E. Schmidt. 2014. Compositional analysis and projected biofuel potentials from common West African agricultural residues. *Biomass and Bioenergy* 63: 210–217

Waghmare, P. R., A. A. Kadam, G. D. Saratale, and S. P. Govindwar. 2014. Enzymatic hydrolysis and characterization of waste lignocellulosic biomass produced after dye bioremediation under solid state fermentation. *Bioresource Technology* 168: 136–141

Wang, Y., W. Huang, N. Sathitsuksanoh, Z. Zhu, and Y. H. P. Zhang. 2011. Biohydrogenation from biomass sugar mediated by in vitro synthetic enzymatic pathways. *Chemistry and Biology* 18(3): 372–380

Wolfrum, E. J. and A. D. Sluiter. 2009. Improved multivariate calibration models for corn stover feedstock and dilute-acid pretreated corn stover. *Cellulose* 16(4): 567–576

Wyman, C. E. 2007. What is (and is not) vital to advancing cellulosic ethanol. *Trends in Biotechnology* 25(4): 153–157

Wyman, C.E. (ed.). 2013. Aqueous pretreatment of plant biomass for biological and chemical conversion to fuels and chemicals. John Wiley & Sons

Xu, F., J. Yu, T. Tesso, F. Dowell, and D. Wang. 2013. Qualitative and quantitative analysis of lignocellulosic biomass using infrared techniques: A mini-review. *Applied Energy* 104: 801–809

Yang, B. and C. E. Wyman. 2008. Pretreatment: The key to unlocking low-cost cellulosic ethanol. *Biofuels, Bioproducts and Biorefining* 2(1): 26–40

Yang, S. J., I. Kataeva, S. D. Hamilton-Brehm, N. L. Engle, T. J. Tschaplinski, C. Doeppke, M. Davis, J. Westpheling, and M. W. Adams. 2009. Efficient degradation of lignocellulosic plant biomass, without pretreatment, by the thermophilic anaerobe "*Anaerocellum thermophilum*" DSM 6725. *Applied and Environmental Microbiology* 75(14): 4762–4769

Zhang, Y. H. P. 2011. Substrate channeling and enzyme complexes for biotechnological applications. *Biotechnology Advances* 29(6): 715–725

Zhang, Y.-H. P., S. Y. Ding, J. R. Mielenz, J. B. Cui, R. T. Elander, M. Laser, M. E. Himmel, J. R. McMillan, and L. R. Lynd. 2007. Fractionating recalcitrant lignocellulose at modest reaction conditions. *Biotechnology and Bioengineering* 97(2): 214–223

Zhang, H., R. Xiao, B. Jin, D. Shen, R. Chen, and G. Xiao. 2013. Catalytic fast pyrolysis of straw biomass in an internally interconnected fluidized bed to produce aromatics and olefins: Effect of different catalysts. *Bioresource Technology* 137: 82–87

Zhong, C., M. W. Lau, V. Balan, B. E. Dale, and Y. J. Yuan. 2009. Optimization of enzymatic hydrolysis and ethanol fermentation from AFEX-treated rice straw. *Applied Microbiology and Biotechnology* 84(4): 667–676

15

Municipal Solid Waste Management in India: Present Status and Energy Conversion Opportunities

Barkha Vaish[a], Pooja Singh[a], Vaibhav Srivastava[a], Prabhat Kumar Singh[b], and Rajeev Pratap Singh[a,*]

[a]*Institute of Environment and Sustainable Development, Banaras Hindu University, Varanasi, Uttar Pradesh 221005*

[b]*Civil Engineering Department, Indian Institute of Technology, Banaras Hindu University, Varanasi, Uttar Pradesh 221005*

[*]E-mail: rajeevprataps@gmail.com

15.1 INTRODUCTION

In emerging economies like India one of the major side effects of urbanization, industrialization, and globalization is the generation of huge amounts of solid waste. It is expected that in the coming decades the total amount of solid waste will increase significantly as the country races towards the status of an industrialized nation by year 2020 (Shekdar, Krishnaswamy, Tikekar, *et al.* 1992; CPCB 2004; Sharma and Shah 2005; Srivastava, Ismail, Singh, *et al.* 2015). Moreover, the country's increasing population is also creating surmounting pressure on its natural resources. Economies of developing countries largely depend on fossil fuels. However, resources for energy production are diminishing at a fast rate and becoming costlier (Asif and Muneer 2007). In keeping with this trend, India's economy is facing the huge challenge of bridging the gap between demand and supply of energy (Figure 15.1). It is projected that the demand of electricity will continuously increase in the near future (Figure 15.2). Therefore, we need to find renewable and non-conventional sources of energy which could hold promise for the future and can help bring down expenditures.

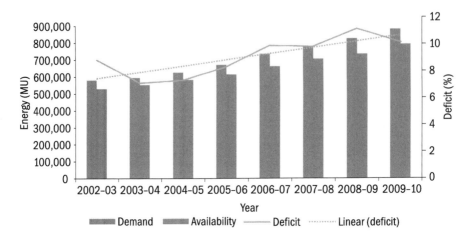

Fig. 15.1 *Gap between demand and supply of energy in India*
Source *Central Electriclity Authority (2012)*

According to the International Energy Agency, about 40% of India's total primary energy demand is consumed by the industrial sector, which was one of the largest consumers in 2009, while the transport sector is fuelled by petroleum, and its products and other traditional biomass that is depleting at a very fast rate. Coal is the main source of energy in India, followed by petrol and other traditional biomass. Estimates suggest that nearly 25% of the population—about 1.4 billion people—lack access to basic electricity (The World Bank 2010; IEA 2013; EIA 2014). Energy (in peta joules) was chiefly consumed in the form of electricity that accounted for nearly about 57.57% during 2011/12, followed by coal and lignite at 19.91% while 18.75% crude petroleum came in at third place (Energy

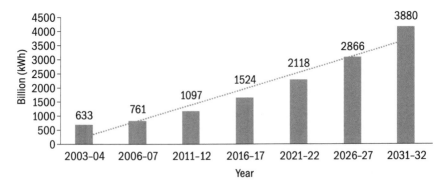

Fig. 15.2 *Projected requirement of energy in India in the near future*
Source *Information and Credit Rating Agency (2010)*

Statistics 2013). A massive imbalance in the demand and supply of energy prevails all across the country and, therefore, the government needs to make serious efforts to meet the challenges of energy supply constraints.

Currently, India and other developing economies are struggling to provide basic facilities, such as proper drainage system, solid waste management (SWM) practices, and the lack of proper facilities lead to an increased burden on local governments, municipalities and other governing bodies that are involved in collection, transportation and disposal of waste. The practice that is most commonly adopted for disposing waste is open dumping in low-lying areas that lead to polluting the environment by emitting greenhouse gases (GHGs) such as carbon dioxide (CO_2) and methane (CH_4), and produce leachate that pollutes the soil and aquifers besides adversely affecting human health (Bagra and Bharti 2014; Srivastava, Ismail, Singh, *et al.* 2015). This shows that waste needs to be disposed in a more technologically advanced manner. Apart from landfilling, open burning of municipal solid waste (MSW) is another common practice that causes air pollution and is a key issue that needs to be dealt with (Mahima and Thomas 2013) while planning proper waste disposal. Municipal solid waste management (MSWM) is one of the most overlooked techniques in the area of policy and strategy making, societal awareness, and research, and is one of the major environmental crises in urban cities. Most metropolitans are still struggling with the challenge of preventing environmental degradation due to disorganized and non-systematic SWM practices (Kumar, Bhattacharyya, Vaidya, *et al.* 2009, Singh, Tyagi, Allen, *et al.* 2011). Fossil fuels are depleting at a fast rate and energy generation is not sufficient to bridge the energy supply and demand gap in the country. Attempts are now being made to come up with technologies that are environmentally friendly, sustainable, and socially acceptable. In this regard, waste-to-energy is an approach that was once neglected but has now been identified as one with strong potential to generate energy from waste and help avert future crises.

15.2 OVERVIEW OF MUNICIPAL SOLID WASTE MANAGEMENT IN INDIA

Management of solid waste is one of the major issues that need to be addressed in Asian countries such as India, Nepal, and Bangladesh. Sources and quantities of solid waste in a region depend on various factors such as lifestyle, culture, attitude, and season. Between 2003 and 2011, per capita waste generation rate in India increased from 0.44 kg per day to approximately 0.5 kg per day (Pattnaik and Reddy 2010; Annepu 2012). Within the next decade, the quantity of solid waste produced is supposed to rise at a rate of 1%–1.33% annually (Bhide and Shekdar 1998; Shekdar 1999; Pappu, Saxena, and Asokar 2007). The total quantity of MSW generated in India is about 188,500 tonnes per day (TPD) (Annepu 2012). This huge quantity of solid waste would place not only a burden on the country's resources in terms of collection, transportation and management, but also be a menace for environment and the health of people. Waste management processes generally followed in Indian cities are not scientific and environmentally friendly. The quantity of solid waste produced is much higher in urban cities than in rural areas (CPCB 2004; Sharholy, Ahmad, Mahmood, et al. 2006; Siddiqui, Siddiqui, and Khan 2006). It is important to know the composition of solid waste in order to decide what kind of technology is to be adopted for its proper management. The composition, and characteristics of solid waste differ in domestic, commercial, and industrial sectors. The composition of waste also varies with location, season, society, and time.

15.2.1   Quantum of Problem

Of the waste streaming from developing countries, one-half comprises organic waste unlike other solid waste materials that include paper, plastic, fabric, glass, metals, ash, stone, hair, and other undefined materials. This is listed in Table 15.1. It has been estimated that the average percentage of organic fraction of solid waste ranges between 50% and 70% (Sharholy, Ahmad, Mahmood, et al. 2008) in developing countries. The high variability in waste composition of developing countries depends on income level, seasonal effects, household fuel supply (Wang and Nie 2001; Metin and Erozturk 2003), living standards, geography, and climate (World Bank

Table 15.1   MSW composition in selected developing countries

Country	Organic (%)	Paper (%)	Plastic (%)	Glass (%)	Metal (%)	Others (%)
Bangladesh	71	5	7	-	-	16
Bhutan	58	17	13	4	1	7
China	38	26	19	3	2	12
India	35	3	2	1	-	59
Indonesia	62	6	10	9	8	4
Iran	43	22	11	2	9	13
Malaysia	62	6	10	9	8	4
Nepal	80	7	3	3	1	7
Pakistan	67	5	18	2	-	7
Sri Lanka	76	11	6	1	1	5

Source World Bank Report (2012)

2003; Buenrostro and Bocco 2003). In most developing countries, dumping waste on the roadside is a very common practice. Streets are swept on an almost daily basis by employees of municipal corporations. Waste is collected by handcarts or other vehicles and put into containers placed on roadsides. Very often, stray animals scatter the waste if it is left out there for long periods. Waste from community bins is collected in open trucks or compactor vehicles and transported to disposal sites, which are mostly open dumps (Figure 15.3). In general, none of the Indian cities completely complies with the Municipal Solid Waste (Management and Handling) Rules, 2000) (Pandey 2011). Open dumping, open burning, and open exposure of waste are widely prastised. This leads to many environmental problems such as soil, air, surface and groundwater pollution, odour nuisance, health problems, impairment of natural aesthetic values and also feed vectors like stray animals (rats, dogs, mosquitoes, monkeys) (Diaz, George, and Eggerth 1997; Kumar 2000). If no timely measures are taken, the situation will get much worse. In growing economies, people

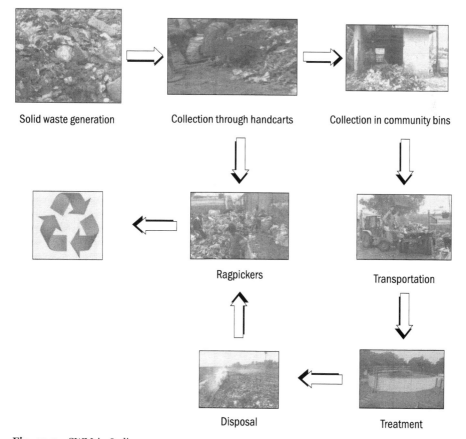

Solid waste generation   Collection through handcarts   Collection in community bins

Ragpickers   Transportation

Disposal   Treatment

Fig. 15.3 *SWM in India*

generate greater amounts of waste and the composition of waste becomes more complex. This is why industrialized cities generate more waste than non-industrialized ones. The quantities of solid waste generated in the megacities of India are listed in Table 15.2.

15.3   OPPORTUNITIES FOR ENERGY CONVERSION

Rapid urbanization, globalization, and economic expansion are creating pressure on every sector, be it food, fuel, or energy which in turn increases the cost of energy and the impending shortage of fuel becomes a more alarming issue. Due to the existing pressure on biomass reserves and the rising fossil fuel prices, generation of energy from solid waste is gaining importance as an alternative source of energy. Therefore, in recent years, there has been an increased interest in treating waste as a valuable

Table 15.2   MSW generation rates in different megacities of India

City	MSW generation (tonnes per day)	
	2004–05[a]	2010–11[b]
Ahmedabad	1302	2300
Bengaluru	1699	3700
Bhopal	574	350
Bhubaneswar	234	400
Chandigarh	326	264
Chennai	3036	4500
Dehradun	131	220
Delhi	5922	6800
Guwahati	166	204
Indore	557	720
Jammu	215	300
Kanpur	1100	1600
Kolkata	2653	3670
Lucknow	475	1200
Mumbai	5320	6500
Patna	511	220
Pune	1175	1300
Shillong	45	97
Srinagar	428	550
Varanasi	425	450

Source [a]CPCB (2012); [b]CIPET (2012)

resource rather than as a refuse, as many beneficial products such as fertilizers and energy can be obtained from it. Research and development around new energy sources from solid waste is attaining great importance these days. Conversion of organic fraction of waste into fuel gas through biochemical and thermochemical treatments is also being explored. Waste-to-energy technologies not only serve in providing alternative sources of energy but also help in solving the problem of waste disposal, lessen environmental problems, and generate wealth and employment (Singh, Tyagi, Allen, *et al.* 2011).

These technologies are still in preliminary stages of development and further technological evolution is needed for converting waste to energy. Currently, biological and thermal processes are frequently used to convert MSW into energy (Figure 15.4). Depending on the type, quantity, and physico-chemical characteristics of solid waste, different techniques, such as direct combustion, gasification, biomethanation, refuse-derived fuel (RDF), are adopted to generate energy from waste. In India, various projects are underway to convert waste into energy. Generally, three technologies, namely RDF, biomethanation, and incineration are used for energy recovery from MSW on a small scale (Unnikrishnan and Singh 2010). These technologies have their own advantages and limitations that will be discussed in the next part of the chapter.

15.3.1 Thermal Conversions

Thermal conversion technologies integrate processes, such as incineration, gasification, and pyrolysis, that create energy in the form of electricity, fuel, or heat from solid waste. Once these processes are completed, their different by-products need to be treated by another set of energy and

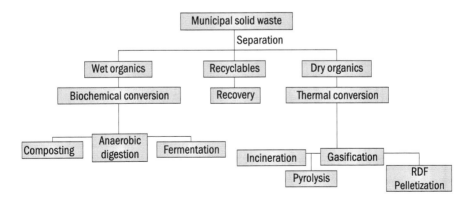

Fig. 15.4 *Different technological options available for SWM*

matter recovery systems (Bridgwater 1994; Singh, Tyagi, Allen, *et al.* 2011). The following are the various kinds of thermal energy conversion processes.

Incineration

Incineration is one of the most basic waste treatment methods that has been used in different parts of the world due to its ability to reduce waste mass by 70% and waste volume by 90% (Singh, Tyagi, Allen, *et al.* 2011; Kalyani and Pandey 2014). In European countries, incineration is the general practice employed in disposing residual waste to avoid landfilling (European Environment Agency 2013). Waste-to-energy conversion through incineration involves three steps, namely, burning (oxidation), energy recovery, and air pollution abatement (Lee, Kwok, Cheung, *et al.* 2007). Incinerators emit air pollutants such as SO_2, NO_x, CO_x, heavy metals, dioxins and toxic flyash, which are proven health hazards (Quina, Bordado, and Ferreira 2010). These air pollutants have the capability to penetrate into the atmosphere and may make the environment toxic (Cordioli, Vincenzi, and De Leo 2014). To control air pollution, incinerators have to be well equipped with air pollution control devices. The incineration process is operated at a temperature range of 750–1000°C, together with steam and electricity generation. At the end, an ash known as Incinerator Bottom Ash (IBA) is produced which must be reused properly as it is generally contaminated with heavy metals, such as Cu, Pb, Mn, Zn (Alhassan and Tanko 2012).

However, there are various challenges that the incineration process still has to overcome before achieving environmentally sound waste management and produce clean energy. The challenges include capital, operating and maintenance costs, corrosion of equipment, and flyash management (Cheng and Hu 2010).

In India, the first incineration plant was installed in 1987 at Timarpur, New Delhi. The unit operated at an initial capital cost of ₹20 crore (US$ 4.4 million), was designed to incinerate a maximum of about 300 tonnes of MSW per day to produce 3.75 MW of electricity. The plant could not run beyond its 21-day trial period and failed due to the poor quality of waste supplied. A net calorific value of at least 1400 kcal/kg was required to run the plant while the incoming waste which was supplied was in the range of 600–700 kcal/kg.

Pyrolysis

Pyrolysis is the thermochemical decomposition of the organic fraction of waste in oxygen-free environment. Pyrolysis works extremely well with organic waste of high heat value (Rhyner, Schwartz, Wenger, *et al.* 1995). During the pyrolysis process, the hydrocarbon present in waste reacts at a temperature range of about 450–500°C in an oxygen-free environment and the end products generated are in the form of pyrolysis gas, pyrolysis coke, char (fixed carbon + ashes), and tar (condensable gas). The accepted equation for pyrolysis is (Tillman 1991).

$$\text{Biomass} + \text{Heat} \rightarrow H_2O + CO_2 + H_2 + CO + CH_4 + C_2H_6 + CH_2 + \text{Tar} + \text{Char} \qquad ...(15.1)$$

Flue gas produced during pyrolysis consists mainly of carbon monoxide (CO) and hydrogen (H_2) (Blanco, Wu, Onwudili, *et al.* 2012), which are suitable either for electricity production or for heat production. The general composition of pyrolysis gas which we get at the end of the process is shown in Figure 15.5 (Singh and Gu 2010). The flue gas has a calorific value in the range of 22–30 MJ/m^3 this is also dependent on the kind of waste material that is being processed.

There are various advantages associated with pyrolysis process. These include the reduction of waste volume by 50%–90%, solid, liquid and gaseous fuel production from waste, transportable fuel or feedstock, and a comparatively less process cost. The process also helps in the management of MSW as a lot of desirable energy is obtained from it. The process is self-sustaining as well, that is, once started the process keeps on operating by itself (Singh and Gu 2010).

Depending on their operational parameters, pyrolysis are of three main kinds, namely, conventional pyrolysis, fast pyrolysis, and flash pyrolysis. Conventional pyrolysis is a slow process that operates under a slow heating temperature (550–900°C) creating solid, liquid, and gaseous end products (Katyal 2007). It is used mainly for producing charcoal.

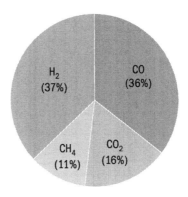

Fig. 15.5 *Composition of pyrolysis gas*
Source *Singh and Gu (2010)*

Fast pyrolysis is associated with tar production and operates at a temperature range of around 850–1250°C, while flash pyrolysis operates at a temperature range of 1050–1300°C. Fast/flash pyrolysis technology, which requires a high-temperature range and has a short residence time, is widely accepted nowadays (Demirbas 2008).

Recently, Tamil Nadu installed a pyrolysis plant of 10-TPD capacity which is operating successfully in Coimbatore. Likewise, other 10-TPD plants have achieved success in Masaipet (Andhra Pradesh) and Navapur (Maharashtra), Madhya Pradesh, Mathura (Uttar Pradesh), Patna (Bihar). 5-TPD plants have been installed in Bichhiwada (Rajasthan), Durg (Chhattisgarh), and Nagpur (Maharashtra).

Gasification

Gasification is a process in which carbonaceous materials including waste and biomass get partially combusted into useful convenient gaseous fuel or chemical feedstock (Bhavanam and Sastry 2011). Gasification is different from incineration, as the latter emits a higher amount of GHGs and acidic gases (HCl, SO_x, HF, NO_x), volatile organic carbon (VOCs), polycyclic aromatic hydrocarbon (PAH), and heavy metals (Wikstrom and Marklund 2000; Wang, Chiang, Tsai, et al. 2001; Liu and Liu 2005). The process of gasification utilizes very little to no oxygen as opposed to combustion or burning associated with incineration.

During gasification process, the product gases such as CO_2 and H_2O are reduced to CO and H_2. The process may also generate some amount of CH_4 and hydrocarbons and various other contaminants such as tar, small char particles, and ash (Appel, Fu, Friedman, et al. 1971; Kalyani and Pandey 2014). The gasification agent used in the process permits the feedstock to be easily converted into gas with the help of different heterogeneous reactions (DiBlasi 2000). Gasification-based power systems also make it possible to achieve high power generation efficiency, which causes CO_2 to be produced in a concentrated form, and it is easier as well as cheaper to capture and sequester it, thereby preventing it from reaching the atmosphere (Guan, Fushimi, Tsutsumi, et al. 2010).

Gasification system contains three main components: (i) the gasifier, which is the heart of the system and converts hydrocarbon feedstock materials into gaseous components; (ii) the clean-up system, which prevents harmful materials from the combustible gases to reach the atmosphere; and (iii) energy recovery systems.

There are three main types of gasifiers: fixed bed, fluidized bed, and indirect. Fixed-bed gasification is a conventional process that operates at a temperature of around 1000°C. The composition of the product gas is typically 40%–50% N_2, 15%–20% H_2, 10%–15% CO, 10%–15% CO_2, and 3%–5% CH_4, with a net calorific value of about 4–6 MJ/Nm^3 (McKendry 2002). In a fluidized bed gasifier, the temperature distribution is uniform in the gasification zone and this uniformity is achieved by using a bed of fine-grained material into which air is introduced (Feo, Belgiorno, Napoli, et al. 2000). Even if the gasification process does not need any oxidizing agent, it requires an external source of energy instead, and this process is known as indirect gasification (Staniewski 1995; Hauserman, Giordano, Lagana, et al. 1997; Kalyani and Pandey 2014). The most commonly used indirect gasification agent is steam because of its easy availability and production, and its capability to increase the H_2 content of the combustible gases (Hauserman, Giordano, Lagana, et al. 1997).

The limitation of the gasification process is that it requires homogeneous carbon-based material for efficient functioning and, therefore, it is not suitable for all types of waste. Industrial waste, such as mixed plastic waste, forest industry waste, paper mills waste and agricultural residues, can be directly used in the gasification process. In spite of all its disadvantages, gasification is the technology for the future due to its ability to convert carbonaceous waste through thermal treatment (Ahrenfeldt, Thomsen, Henriksen, et al. 2013). According to Belgiorno, Feo, Rocca, et al. (2003), RDF is utilized by gasification technology to produce energy. The new emerging technology in the waste-to-energy world is plasma arc gasification (PAG).

There are several gasification plants of various power capacities operating in India, such as a 150 kW capacity plant installed by ENERGREEN at BERI, Tumkur, in Karnataka, and the West Bengal Renewable Energy Development Agency (WBREDA) installed by SYNERGY. A 415 kW power capacity plant was installed in Jammu at Hindustan Pencils by BETEL. A 500 kW plant was installed at Bethmangla, Karnataka by ENERGREEN as well. At Arashi and Gomathi in Tamil Nadu, a 1 MW power capacity plant is running successfully.

Plasma arc gasification

This new technology uses a plasma arc torch that passes exceptionally high voltage electric current via two electrodes. During this process temperature reaches around 3,000–10,000°C by creating an arc in the space between them in an oxygen-free environment (Moustakas, Fatta, Malamis, et al. 2005). PAG is a two-stage process that can work with any kind of feedstock. In the first stage, the feedstock is gasified, that is, partially oxidized. In the second stage, the gas is then cleaned and completely oxidized to generate electricity. The extreme heat produced by this technology breaks down the waste and forms syngas (H_2 and CO) which is used as a primary fuel for power generation, and inert slag is released as a by-product, which finds utilization in construction, tiles, bricks, and road asphalt (Fabry, Rehmet, Rohani, et al. 2013). A variety of marketable fuels can be generated from this syngas, including natural gas (CH_4), ethanol and H_2; electricity can also be produced directly. To eliminate volatile metals and pollutants from the process, a high-efficiency scrubber is used.

The advantage of using PAG is that it can work with any kind of waste, be it steel, concrete or toxic chemicals, and the waste volume is reduced by up to 98%, depending on the kind of feedstock and how the process is operated (Huang and Tang 2007). Like incineration, this is also an excellent way of dealing with toxic and hazardous waste. However, the main disadvantage associated with the process is that it requires a large input of energy and capital cost. This technology is yet to be fully proven, and as a result has not been implemented on a large scale.

Two plants using PAG technology for disposing hazardous waste with a facility to treat 72 tonnes of waste per day to produce electricity are under construction in Pune and Nagpur by SMS Infrastructure Ltd. It is expected that when they are functional, they will produce 5 MW of electricity (Bhasin 2009).

Refuse-derived fuel pelletization

Refuse-derived fuel is an effective process of converting waste to energy. Conversion of MSW to RDF involves various steps such as drying, separation of combustibles, size reduction, and pelletization (Chu, Zhu, Tompsett, et al. 2015) (Figure 15.6). After collection, MSW is heated to remove about 30%–60% moisture content so that the pellets have a reasonable heating value. The solid waste is dried by spreading in an open yard and allowing it to sun-dry. Sun-drying continues for 1–2 days depending on the moisture present in the waste as well as the climate of the region. To get rid of large quantity of debris, tree cuttings, and other undesirable material unfit for the process, manual separation is employed during the sun-drying period. After drying, the waste is passed through a screen for removing

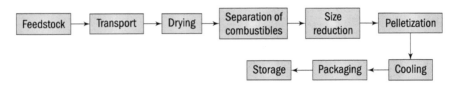

Fig. 15.6 *Flowchart for various steps involved in RDF pelletization*

sand and grit. Thereafter, it is passed through separators in which the light combustibles and dense fractions are separated over an air barrier. Magnetic separators are used to remove all ferrous particles which are then ground to a 20–25 mm particle size. This feed, which may be made of glues, natural starch and celluloses, calcium hydroxide, free lime, cement kiln dust, lignite flyash, is then densified in the pelletizer with a 5%–10% binder in order to improve the calorific value of the waste (Sudhir 2000). RDF pelletization also improves the effective bulk density of the waste (Marsh, Griffiths, Williams, *et al.* 2007). Fuel pellets obtained after the process are homogenized. The pellets can be stored easily and transported, and can serve as a substitute for combustion, pyrolysis, or gasification. The major difficulty with this technique is that pelletization and operation become difficult in the rainy season. Also more energy is consumed while drying the moisture contained in the waste. However, it is still considered a high-potential mechanism for transforming waste to energy.

In Chandigarh, India an RDF plant has been set-up to treat MSW into fuel. Operators estimated that 350 TPD of MSW were available for processing by the plant (Ravindra, Kaur, and Mor 2015) and it is expected to treat waste up to 500 TPD in the coming years. However, presently the plant is not in a working condition (Annepu 2012). Finding suitable uses for RDF continues to be a challenge.

15.3.2 Biochemical Conversion

Biochemical conversion of organic waste into energy is more environmentally friendly, economical, and sustainable in relation to other techniques. It involves the use of bacterial action and of other microorganisms that help in degrading the organic fraction of waste into useful end products. Two processes that belong to this category are anaerobic digestion and composting. The process of fermentation also falls under biochemical conversion.

Anaerobic digestion

It is a biochemical process which is carried out in an oxygen-free environment where microorganisms act on the organic feedstock liberating biogas, and reducing the amount of waste in the process. The biogas, thus, produced can be utilized as combined heat and power (CHP) or may be used as fuel. During the decomposition process, the temperature initially rises as high as 65°C and then begins to fall gradually over a couple of months (Singh, Tyagi, Allen, *et al.* 2011). At the end of the process, CH_4 gas is produced, which has a high-heat value and can also be converted to methanol. A series of metabolic reactions such as hydrolysis, acidogenesis, and methanogenesis (Themelis and Ulloa 2007) occurs during anaerobic digestion. Through this process, under controlled environmental conditions, useful products such as biogas and digestate, which can be utilized as organic amendment (soil conditioner), can be produced and it is possible without any oxygen supply (Chanakya, Ramachandra, and Vijayachamundeeswari 2007; Guermoud, Ouagjnia, Avdelmalek, *et al.* 2009). CH_4 and H_2 as potential fuels are considered comparatively cleaner than fossil fuels (Khalid, Arshad, Anjum, *et al.* 2011). Estimates indicate that by controlled anaerobic digestion, 1 tonne of MSW can produce that amount of CH_4 in three weeks which a landfill will take 6–7 years to produce (Saxena, Adhikari, and Goyal 2009). Therefore, this treatment provides a good opportunity to solve the problem of solid waste disposal and environmental pollution on the one hand and produce biogas and digestate that can be used as organic amendment or carrier material for biofertilizers on the other hand. However, the process needs a long time for biostabilization (Fernandez, Perez, and Romero 2008). Specific moisture and temperature are required by methanogenic bacteria as they are sensitive and can survive only under certain conditions (Ward, Hobbs, Holliman, *et al.* 2008).

In India, Gujarat is the only state that is successfully running waste-to-electricity plants through anaerobic digestion. M/S Kanoria Chemicals Ltd, in Ankleshwar generates 2 MW of power using this treatment, while 4800 nm^3 of biogas is generated by M/S Anil Starch Products Ltd (Kalyani 2003). Similarly, Maharashtra is also taking the initiative to utilize waste and convert it into energy by installing many pilot projects at Mumbai, Pune, Nasik, among other places (Ramchandra 2006). The Ministry of New and Renewable Energy (MNRE) has given subsidies to 3 demonstration projects at Hyderabad (6.6 MW), Vijaywada (6 MW), and Lucknow (5MW). However, these projects were not successful in turning waste to energy (Annepu 2012).

Composting

Composting is the biological decomposition of biodegradable organic waste by bacteria and microorganisms under aerobic conditions, to a state that is adequately stable for odour- and pathogen-free storage and safe for use as biofertilizers in agriculture (United Nations 1993; Owamah, Dahunsi, Oranusi, *et al.* 2014). It has been evaluated and proven that compost from waste is very effective in soil remediation and increases the nutrient availability in the soil, aids in plant growth, and, ultimately, increases the plant yield (Tejada, Garcia, Gonzalez, *et al.* 2008; Walker and Bernal 2008; Alburquerque, Gonza´lvez, Garci´a, *et al.* 2007). Composting is not popular in some countries because of its high investment cost, possibility of secondary pollution, large area requirement, and low value of the compost product (Sakawi 2011; Ng, Lam, Varbanov, *et al.* 2014).

In India, under the National Scheme of Solid Waste Disposal (1975–80), cities with a population of more than 300,000 were set up with 10 operational composting plants which had composting capacities to treat 150–300 tonnes of MSW per day (Government of India 1995; Hoornweg, Laura, and Lambert 2000). Currently, low quality of waste, wrong selection of equipment, poor maintenance, high production costs, financial losses, low priority at the top level, and poor marketing efforts are some of the key challenges that the composting process has to overcome so it can be successfully implemented on a large scale (Selvam 1996).

Fermentation

Fermentation is a biological process where microorganisms convert sugars to acids, gases or alcohols, and a mixture of gases that include CO and H_2S. It is carried out in the absence of oxygen while the fermentation rate depends on several factors such as number of cells and cellular components, concentration of microorganisms, enzymes, temperature, and pH. Fermentation is a time-consuming process but the US and Brazil still use this process on a large scale for the production of ethanol (MacDonald, Yowell, and McCormack 2001).

This technology has attained much importance in the last few decades as ethanol can be produced from non-food plants and agronomic remains that are collectively called biomass (Urbaniec and Bakker 2015). The production of ethanol through fermentation requires three important types of raw materials: sugars, starches, and cellulose. Sugars can be obtained from sugarcane, sugarbeet, fruits, and molasses and they can be directly transformed to ethanol. Corn, potatoes, cassava, and root crops are used to derive starches. But they need to be hydrolysed first into fermentable sugars with the help of enzymes. Agricultural residues, wood, waste material from paper and the pulp industry are used to derive cellulose. With the help of mineral acids they are first transformed into sugars. As soon as the sugars are formed, the enzymes secreted by microorganisms ferment them into ethanol (Lin and Tanaka 2006).

The use of this technology is a promising approach, but its high cost is the main hindrance preventing it from moving out of the laboratory and into the commercial world. Table 15.3 compares different waste treatment methods.

Table 15.3   Comparison between different waste treatment methods

Waste treatment method	Advantages	Disadvantages
Thermochemical conversion		
(a) Incineration	• Most suitable for high calorific value waste, pathological wastes • Thermal energy recovery for direct heating or power generation. • Relatively noiseless and odourless. • Low land area requirement. • Can be located within city limits, reducing the cost of waste transportation. • Hygienic	• Least suitable for high moisture content/low calorific value and chlorinated waste. • Varying degree of pollutants such as dioxins, PCBB, SO_x, NO_x, chlorinated compounds are released into the atmosphere. • High capital and operation and maintenance costs. • Skilled personnel required for operation and management.
(b) Pyrolysis/gasification	• Production of fuel gas/oil, which can be used in a variety of applications. • Compared to incineration, control of atmospheric pollution can be dealt with in a superior way, in a techno-economic sense. • The reducing environment of the gasifier: (i) improves the quality of solid residues, particularly metals; (ii) strongly reduces the generation of some pollutants (dioxins, furans, and NO_x). • Syngas could be used, after proper treatment, to generate high-quality fuels (diesel, gasoline, or hydrogen) or chemicals.	• Net energy recovery may suffer in the case of wastes with excessive moisture. • High viscosity of pyrolysis oil may be problematic for its transportation and burning. • The actual production of pollutants depends on how syngas is processed downstream of the gasifier; if syngas is eventually oxidized, dioxins, furans, and NO_x potentially may still be an issue. • Syngas is toxic and potentially explosive; its presence raises major security concerns and requires reliable control equipment.

Contd...

Table 15.3 *Contd...*

Waste treatment method	Advantages	Disadvantages
(c) RDF pelletization	• High and constant heating value (3000–6000 kcal/kg). • Low level of pollution, and it does not produce a foul smell. • Waste is reduced to one-tenth of its original volume after processing; it can be easily transported or stored. • Convenient to use as it can be stored under normal temperatures for 6–12 months without decaying. • No need for expensive equipment; land costs are irrelevant.	• Low gas productivity • Generation of heavy tarry compounds, causing corrosion problems for the gas collection equipment.
Biochemical conversion		
(a) Anaerobic digestion	• Biogas is renewable. • Biogas is relatively inexpensive to produce. • Low calorific organic waste can be digested. • Viable technology for the competent treatment of organic waste.	• Long retention times and low removal efficiencies of organic compounds. • Temperature- and pH-sensitive. • High moisture is required throughout the digestion cycle.
(b) Composting	• Helps in reducing the amount of waste generated. • Compost helps improve soil health. • Local employment opportunities are generated. • Low cost, simple methodology, and eco-friendly. • It has a large advantage over inorganic fertilizers in long-term application.	• There is always a potential threat of heavy metal contamination; usually depends on type of substrate. • Slow rate of nutrient release compared to chemical fertilizers. • GHGs are released if not properly processed.
(c) Fermentation	• Low-water demand and high concentration of the end product. • High-volume productivity. • Easy aeration and no anti-foam chemicals. • Utilization of otherwise unusable carbon sources.	• Microorganisms may not be compatible with the type of waste being processed. • Chances of contamination with foreign microorganisms. • Careful process control is required, such as pH, temperature.

Source Singh, Tyagi, Allen, *et al.* (2011); Branconnier, Cumming, Jackman, *et al.* (1998); Bhide and Monroy (2011); Bhattacharyya (2010); Mata-Alvarez, Mace, and Llabres 2000; Holker, Hofer, and Lenz (2004); Khalid, Arshad, Anjum, *et al.* (2011)

15.4   CHALLENGES

India generates about 127,486 tonnes of garbage each day (SPCB's/PCB's 2009–12), but in spite of the availability of such a huge amount of waste, efforts to convert waste to energy are not very successful. The possible reasons are as follows (Bhattacharyya 2010):

- High moisture content in the waste makes certain technologies unsuitable for conversion of waste into electricity, the only exception being biochemical conversion.
- The Indian Government provides subsidies only for the construction and not for the running of plants. As a result, the plants become dormant after a while due to lack of funding.
- Initial establishment cost is high in setting up of these plants.
- The general public is unwilling to pay for renewable energy technologies which come with a high initial investment cost.
- Due to inadequate legal frameworks, private companies usually show unwillingness or are unable to participate in renewable energy technology programmes.
- There is a lack of institutional capacity for and technical knowledge in waste-to-energy conversion technology programmes. There are not enough technical experts in the field either.
- There is a lack of promotion of waste-to-energy programmes at a decentralized level.
- There is a lack of scientific decision-making among government officials when it comes to implementing cleaner and energy-efficient conversion technologies.
- Instead of motivating, the government makes unrealistic commitments and demotivates the programme implementers as they believe that their goals are too unrealistic.

15.5   THE WAY FORWARD

India should make every possible effort to develop and implement Integrated Waste Management technologies by strategically choosing a set of technologies that can work together in a coordinated way (Branconnier, Cumming, Jackman, *et al.* 1998). The following strategic points should be taken into consideration for sustainable waste management:

- Promoting R&D in the waste conversion sector will not only reduce the huge amount of waste the country produces but also open the gate for energy generation, thus lessening the gap between demand and supply of energy.

- Economic and social analyses of various waste management techniques need to be made before converting waste into energy.
- A decentralized system for waste-to-energy generation needs to be promoted.
- India should strive to adopt internationally approved practices that facilitate waste-to-energy conversion.
- Indigenous resources should be efficiently used.
- As India is in the early stages of economic transformation, it is also in an advantageous position as it can still adopt highly efficient and internationally approved waste-to-energy conversion technologies which can provide long-term benefits.
- Scientific decisions need to be made by government officials for the proper implementation of cleaner and energy-efficient conversion technologies.
- Proper changes in planning and policies from the regional to the national level should be made for better functioning of the judiciary.
- The Indian Judiciary needs to play a positive role in achieving these goals as their strong and independent decisions hold the power to improve the condition of waste management in the country.
- A suitable policy framework should be developed to promote the involvement of private companies in such programmes.
- Public awareness programmes should also be popularized.

15.6   CONCLUSION

The enormous amount of solid waste generated in developing countries is primarily associated with the economic conditions of these societies. SWM poses a big challenge mainly due to the increasing amount of waste generated, inappropriate handling, and the financial constraints of municipal authorities. There appear to be no immediate solutions to the problem of SWM while the health of the public as well as the environment continues to deteriorate. Most municipal officials report a scarcity of land for waste disposal. Therefore, the need of the hour is to integrate waste-to-energy conversion technologies with SWM in an efficient way. This will not only increase energy and aid material recovery, but will also solve the problem of environmental pollution. Cost-effective technologies in waste management together with a concrete and science-based national policy needs to be implemented. It is well understood that the organic component of waste is an important resource that could help drive growth in the form of energy. Therefore, it would be wise to bridge the gap between the demand and supply of energy in India.

ACKNOWLEDGEMENTS

The Authors are thankful to DST (P-45/18) for providing the funds and to the Director, Dean and Head, Institute of Environment and Sustainable Development, Banaras Hindu University, for providing necessary facilities.

REFERENCES

Ahrenfeldt, J., T. P. Thomsen, U. Henriksen, and L. R. Clausen. 2013. Biomass gasification cogeneration: A review of state-of-the-art technology and near future perspectives. *Applied Thermal Engineering* 50: 1407–1417

Alburquerque, J. A., J. Gonza'lvez, D. Garcı'a, and J. Cegarra. 2007. Effects of a compost made from the solid by-product ("alperujo") of the two-phase centrifugation system for olive oil extraction and cotton gin waste on growth and nutrient content of ryegrass (*Lolium perenne* L). *Bioresource Technology* 98: 940–945

Alhassan, H. M. and A. M. Tanko. 2012. Characterization of Solid Waste Incinerator Bottom Ash and the Potential for its Use. *International Journal of Engineering and Research Application* 2(4): 516–522

Annepu, R. K. 2012. *Sustainable Solid Waste Management in India*. New York: Department of Earth and Environmental Engineering, Columbia University, Department of Earth and Environmental Engineering

Appel, H. R., Y. C. Fu, S. Friedman, P. M. Yavorsky, and I. Wender. 1971. Converting organic wastes to oil. *US Bureau of Mines Report of Investigation No.* 7560

Arena, U. 2012. Process and technological aspects of municipal solid waste gasification:. A review. *Waste Management* 32: 625–639

Asif, M. and T. Muneer. 2007. Energy supply, its demand and security issues for developed and emerging economies. *Renewable and Sustainable Energy Reviews* 11: 1388–1413

Bagra, K. and A., Bharti. 2014. Quantitative and Qualitative Study of MSW from Papumpare District in Arunachal Pradesh. *International Journal of Innovative Research in Science Engineering and Technology* 3: 200–207

Belgiorno, V., G. D. Feo, C. D. Rocca, and R. M. A. Napoli. 2003. Energy from gasification of solid wastes. *Journal of Waste Management* 23: 1–15

Bhasin, K. C. 2009. Plasma Arc arc Gasification gasification for Waste waste management. Details available at http://www.electronicsforu.com/EFYLinux/efyhome/cover/February2009/Plasma-Arc-2.pdf. 2009

Bhattacharyya, S. C. 2010. Shaping a sustainable energy future for India: Management challenges. *Energy Policy* 38: 4173–4185

Bhavanam, A. and R. C. Sastry. 2011. Biomass gasification processes in downdraft fixed bed reactors: a review. *International Journal of Chemical Engineering and Applications* 2(6): 425

Bhide, A. and A. V. Shekdar. 1998. Solid waste management in Indian urban centres. *International Solid Waste Association Times* 1: 26–28

Bhide, A. and C. R. Monroy. 2011. Energy poverty: A special focus on energy poverty in India and renewable energy technologies. *Renewable and Sustainable Energy Reviews* 15: 1057–1066

Blanco, P. H., C. Wu, J. A. Onwudili, and P. T. Williams. 2012. Characterization of tar from the pyrolysis/ gasification of refuse derived fuel: Influence of process parameters and catalysis. *Energy and Fuel* 26: 2107–2115

Branconnier, J. R, B. D. Cumming, and L. R. Jackman. 1998. Process for thermophilic aerobic fermentation of organic waste. *American Patent* 5: 810–903

Bridgwater, A. V. 1994. Catalysis in thermal biomass conversion. *Applied Catalysis* 116: 5–47

Buenrostro, O. and G. Bocco. 2003. Solid waste management in municipalities in Mexico: goals and perspectives. *Resource, Conservation and Recycling* 39: 251–263

Central Electrical Authority. 2011. *Annual Report (2010-11)*. Details available at http://www.cea.nic.in/reports/powersystems/nep2012/generation_12.pdf

Central Institute of Plastics Engineering and Technology. 2012. Details available at http://www.cipetlibrary.gov.in/pdf/CIPET_Times_April_October_2012.pdf

Central Pollution Control Board (CPCB). 2004. *Management of Municipal Solid Waste*. New Delhi: Ministry of Environment and Forests, New Delhi, India

Central Pollution Control Board (CPCB). 2011. *Annual Report* 20102011. Details available at. http://www.cpcb.nic.in/upload/AnnualReports/AnnualReport_41_Annaul_Report_2010_11.pdf

Chanakya, H. N., T. V. Ramachandra, and M. Vijayachamundeeswari. 2007. Resource recovery potential from secondary components of segregated municipal solid wastes. *Environment Monitoring Assessment* 135: 119–127

United Nations. 1993. Changing consumption patterns. 1993. Agenda 21: United Nations Conference on Environment and Development, Rio de Janeiro, Brazil. New York: United Nations Department of Public Information

Cheng, H. and Y. Hu. 2010. Municipal solid waste (MSW) as a renewable source of energy: Current and future practices in China. *Bioresource Technology* 101: 3816–3824

Chu, S., C. Zhu, G. A. Tompsett, T. J. Mountziaris, and P. J. Dauenhauer. 2015. Refuse-Derived Fuel and Integrated Calcium Hydroxide Sorbent for Coal Combustion Desulfurization. *Industrial and Engineering Chemistry Research* 54: 3136–3144

Cordioli, M., S. Vincenzi, and G. A. De Leo. 2014. Effects of heat recovery for district heating on waste incineration health impact: A simulation study in Northern Italy. *Science of Total Environment* 444: 369–380

Demirbas, A. 2008. Producing bio-oil from olive cake by fast pyrolysis. *Energy Sources* Part A 30: 38–44

Diaz, L. F., M. S. George, and L. L. Eggerth. 1997. Managing solid wastes in developing countries. *Journal of Waste Management* 43–45

DiBlasi, C. 2000. Dynamic behaviour of stratified downdraft gasifier. *Chemical Engineering and Science* 55: 2931–44

Energy Information Administration United States. 2014. Independent Statistics and Analysis. Details available at http://www.eia.gov/countries/cab.cfm?fips=in.

Energy Statistics. 2013. *Twentieth Issue.* New Delhi: Central Statistics Office, Ministry of Statistics and Programme Implementation, Government of India, New Delhi.

European Environment Agency. 2013. Municipal waste management in the Netherlands. Prepared by Leonidas Milios.

Fabry, F., C. Rehmet, V. Rohani, and L. Fulcheri. 2013. Waste gasification by thermal plasma: a review. *Waste and Biomass Valorization* 3(4): 421–439

Feo, D. G., V. Belgiorno, R. M. A. Napoli, and U. Papale. 2000. Solid Wastes Gasification. *SIDISA International Symposium on Sanitary and Environmental Engineering.*

Fernandez, J., M. Perez, and L. I. Romero. 2008. Effect of substrate concentration on dry mesophilic anaerobic digestion of organic fraction of municipal solid waste (OFMSW). *Bioresource Technology* 99: 6075–6080

Government of India (GoI). 1995. Urban solid waste management in India. *Report of the High Power Committee.* New Delhi: Planning Commission, Government of India

Guan, G., C. Fushimi, A. Tsutsumi, M. Ishizuka, S. Matsuda, H. Hatano, and Y. Suzuki. 2010. High-density circulating fluidized bed gasifier for advanced IGCC/IGFC—Advantages and challenges. *Particuology* 8(6): 602–606

Guermoud, N., F. Ouagjnia, F. Avdelmalek, F. Taleb, and A. Addou. 2009. Municipal solid waste in Mostagnem city (Western Algeria). *Journal of Waste Management* 29: 896–902

Hauserman, W. B., N. Giordano, M. Lagana, and V. Recupero. 1997. Biomass gasifiers for fuel cells systems. *La Chimica & L' Industria* 2: 199–206

Holker, U., M. Hofer, and J. Lenz. 2004. Biotechnological advantages of laboratory-scale solid-state fermentation with fungi (Mini review). *Applied Microbiology Biotechnology* 64: 175–186

Hoornweg, D., T. Laura, and O. Lambert. 2000. Composting and its applicability in developing countries. *Working Paper Series* No. 8., Urban Development Division, The World Bank, Washington, DC. Details available at http://www.no-burn.org/downloads/Timarpur.pdf

Huang, H., and L. Tang. 2007. Treatment of organic waste using thermal plasma pyrolysis technology. *Energy Conservation and Management* 48: 1331–1337

ICRA Management Consulting Services (IMaCS). 2010. Details available at http://www.imacs.in/store/energy/IMaCS%20Capability%20Statement%20-%20Power%20Sector%202010.pdf.

International Energy Agency. 2013. *World Energy Outlook*. Details available at http://www.worldenergyoutlook.org/publications/weo-2013/

Jana, K. and S. De. 2015. Techno-economic evaluation of a polygeneration using agricultural residue: – A case study for an Indian district. *Bioresource Technology* 181:163–173

Kalyani, A. K. and K. K. Pandey. 2014. Waste to energy status in India: A short review. *Renewable and Sustainable Energy Reviews* 31:113–120

Kalyani, K. 2003. *Alternative Energy Supply Option from Household Waste Gasification Process: -A Feasibility Analysis on Rajkot City*. Dehradun: University of Petroleum and Energy Studies, College of Management and Economic Studies

Katyal, S. 2007. Effect of carbonization temperature on combustion reactivity of bagasse char. *Energy Sources* Part A. 29: 1477–85

Kaushal, R. K., G. K. Varghese, and M. Chabukdhara. 2012. Municipal Solid Waste Management in India—Current State and Future Challenges: A Review. *International Journal of Engineering Science Technology* 4(4): 1473–89

Khalid, A., M. Arshad, M. Anjum, T. Mahmood, and L. Dawson. 2011. The anaerobic digestion of solid organic waste. *Journal of Waste Management* 31: 1737–1744

Kumar, S. 2000. Technology options for municipal solid waste-to-energy project. *TERI Information Monitor on Environmental Science* 5(1): 1–11

Kumar, S., J. K. Bhattacharyya, A. N. Vaidya, T. Chakrabarti, S. Devotta, and A. B. Akolkar. 2009. Assessment of the status of municipal solid waste management in metro cities, state capitals, class I cities, and class II towns in India: An insight. *Waste Management* 29: 883–895

Lee, V., K. Kwok, W. Cheung, and G. McKay. 2007. Operation of a municipal solid waste co- combustion pilot plant. *Asia-Pacific Journal of Chemical Enginering* 2: 631–639

Lin, Y. and S. Tanaka. 2006. Ethanol fermentation from biomass resources: current state and prospects. *Applied Microbiology and Biotechnology* 69: 627–642

Liu, Y. and Y. Liu. 2005. Novel Incineration Technology Integrated with Drying, Pyrolysis, Gasification, and Combustion of MSW and Ashes Vitrification. *Environmental Science and Technology* 39: 3855–3863

MacDonald, T., G. Yowell, and M. McCormack. 2001. US ethanol industry production capacity outlook. California energy commission. Staff report. Details available at http://www.energy.ca.gov/reports/2001-08-29_600-01-017

Mahima, S. and S. Thomas. 2013. Estimating households' willingness to pay for solid waste management with special reference to Palakkad district in Kerala. *International Journal of Social Science and Interdisciplinary Research* 2(1): 73–80

Marsh, R., A. J. Griffiths, K. P. Williams, and S. J. Wilcox. 2007. Physical and thermal properties of extruded refuse derived fuel. *Fuel Process Technology* 88: 701–706

Mata-Alvarez, J., S. Mace, and P. Llabres. 2000. Anaerobic digestion of organic solid wastes: An overview of research achievements and perspectives. *Bioresource Technology* 74: 3–16

McKendry, P. 2002. Energy production from biomass (Part 3): gasification technologies. *Bioresource Technology* 83: 55–63

Metin, E. A. and C. Eroztürk. 2003. Solid waste management practices and review of recovery and recycling operations in Turkey. *Journal of Waste Management* 23: 425–432

Moustakas, K., D. Fatta, S. Malamis, K. Haralambous, and M. Loizidou. 2005. Demonstration of plasma gasification/vitrification system for effective hazardous waste treatment. *Journal of Hazardous Matter* B123: 120–126

Ng, W. P. Q., H. L. Lam, P. S., Varbanov, and J. J., Klemeš. 2014. Waste-to-Energy (WTE) network synthesis for Municipal Solid Waste (MSW). *Energy Conservation and Management.* 85: 866–874

Owamah, H. I., S. O. Dahunsi, U. S. Oranusi, and M. I. Alfa. 2014. Fertilizer and sanitary quality of digestate biofertilizer from the co-digestion of food waste and human excreta. *Journal of Waste Management* 34: 747–752

Pandey, P. K. 2011. Management of Municipal Solid Waste in India: A Legal Study. *Journal of Science Forum* 2(1): 182–196

Pappu. A., M. Saxena, and S. R. Asokar. 2007. Solid waste generation in India and their recycling potential in building materials. *Journal of Building Environment* 42 (6): 2311–2324

Pattnaik, S. and M. V. Reddy. 2010. Heavy metals remediation from urban wastes using three species of earthworm (*Eudrilus eugeniae, Eisenia fetida* and *Perionyx excavatus*). *Journal of Environmental Chemistry and Ecotoxicology* 3(14): 345–356

Phillip, N. P., T. N. Aziz, J. F. DeCarolis, M. A. Barlaz, F. He, F. Li, and A. Damgaard. 2014. Municipal solid waste conversion to transportation fuels: a life-cycle estimation of global warming potential and energy consumption. *Journal of Cleaner Production* 70: 145–153

Quina, M. J., J. C. M. Bordado, and R. M. Q. Ferreira. 2010. Chemical stabilization of air pollution control residues from municipal solid waste incineration. *Journal of Hazardous Matter* 179: 382–392

Ramachandra, T. 2006. *Management of Municipal Solid Waste*. New Delhi: Capital Publishimg Publishing Company

Ravindra, K., K. Kaur, and S. Mor. 2015. System analysis of municipal solid waste management in Chandigarh and minimization practices for cleaner emissions. *Journal of Cleaner Production* 89: 251–256

Rhyner, C. R., L. Z. Schwartz, R. B. Wenger, and M. G. Kohrell. 1995. *Waste Management and Resource Recovery*. New York: CRC Press, Inc.

Sakawi, Z. 2011. Municipal solid waste management in Malaysia: solution for sustainable waste management. *Journal of Applied Science and Environmental Sanitation* 6: 29–38

Saxena, R. C., D. K. Adhikari, and H. B. Goyal. 2009. Biomass-based energy fuel through biochemical routes: A review. *Renewable and Sustainable Energy Reviews* 13: 167–178

Selvam, P. 1996. A review of Indian experiences in composting of municipal solid wastes and a case study on private sector on private sector participation. *Conference of Recycling Waste for Agriculture: The Rural-Urban Connection.* Washington, DC, USA, 23–24

Sharholy, M., K. Ahmad, G. Mahmood, and R. C. Trivedi. 2008. Municipal solid waste management in India: A review. *Journal of Waste Management* 28: 459–67

Sharholy, M., V. Ahmad, G. Mahmood, and R.C. Trivedi. 2006. Development of prediction models for municipal solid waste generation for Delhi city. In *Proceedings of National Conference of Advanced in Mechanical Engineering (AIME-2006),* Jamia Millia Islamia, New Delhi, India, pp. 1176–86

Sharma, S. and K. W. Shah. 2005. Generation and disposal of solid waste in Hoshangabad. In: *Book of Proceedings of the Second International Congress of Chemistry and Environment*, Indore, India, 749–751

Shekdar, A. V. 1999. Municipal solid waste management the Indian perspective. *Journal of Industrial Association Environment Management* 26(2): 100–108

Shekdar, A. V., K. N. Krishnawamy, V. G. Tikekar, and A. D. Bhide. 1992. Indian urban solid waste management systems—jaded systems in need of resource augmentation. *Journal of Waste Management* 12 (4): 379–387

Siddiqui, T. Z., F. Z. Siddiqui, and E. Khan. 2006. Sustainable development through integrated municipal solid waste management (MSWM) approach—a case study of Aligarh District. In *Proceedings of National Conference of Advanced in Mechanical Engineering* (AIME-2006), Jamia Millia Islamia, New Delhi, India, pp. 1168–1175

Singh, J. and S. Gu. 2010. Biomass conversion to energy in India—A critique. *Renewable and Sustainable Energy Reviews* 14: 1367–1378

Singh, R. P., V. V. Tyagi, T. Allen, M. H. Ibrahim, and R. Kothari. 2011. An overview for exploring the possibilities of energy generation from municipal solid waste (MSW) in Indian scenario. *Renewable and Sustainable Energy Reviews* 15: 4797– 4808

Srivastava, V., S. A. Ismail, P. Singh, and R. P. Singh. 2015. Urban solid waste management in the developing world with emphasis on India: Challenges and opportunities. *Rev. Environ. Sci. Bio/Technol.* 14: 317–337

Staniewski, E. 1995. Gasification: the benefits of thermochemical conversion over combustion. *Hazardous Matter Management October/November*

Sudhir, K. 2000. Technology options for municipal solid waste-to-energy project. *TERI Information Monitor on Environmental Science* 5(1): 1–11

Tejada, M., C. Garcia, J. L. Gonzalvez, and M. T. Hernandez. 2008. Use of organic amendment as a strategy for saline soil remediation: Influence on the physical, chemical and biological properties of soil. *Soil Biology and Biochemistry* 38: 1413–1421

Themelis, N. J. and P. A. Ulloa. 2007. Methane generation in landfills. *Renewable Energy* 32: 1243–1257

Tillman, D. A. 1991. *The Combustion of Solid Fuels and Wastes.* San Diego: Academic Press,

Unnikrishnan, S. and A. Singh. 2010. Energy recovery in solid waste management through CDM in India and other countries. *Resource, Conservation and Recycling* 54: 630–640

Urbaniec, K. and R. R. Bakker. 2015. Biomass residues as raw material for dark hydrogen fermentation: A review. *International Journal of Hydro Energy* 40(9): 3648–3658

Walker, D. J. and P. M. Bernal. 2008. The effects of olive mill waste compost and poultry manure on the availability and plant uptake of nutrients in a highly saline soil. *Bioresource Technology* 99: 396–403

Wang, H. and Y. Nie. 2001. Municipal solid waste characteristics and management in China. *Journal of Air and Waste Management Association* 51: 250–263

Wang, K. S., K. Y. Chiang, C. C. Tsai, C. J. Sun, C. C. Tsai, and K. L. Lin. 2001. The effects of $FeCl_3$ on the distribution of the heavy metals Cd, Cu, Cr, and Zn in a simulated multimetal incineration system. *Environment International* 26: 257–263

Ward, A.J, P. J. Hobbs, P. J. Holliman, and D. L. Jones. 2008. Optimization of the anaerobic digestion of agricultural resources. *Bioresource Technology* 99: 7928–7940

Waste-to-Energy Research and Technology Council (WTERT) 2009. Details available at http://www.wtert.eu/default.asp?Menue=12&ShowDok=14

Wikstrom, E. and S., Marklund. 2000. Secondary Formation of Chlorinated Dibenzo-p-dioxins, Dibenzofurans, Biphenyls, Benzenes, and Phenols during MSW Combustion. *Environmental Science and Technology* 34: 604–609.

World Bank IBRD-IDA. 2010. Details available at http://www.worldbank.org/en/news/feature/2010/04/19/india-power-sector

World Bank. 2003. *Indonesia Environment Monitor*, pp. 33–41. Jakarta: World Bank Indonesia Office, Jakarta, Indonesia

World Bank. 2012. What a Waste: A Global Review of Solid Waste Management. Details available at http://siteresources.worldbank.org/INTURBANDEVELOPMENT/Resources/336387-1334852610766/What_a_Waste2012_Final.pdf

16

Organic Superfluous Waste as a Contemporary Source of Clean Energy

Yogita Basene[*] and Ragini Gothalwal

Department of Biotechnology and Bioinformatics Centre, Barkatullah University, Bhopal, Madhya Pradesh 462026

[*]E-mail: yogita.basene@gmail.com

16.1 INTRODUCTION

Every day, a substantial amount of waste gets produced that has a deteriorating effect on environment. There is a lot of challenges in the urban solid waste management. Different types of waste are responsible for producing harmful effects on living organisms and environment. Both human and animal are suffering from various kinds of diseases, the quality of air and soil are affected, and, therefore, the entire ecosystem is disturbed. In step with Environment Protection Act, 1989, 80% of the municipal solid waste (MSW) is deposited in landfills, 10% is incinerated, and only 10% of it is recycled. In developing countries, the solid waste production rate is 0.4–0.6 kg/person/day as opposed to 0.7–0.8 kg/person/day in developed countries.

Nowadays, a large amount of industrial, agricultural, and municipal waste are being used because their utilization process rate is high. The bulk of the organic wastes from different plenary industries is mostly paper mill industrial waste. It contains a substantial amount of solids with a high amount of lignocelluloses (Kim, Gil, Chang, *et al.* 2003).Lignocellulose waste can be different types including citrus peel waste, sawdust, paper pulp, industrial waste, MSW, and paper mill sludge (Greene, Celik, Dale, *et al.* 2004). The organic wastes produced in cafeterias, restaurants, halls, aliment process plants and domestic kitchens contain a large amount of carbohydrate constituent like cellulose and starch (Shiratori, Hironori,

Shohei, *et al.* 2006). If these sugars present in the wastes can be recycled, then the alcohol price can be kept at a check (Cekmecelioglu and Uncu 2013). In Asia each year approximately 4.4 billion tonnes of solid waste and 790 million tonnes of MSW are produced. India alone produces 48 million tonnes of waste. By 2047, it is predicted that India will produce 300 million tonnes of solid waste, and 169.6 km^2 of land will be required for disposing of this waste. When waste is disposed of unsystematically, it has a harmful effect on both the environment and the living beings (Gautam, Bundela, Pandey *et al.* 2012). Agricultural crop residues contain biochemical energy in the form of cellulose and lignin (Gregoire and Becker 2012). During municipal waste water treatment, the resultant primary sludge contains an abundant amount of cellulosic polysaccharides which can be a good resource to yield-reducing sugars (Champagne and Li 2009).

Renewable energy offers effective technologies that can be a help to tackle the rising energy challenges; climate change and the increasing demand for energy. Every country in the world has set renewable energy target and is forming policies to reach the goals. The countries are of the view that apart from creating a trade that will boost their economic development, it will also create energy independence. Renewable energy can be produced from solar, wind, water, biomass and geothermal heat flows. Energy produced through fossil fuels, or atomic power is not renewable (International Energy Agency 2002). In 2006 about 18% of the world's energy requirement was from renewable source that led research being conducted on biomass fuel, hydropower, and other forms of renewable energy that are at a nascent stage.

16.2   CURRENT STATUS OF ENERGY

Worldwide, India is the fifth largest consumer of energy. Approximately 12000 million tonnes of energy produced from petroleum is consumed globally of which India's share is 4.4%. In the last ten years, the worldwide consumption of energy produced from coal, oil, gas, nuclear has grown approximately 2.6%. In India it is around 6.8% while the supply increased at a compounded annual rate of only 1% of all the primary energy produced, 45% is generated from oil and gas. Between 1990 and 2007 the gross domestic product worldwide multiplied by 156% and the world energy demand rose by 39%. As a result, the greenhouse gas(GHG) emission increased at an alarming rate of 38%. In the last decade, the percentage of energy produced through renewable technologies, such as solar and wind are in double digits. From 2003 to 2008 energy produced from hydropower increased from 5% to 23%. United States, Germany, Sweden, and Spain have been successful in reducing carbon emissions due to the utilization of renewable energy technologies. As on 2012, India produced 89,774 MW of energy of which from wind 49,130 MW, small hydropower 15,399 MW, from biomass 17,538 MW and bagasse-based sugar mills 5000 MW (Energy Statics 2013, 20th issue). In 1970/71, the all-India gross electricity production from utilities, omitting that from captive generating plants, was 55,828 GigaWatt-Hours (GWh). In 1980/81, it rose to 110,844 GWh. In 1990/91, it was 264,329 GWh. The total electricity demand in 2010/11 was 844,846 GWh and reached 923,203 GWh in 2011/12, registering an additional need by about 9.27%. The annual increase in Per-capita Energy Consumption (PEC) from 2010/11 to 2011/12 was 3.36% (Energy statics 2013 20th issue) (www.mospi.gov.in).

To be one of the developed countries, India needs to make more parts of the country electrified and meet the growing electricity demand too. At present, the state of electricity services across India is said to be acute if not in a critical mode. The poor and the rural people do not have access to electricity because the country is unable able to meet the peak demand. It is the country's responsibility to provide sufficient amount and quality of power supply (EA 2003).

As per national statistics, the major issues include generation, transmission, and distribution of electricity. During 2000 and 2003 the scarcity in energy need and peak power demand was 8% and 12%, respectively. India's electricity consumption with a rough year per capita was 400 kWh, which was way below China's 900 kWh, Malaysia's 2500 kWh, and Thailand's 1500 kWh (Modi 2005).

Different renewable technologies are suitable for fuel generation, electricity production and distribution. It is also reducing transmission and transportation losses so that little amount of primary energy is needed to supply the same energy services (Sawin and Moomaw 2009).

Bioenergy can be produced from global biomass systems which contain enormous amount of biomass in the form of food, fodder and fibre production and forest wastes and residues management.

16.3   BIOENERGY

Bioenergy means biomass energy system that provides heating, cooling, and electricity. Biofuels mean liquid fuels that are used for transportation. The basic biomass resources and key components are identified by the biomass supply chain (Department for Environment Food and Rural Affairs 2007). Bioenergy can be produced from organic waste as the latter is a promising natural resource for sustainable energy production. Organic waste has a substantial commercial use because of effective utilization and treatment of polysaccharide residues as frugal carbon substrate offer renewable and sundry alternative for transportation fuel (Coughlan 1990; Geddes, Nieves, Ingram, *et al.* 2011). Commercially, bioenergy can be produced from agricultural residues, industrial wastewater and products. Biofuel produced from industrial waste is only a fraction of liquid fuels that are generated from petroleum (IEA Bioenergy 2009). Even today developing countries are struggling with simple energy grids. Lack of constant electricity makes it tough for industrial development apart from causing appalling living conditions for people as they cannot afford a supersession supply for energy like a generator that converts chemical fuel into electric power. A continuous supply of energy is paramount for any country permitted the wastewater treatment system to function properly (Nwogu 2007).

Biohydrogen production from organic waste has potential source, as it is not only abundant and readily available, but also cheap and degradable (Guo, Table, Latrille, *et al.* 2010). Organic waste may serve not only as a source of hydrogen gas production but also as a source of hydrogen producing microorganisms (Favaro, Alibardi, Lavagnolo, *et al.* 2013). The efficiency of energy generation can be increased by screening microbial diversity and simply possible fermentable organic substrate (Kalia and Purohit 2008). Some of the pure cultures of microorganisms such as *Enterobacter, Bacillus,* and *Clostridium* species produce hydrogen gas via complicated fermentative processes. The latter two groups are characterized by the formation of spores in response to adverse environmental conditions such as scarcity of nutrients and rising temperature (Hawkes, Dinsdale, Hawkes, *et al.* 2002).

Microbial fuel cells (MFCs) is a novel way of current generation or bioelectricity generation by using various types of organic substrate including wastewater (Prasad, Sivaram, Berchmans *et al.* 2006; Ghangrekar and Shinde 2007).

16.4   ORGANIC WASTE

Biomass is a combination of different organic substances, raw or processed, with the natural form of energy. Based on its end uses, biomass can be divided into various categories, such as heating/cooling, power (electricity) generation, or transportation. A sizable quantity of agricultural waste (example, corn stalks, straw, sugarcane waste, mill scrap, municipal waste) is generated during harvest, transport, storage, and promotion and processing. Owing to their nature and composition, these wastes easily decompose and produce a foul odour. The vegetable waste generated in markets are dumped in municipal landfills and dumping ground. These wastes can be reused for compost preparation, biogas generation, and electricity production, and so on. Paper is widely used as packing materials, particularly in the form of cardboard. Currently, 49% of paper and board packaging waste is recycled (UNIDO 2007).

The essential biomass resources and key components are penned within the biomass supply chain (DEFRA 2007). There is scope to convert the energy potential of biomass supplemental expeditiously through additional refined processes. Biomass resources are a combination of various residues of an industrial process, agricultural resources, municipal waste and specific forest residues. It can be utilized for producing heat and power, fuel, as well as liquid fuels, or fossil fuel-based material and products (DEFRA 2007). Biomass, a suitable form of renewable energy, provides 10% of the worldwide energy (50 EJ^{-year}). Biomass is used in developing countries for heating and domestic needs such as cooking. A threefold raise within the utilization of bioenergy, to 150 EJ/year would require approximately the entire current world biomass produce (IPCC).

By 2050 the demand for biomass may increase 100 EJ to 300 EJ. Since transportation fuels are less easy to exchange than heat and power, future use of biomass for energy is to focus on the former. Today's applications of bioenergy are based on pure solid, liquid and gaseous energy carriers, example bioethanol and methane. The annual production of bioethanol is 84 billion litres (2010) projected to reach 125 billion litres in 2017 (Walker 2011).

Table 16.1   Organic waste converted into energy (biomass supply chain)

Bio resource	Supply system	Conversion	End use
Convention forestry	Harvesting	Biochemical	Transportation fuel,
Short rotation forestry		Thermochemical	Heat
	Collection	Physical/chemical process	Electricity
Oil bearing plant		Example:	Solid fuels
Animal products	Delivery	Deoxygenation	Renewable construction material
Muncipal solid waste		Depolymerization	
Industrial waste	Storage	Pyrolysis,	Plant-based pharmaceutical
		Gasification	Renewable

Source DEFRA (2007)

16.4.1   Major Organic Waste

Cellulose and hemicellulose are superior material that can be broken down chemically into glucose units by treating it with concentrated acids at high temperature. Cellulose is the long-chained polymer of glucose monomers joined by β- (1→4)-glycosidic bond. Cellulose is different from starch and glycogen found in animals. This is because polysaccharides such as starch and glycogen contain α- (1→4)-glycosidic bonds. Unlike starch, cellulose is a straight chain polymer where no coiling or branching occurs, and the molecule adopts an extended and rather stiff-rod conformation, aided by the equatorial conformation of the glucose residues. Cellulose is mostly found in plants. Cellulose chains consist of about 300–10,000 units that vary depending on the plant species (Demers, Doane, Guzman, *et al.* 2009).

Cellulose is the most abundant renewable natural biological resource. Therefore, various cellulosic materials along with modified cellulose of the polymer are used currently in the world (Hon 1992). Gas produced as a result of microbial degradation of transuranic waste contain cellulosic material (Francis, Gillow, Giles, *et al.* 1997). Around 30 billion tonnes of terrestrial cellulose is produced annually (Cox, Betts, Jones, *et al.* 2000). Cellulosic waste materials include sawdust, sugarcane bagasse, waste paper (example, printing papers and cartons), and corncob. Agricultural wastes are decomposed by physical, chemical, and biological processes because they largely contain cellulosic material. Rice straw and banana plant waste are primary sources of polysaccharides (Dhegiha, Waleed, Zawawy, *et al.* 2013).

Primarily cellulosic wastes (wood, cardboard, paper, and tissue) mediate by a variety of microbial populations. Agricultural residues are a good example of lignocellulosic biomass that is renewable, primarily underutilized and produced at a significantly low cost. Lignocellulosic biomass is composed of hemicellulose, cellulose, and a polymer and includes various parts of plants (leaves, stems, and stalks) such as corn fibre, sugarcane pulp, crops residues. Agriculture waste consists of 70% moisture and 30% solid, of which cellulose accounts for 65.5%, lignin 21.2%, ash 3.5%, hot water-soluble substances 5.6% and alcohol–benzene soluble 4%–1% (Thambirajah, Zulkafli, Hashim, *et al.* 2005).

At present, abundant polymers and lignin-related compounds are released as waste effluents from pulp and paper industry. The presence of a group of efficient cellulose digesting clostridia was confirmed by molecular ecological analysis (Shiratori, Hironori, Shohei, *et al.* 2006). Methane production takes place where microbial consortium MC1 is used to degrade or pretreat filter paper, newspaper, and cardboard (Yuan, Cao, Li, *et al.* 2012).

Pulp and paper industry is amongst the foremost energy utilizing industry in the world. This industry is the fifth-largest energy consumer in the world; utilizing approximately 4% of the global energy. During pulp and paper process, the considerable amount of waste is generated. Three different raw materials are utilized in the industry: non-wood fibres, soft wood, and hard wood. It has been calculated that 500 million tonnes of paper will be produced by 2020 (Cheremisinoff and Rosenfeld 2010).

Unwanted cellulosic components are disposed of as waste. Processed MSW contains 40%–50% cellulose, 11%–12% hemicelluloses, and 10%–15% lignin by dry weight. The sludge pumped from effluent treatment plants consists of large quantities of cellulosic material that can be efficiently used for hydrogen gas production with the help of dark fermentation (Chairattanamanokorn et al. 2012). The treatment of microcrystalline cellulose effluent was utilized in a two-stage UASB and two-stage BAF series bioreactor (Wang, Ji, Wu, et al. 2012). Milling of cane stalks to extract raw sugar yields by-products of various quality. These by-products contain effluent, molasses, bagasse, mill mud, and boiler ash. On average, each metric weight unit of raw sugar produced yields 7 tonnes of waste product (Qureshi, Wegener, and Mallawaarachchi 2001).

16.4.2 Degradation of Cellulose

Eubacteria and fungi mostly act as cellulolytic microorganisms. Cellulolytic microorganisms are able to establish synergistic relationships with non-cellulolytic species in cellulosic wastage. The interactions between both the population lead to complete degradation of cellulose, releasing carbon dioxide and water under aerobic conditions and carbon dioxide, methane and water under anaerobic conditions (Leschine 1995).

Various anaerobic thermophiles are shown to use polysaccharide, such as *Clostridium thermocellum, Clostridium straminisolvens,* and *Clostridium stercorarium* (Madden 1983). Several microorganisms

Table 16.2 Percentage of (C, N, and H) in various cellulosic materials

Residues	Weight of substrate (mg)	Carbon (%)	Hydrogen (%)	Nitrogen (%)	Molar ratio (C:H)
Corncob	2.9	42.5	6.56	0.334	6:10.9
Sawdust components	2.6	48.6	5.96	0	6:8.8
Sugarcane bagasse	2.8	44.4	6.08	0	6:9.8
Waste paper	2.7	39.1	5.96	0	6:10.9

Source Pang, Chin, and Yih (2011)

are engaged with cellulose-degrading activities as well as several cellulolytic microorganism species including *Trichonympha, Clostridium, Actinomycetes, Bacteroides succinogenes, Butyrivibrio fibrisolvens, and Ruminococcus albums* (Schwarz *et al.* 2001; Milala, Shugaba, Gidado, *et al.* 2005). Aerobic and anaerobic fungal strains, such as *Chaetomium, Fusarium, Myrothecium, Trichoderma, Penicillium, Aspergillus* and so forth are some of the reported fungal species for cellulosic biomass hydrolysis. Two model organisms *Thermobifida fusca* and *Cellulomonas fimi* have the ability of cellulose degradation (Wilson 2004).

In natural and designed anaerobic environments, microorganisms use cellulose as the primary carbon source of energy, especially in rumina (Whitford, Forster, Beard, *et al.* 1998), decomposing wood (Borsodi, Micsinai, Rusznyák, *et al.* 2005), sulphate-reducing and methanogenic sediments (Heijs, Haese, Forney, *et al.* 2007; Cardenas, Wu, Leigh, *et al.* 2008), insect guts (Warnecke, Luginbuhl, Ivanova, *et al.* 2007), wetlands (Ibekwe, Grieve, Lyon, *et al.* 2003) and sulphate-reducing bioreactors (Hiibel, Pereyra, Inman, *et al.* 2008). In cellulosic wastes synergistic relationship can be established between cellulolytic microorganisms and non-cellulolytic species. Cellulose biomass may be an upgrade strategy for biomass process due to microbial conversion (Lynd, Laser, Bransby, *et al.* 2008). Cellulases are enzymes that catalyse the β-1,4- glycosidic linkage in cellulose chains and are synthesized by microorganisms during their metabolic activity and use cellulose as a substrate (Lee and Koo, 2001). The enzymatic hydrolysis of cellulose needs different types of cellulases; namely endoglucanase, (1,4-D-glucan-4-glucanohydrolase; EC 3.2.1.4), exco cellobiohydrolase (1,4-D-glucan glucohydrolase) and glucosidase (D-glucoside glucohydrolase) (Yi, Sandra, John, *et al.* 2009). Mostly aerobic microorganisms follow independent cellulase mechanism within which they produce a group of individual cellulases, most of which contain a carbohydrate binding module (CBM) that is connected via a flexible linker to one end of the catalytic domain. The cellulase within the mixture act synergistically to degrade solid or crystalline cellulose hydrolysis (Wilson 2008). Cellulase synergism can result in increase in the particular activity of appropriate mixtures which are more than fifteen-fold higher than that of any individual cellulase (Irwin, Spezio, Walker, *et al.* 1993). Anaerobic microorganisms produce multi-enzyme complexes called cellulosomes that degrade carbon sources (Ding, Xu, Crowley, *et al.* 2008).

Only a few enzymes in cellulosomes contain a CBM, however, the protein scaffolding contains a family of 3 CBMs, which bind the cellulosomes to cellulose. Aerobic and anaerobic microorganisms secrete the cellulase which belongs to the same families, except that only aerobic fungi manufacture GH-7 cellulases and cellulosomes do not contain family 6 exocellulase. Some anaerobic cellulolytic thermophilic bacteria such as *Caldicellulo disruptor* species release multidomain cellulases that contain CBMs. These microorganisms have an effective system of plant cell wall degradation that can hydrolyse untreated plant biomass in contrast to most other cellulose-degrading microorganisms (Blumer-Schuette, Lewis, and Kelly 2010).

16.5   TECHNOLOGY FOR ENERGY

Biotechnological processes of cellulosic biomass for the generation of hydrogen gas, methane, and grain alcohol are well studied (Bridgwater 2006). The advantages of the bioprocess are partially offset by the high cost of required cellulose degrading enzymes, generation of toxic compounds during pretreatment of cellulose, and lack of fermentative microorganisms that may efficiently utilize all the by-products. Moreover, methane and grain alcohol still need to be combusted which reduces their energy efficiency, and separation, purification, and storage of hydrogen gas is extravagant. Hence, there is a need for cost-effective, energy-efficient, and environmentally friendly technique to treat cellulosic biomass. For this, scientists need to develop a breakthrough technology.

Bioenergy productions start with organic substrates such as forest or agriculture crops, industrial and municipal waste and by-products.

Table 16.3   Percentage of lignin, hemicellulose, and cellulose in lignocellulosic material

Lignocellulose material	Lignin (%)	Hemicellulose (%)	Cellulose (%)
Bagasse	20	25	42
Sorghum	21	25	45
Hardwood	18–25	24–40	40–55
Softwood	25–35	25–35	45–50
Corncobs	15	35	45
Corn stover	19	26	38
Rice straw	18	24	32.1
Nut shells	30–40	25–30	25–30
Newspaper	18–30	25–40	40–55
Grasses	10–30	25–50	25–40
Wheat straw	16–21	26–32	29–35

Sources Iqbal, Kyazze, Keshavarz, *et al.* (2013); Anwar, Gulfraz, and Irshad (2014)

Now in a period of advanced technology and research, developing a gravid method to complete degradation of waste and produce energy. In microbial fuel cell, microorganisms act as a catalyst for oxidation of organic compound (Salgado 2009).

16.6   MICROBIAL FUEL CELL

M. C. Potter, a professor of Botany, was first to perform the work of obtaining energy by bacteria. He understood that this discovery will be beneficial for human beings. However, no significant development occurred on his basic designs till the 1980s.

Microbial fuel cell converts chemical energy into electric power throughout oxidation of substrate or organic material with the assistance of microorganisms (Allen and Bennetto 1993) See Figure 16.1. The microbial fuel cell has been used to explain the diversity of systems for electricity generation with the help of microbes (Shukla, Suresh, Berchmans, et al. 2004; Lovely, 2006). Oxidative metabolic pathway of microorganisms plays an important role. Electrons and protons will be transferred through the ubiquinone, coenzyme Q or cytochrome, and

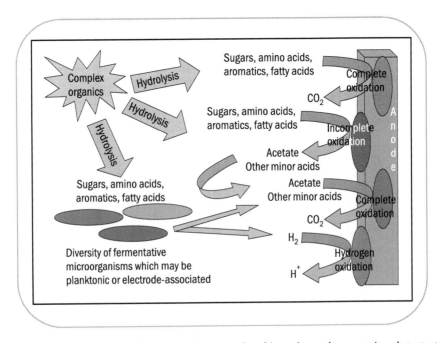

Fig. 16.1 *Representation of the degradation or breaking of complex organic substrate to electric power*

Source *Lovley (2008)*

NADH dehydrogenase. They act as electron carriers in electron transport chain in the microbial fuel cell (Chaudhuri and Lovley 2003).

Many microorganisms have the ability to transfer an electron to the anode via oxidation of organic materials. In a microbial fuel cell assembly, both pure and mixed culture of microorganisms play a significant role (Cheng, Moon, Jank, *et al.* 2005). This indicates that the diverse cultures show high resistance to process disturbance, substrate utilization and high efficiency of power output. Pure cultures of bacterial strain belonging to Firmicutes (Park, Kim, Kim, *et al.* 2001) and Proteobacteria (Holmes, Bond, Lovely, *et al.* 2004; Zhang, Cui, Chen, *et al.* 2008) as well as yeast strains *Saccharomyces cerevisiae* (Walker and Walker 2006) and *Hansenula anomala* (Prasad, Arun, Murugesan, *et al.* 2007) are capable of generating current in an MFC. Electrochemically active and iron-reducing bacteria *Shewanella* and *Geobacter, Klebsiella pneumonia, Rhodopseudomonas palustrine, Dessulfobulbus propionicus* screened from waste product showed higher efficiency to power generation in a microbial fuel cell (Sharma and Kandu 2010). Biofilm indicated that not a single microorganism has the capacity to transfer an electron that develops on the anode. A wide range of microorganisms are found in association with electrodes in MFC systems, especially when a natural inoculum is used to seed the MFC (Aelterman, Rabaey, Pham, *et al.* 2006). A *Enterobacter aerogenes* XM02, a hydrogen producing strain with Fe (III) reducing activity was selected as a biocatalyst for oxidation of carbohydrates source such as starch for electricity generation (Zhang, Zhuang, Zhou, *et al.* 2009). The anode compartment of an microbial fuel cell can be enriched with electrochemically active chromium reducing and β- Proteobacteria, Actinobacteria and Acinetobacter species (Ryu, Kim, Lee, *et al.* 2011).

16.6.1 Microbes–Electrode Interaction

It was observed that electrons were transferred from inside the cell to extracellular acceptors via c- type cytochromes, biofilms and conductive pili (called nanowires) in cultures of *Shewanella putrefaciens, Geobacter sulferreducens, Geobacter Metallireducens* and *Rhodoferax ferrireducens* (Lovley 2008). These microbes may form biofilms on the surface of the anode and play a role as electron acceptor and transfer electron on to the anode under high Coulombic efficiency (Chaudhuri and Lovley 2003).

In another mechanism, electrons from microbial carriers are transferred onto the conductor or anode surface either by a microorganism's (*Shewanella oneidensis, Geothrix fermentans*) own mediator that successively facilitates extracellular electron transfer or by additional mediators. The use of mediator in microbial fuel cell as electron shuttles is referred to as mediator

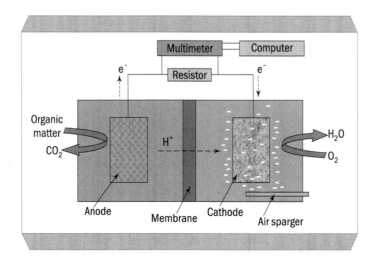

Fig. 16.2 *Design of microbial fuel cells*
Source *Logan (2008)*

MFC. It provides a base for the microbes to come up with electrochemically active reduced product (Figure 16.2).

Table 16.4 Biochemical reaction in MFC

Anode (oxidation reaction): $C_6H_{12}O_6 + 6H_2O \rightarrow 6CO_2 + 24H^+ + 24\ e^-$
Cathode (reduction reaction): $6O_2 + 24H^+ + 24\ e^- \rightarrow 12\ H_2O$
Over all reaction: $C_6H_{12}O_6 + 6O_2 \rightarrow 6CO_2 + 6H_2O$ + electrical energy

Source Du, Li, Gu, *et al.* (2007)

Table 16.5 Several substrates and production

Substrate	Anode	Current (mA)	Microorganisms	References
Lactate	Woven graphite	0.31	*Shewanella putrefaciens*	Kim, Park, Hyun, *et al.* (2002)
Acetate	Graphite	0.40	*Geobacter sulfurreducens*	Bond and Lovely (2003)
Glucose	Graphite	0.2	*Rhodoferax ferrireducens*	Chaudhuri and Lovley (2003)
Glucose	Woven graphite	0.7	*Rhodoferax ferrireducens*	
Glucose	Porous graphite	74	*Rhodoferax ferrireducens*	
Acetate	Graphite	0.23	Mixed culture	Bond, Holmes, Tender, *et al.* (2002)
Sulphide/ acetate	Graphite	60	Mixed culture	Tender, Reimers, Stecher, *et al.* (2002)

The electron donor can be a reduced product of microbial metabolism that transfers electron from the electron carrier to the electrode surface (Lovely 2006). Mostly neutral red, thionine, methylthionine chloride, iron chelates are (Chaudhuri and Lovley 2003) added to the system as reaction mediators (Du, Li, Gu, *et al.* 2007) (Tables 16.4 and 16.5). *Proteus vulgaris, Escherichia coli, Streptococcus lactis* and *Pseudomonas* species transfer electrons outside the cell through mediators. The mediator should be capable of penetrating the cell membranes, transport electron from carriers of the electron transport chain, the rate of electron transfer from the metabolite should be high, stable during long periods of redox cycling and non-toxic to microbes (Ieropoulos, Greenman, Melhuish, *et al.* 2005; Osman, Shah, Walsh, *et al.* 2010).

16.6.2   Mechanisms

In an MFC, the microorganisms present within the anode compartment use organic substrates as fuels to produce electrons and protons through biological process (Rabaey and Verstaete 2005). These electrons are accepted by nicotinamide adenine dinucleotide in the electron transport chain transferred to terminal electron acceptors such as nitrate, sulphate and oxygen and reaching the outer membrane proteins (Logan and Regan 2006; Salgado 2009). Microorganisms then transfer electrons to the anode from where they reach the cathode via an external electrical circuit. The proton migrates to cathode and fuses with the electron and catholyte, a chemical like oxygen, is reduced at the cathode surface. Electricity is generated which is measured by a voltmeter or ammeter. Microbe acts as an enhancer on the anode surface (Lovely 2006). The anode compartment is maintained under anaerobic condition as oxygen inhibits electricity whereas while the the cathode is exposed to oxygen (Mostafa 2009).

16.6.3   Key Player of Microbial Fuel Cells

Geobacter sulfurreducens is a novel microorganism for microbial fuel cell technology because it has a high potential for current generation. *Geobacter sulfurreducens* has become a model for a microbial process for electricity production (Holmes, Bond, Lovely, *et al.* 2004). The complete genome sequence reveals evidence of aerobic metabolism, one-carbon and complex carbon metabolism, motility and chemotactic behaviour (Methé, Nelson, Eisen, *et al.* 2003) and is amenable to genetic changes (Coppi, Leang, Sandler, *et al.* 2001). The study of expression of genes have steered that the assembly of nanowires in microbes is vital for electron transfer via the biofilm (Reguera, Nevin, Nicoll, *et al.* 2006). See Figure 16.3.

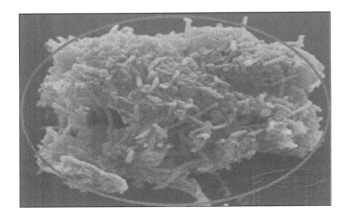

Fig. 16.3   *Geobacter sulfurreducens for electricity generation*

16.7   COMPONENTS OF MICROBIAL FUEL CELLS

Engineers have developed different kinds of MFC designs, however, all of them share the same operation principles. They are operated under various conditions to boost up current production and scale back the value (Logan 2006). See Table 16.6.

16.7.1   Cost Effectiveness

Microbial fuel cell can be analysed with the novel parameter of "Power-to-cost Ratio" (PCR). The municipal waste material contains microorganisms that can function to treat wastewater and produce electricity in an MFC. Functioning MFCs can be constructed from inexpensive, readily available components. That technique of bonding the anode directly to the membrane

Table 16.6   Exhibiting components and material for MFC construction

Components	Materials
Anode electrode	Carbon cloth, carbon felt, graphite felt, carbon mesh, graphite fiber brush (Logan 2006)
Cathode electrode	Platinum black , activated carbon (AC), graphite based cathode (Chen, Choi , Lee, *et al.* 2008)
Proton exchange membrane	Polymer like Nafion, Ultrex (Schwartz 2007)
Substrate	Organic material Carbohydrates, protein, cellulose & waste water (Lee, Chen, Choi, *et al.* 2008)
Mediators	Neutralred, thionine, methyleneblue, anthraquinone-2,6-disulfonate, phenazine and iron chelates (Lovley 2006)

achieves the principal objective increasing the power output and reducing the internal resistance of the MFC. This method led to MFC design that approximately four times the (estimated) PCR efficiency of existing MFC designs. The PCR metric was used to compare the low-cost MFC to more expensive MFC that produces higher amount of current (Patra 2008). See Table 16.7.

16.8 BIO-BATTERY

Bio-battery is an energy storing device that is powered by organic substrate. It produces electricity from renewable fuels and provides sustained, on-demand portable power supply. These batteries can store this energy for later use. Bio-battery uses expeditor, either biomolecules like enzymes or maybe an entire living organism to oxidate biomass for generating electrical charge.

Microbial batteries (Scholz and Schroder 2003) or bio-batteries produce direct current like different electrochemical cells once connected to a circuit. Bacterial batteries will probably operate for years, providing a less expensive and excellent environmentally friendly option to chemical batteries (Ajayi and Weigele 2012). The bio-batteries are more advantageous in comparison to other cells because of their ability to recharge fast. In other words through a continuous supply of organic substrate such as

Table 16.7 Components for MFC construction

Components	Conventional MFC	Low-cost MFC
Chamber	MFC chamber is customized machined cylinders with end caps.	An MFC chamber ought to be a bottle. The supersession was a 2″ PVC pipe that purchased for $ 0.50 (₹ 24.29)
Anode	Most typical MFC anodes use carbon artifact (cloth) 2500 $ (₹ 1,21450)	The MFC used anode engendered by carbon cloth carbon artifact that price $620/m^2 (₹ 30119.6/m^2)
Membrane	Membrane Most MFCs have Nafion® membranes that price $2500/ m^2 (₹ 97160/ m^2)	Agar $165/m^2 (₹ 8015.70/ m^2) and Gore-Tex $82.5/m^2 (₹ 4007.85/m^2) were evaluated as membrane materials
Cathode	Typical MFC cathodes are composed of platinum coated carbon artifact cost accounting regarding costing about $2000/m^2 (₹ 97160/m^2).	The supersession material was galvanized aluminum grating, additionally referred to as Kiwi -Mesh®. It was $12.00/m^2 (₹ 582.96/m^2). It allowed to air to flow through it so that good contact was obtained between the two reactant in the half reaction.

Source Patra (2008)

sugar, it can continuously keep themselves charged without an external power supply with high-fuel flexibility such as sugar, alcohol, diesel, ethanol, blood and so on (Siddiqui and Pathrikar 2013).

16.9   REMOTE POWER SOURCE

One primary advantage of an MFC in remote sensing over traditional battery is that bacteria can reproduce, providing the MFC a considerably long lifetime than traditional batteries.

16.10   EFFECTIVE FOR SEA WATER

Desalination of ocean water and brackish water to be utilized as drinking water is a serious problem because of the enormous amount of energy needed to remove the dissolved salts from the water. An MFC can remove salt by up to 90%. However, higher extraction efficiencies are needed in order to get drinking-quality water.

16.11   CONCLUSION

In the current situation of energy crisis, scientists are forced to alternate forms of possible form of energy. Moreover fossil fuels exploitation have led to environmental pollution. Clean fuels, such as microbial fuel cells and biofuels, canbe used instead of traditional fossil fuels. This chapter summarizes the different types of waste that are used in microbial fuel cells for electricity production in cost-efficient manner. Organic wastes are a good source of bioenergy as compared to another forms of energy.

ACKNOWLEDGEMENTS

The authors are thankful to DBT-IPLS builder programme, Barkatullah University Bhopal for providing junior research fellowship to Yogita Basene.

REFERENCES

Aelterman, P., K. Rabaey, H. T. Pham, N. Boon, and W. Verstraete. 2006. Continuous electricity generation at high voltages and currents using stacked microbial fuel cells. *Environmental Science and Technology* 40(10): 3388–94

Ajayi, F. and P. R. Weigele. 2012. A terracotta bio-battery. *Bioresource Technology* 116: 86–91

Allen, R. M. and H. P. Bennetto. 1993. Microbial Fuel-Cells: electricity production from carbohydrates. *Applied Biochemstry Biotechnology* 39(40): 27–40

Blumer-Schuette, S. E., D. L. Lewis, and R. M. Kelly. 2010. Phylogenetic, microbiological, and glycoside hydrolase diversities within the extremely thermophilic, plant biomass degrading genus Caldicellulosiruptor. *Applied and Environmental Microbiology* 76: 8084–8092

Bond, D. R. and D. R. Lovley. 2003. Electricity production by *Geobacter sulfurreducens* attached to electrodes. *Applied Environmental Microbiology* 69: 1548–1555

Bond, D. R., D. E. Holmes, L. M. Tender, and D. R. Lovley. 2002. Electrode-reducing microorganisms that harvest energy from marine sediments. *Science* 295: 483–485

Borsodi, A. K., A. Micsinai, A. Rusznyák, P. Vladár, G. Kovács, E. M. Tóth, and K. Márialigeti. 2005. The diversity of alkaliphilic and alkali tolerant bacteria cultivated from decomposing reed rhizomes in a Hungarian soda lake. *Microbial Ecology* 50: 9–18

Cardenas, E., W. M. Wu, M. B. Leigh, J. Carley, S. Carroll, T. Gentry, J. Luo, D. Watson, B. Gu, M. Ginder-Vogel, P. K. Kitanidis, P. M. Jardine, J. Zhou, C. S. Criddle, T. L. Marsh, and J. A. Tiedje. 2008. Microbial communities in contaminated sediments, associated with bioremediation of uranium levels. *Applied Environmental Microbiology* 74: 3718–3729

Cekmecelioglu, D. and O. N. Uncut. 2013. Kinetic modeling of enzymatic hydrolysis of pretreated kitchen wastes for enhancing bioethanol production. *Waste Management* 33 (3): 735–739

Chairattanamanokorn, P., S. Tapananont, S. Detjaroen, J. Sangkhatim, P. Anurakpongsatorn, and P. Sirirote. 2012. Additional paper waste in pulping sludge for biohydrogen production by heat-shocked sludge. *Applied Biochemistry Biotechnology* 166(2): 389–401

Champagne, P. and C. Li. 2009. Enzymatic hydrolysis of cellulosic municipal wastewater treatment process residuals as feedstocks for the recovery of pure sugars. *Bioresource Technology* 100(23): 5700–5706

Chaudhuri, S. K. and D. R. Love. 2003. Electricity generation by direct oxidation of glucose in mediatorless microbial fuel cells. *Nature Biotechnology* 21: 1229–1232

Chen, G. W., S. J. Choi, T. H. Lee, G. Y. Lee, J. H. Cha, and C. W. Kim. 2008. Application of biocathode in microbial fuel cells: Cell performance and microbial community. *Applied Microbiology and Biotechnology* 79: 379–388

Cheremisinoff, N. P. and P. E. Rosenfeld. 2010. Best Practices in the Agrochemical industry Amsterdam: Elsevier

Coppi, M. V., C. Leung, S. J. Sandler, and D. R. Lovley. 2001. Development of a genetic system for *Geobacter sulfurreducens*. *Applied and Environmental Microbiology* 67: 3180–3187

Coughlan, M. P. 1990. Cellulose degradation by fungi. *Microbial Enzymes and Biotechnology*, 2nd edn, pp. 1–36. New York: Elsevier

Cox, P. M., R. A. Betts, C. D. Jones, S. A. Spall, and I. J. Totterdell. 2000. Acceleration of global warming due to carbon-cycle feedbacks in a coupled climate model. *Nature* 408: 184–187

DEFRA (Department for Environment Food and Rural Affairs). 2009. *The 2007/08 Agricultural Price Spikes: Causes and Policy Implications*. London: HM Government

Demers, A., R. Doane, S., Guzman, and R. Pagan. 2009. Enzymatic hydrolysis of cellulosic biomass for the production of second generation biofuels. *Project Report Submitted to the Faculty of Worcester Polytechnic Institute US*. Details available at https://www.wpi.edu/Pubs/E-project/Available/E-project-043009-114037

Ding, S. Y., Q. Xu, M. Crowley, Y. Zeng, M. Nimlos, R. Lamed, E. A. Bayer, and M. Himmel. 2008. A biophysical perspective on the cellulosome: new opportunities for biomass conversion. *Current Opinion in Biotechnology* 19: 218–227

Du, Z., H. Li, and T. Gu. 2007. A state of the art review on microbial fuel cells: A promising technology for wastewater treatment and bioenergy. *Biotechnology Advances* 25: 464–482

Favaro, L., L. Alibardi, M. C. Lavagnolo, S. Casella, and M. Basaglia. 2013. Effects of inoculum and indigenous microflora on hydrogen production from the organic fraction of the municipal solid waste. *International Journal of Hydrogen Energy* 38(27): 11774–11779

Feng, Y., X. Wang, B. Logan, and H. Lee. 2008.Brewery wastewater treatment using air-cathode microbial fuel cells. *Applied Microbiology and Biotechnology* 78: 873–880

Francis, A. J., J. Gillow, and B. Giles. 1997. The microbial gas generation under normal waste isolation pilot plant repository conditions. *SAND 96-2582.* 28(24): 243

Gautam, S. P., P. S. Bundela, A. K. Pandey, Awasthi M. K. Jamaluddin, and S. Sarsaiya. 2012. The diversity of cellulolytic microbes and the biodegradation of municipal solid waste by a potential strain. *International journal of Microbiology* 3(2): 59–67

Geddes, C. C., I. U. Nieves, and L. O. Ingram. 2011. Advances in ethanol production. *Current Opinion in Biotechnology* 22: 312–319

Ghangrekar, M. M. and V. B. Shinde. 2007. The performance of membrane-less microbial fuel cell is treating wastewater and effect of electrode distance and area of electricity production. *Bioresource Technology* 98: 2879–2885

Gil, G. C., I. S. Chang, B. H. Kim, M. Kim, J. K. Jang, H. S. Park, and H. J. Kim. 2003. Operational parameters are affecting the performance of a mediator-less microbial fuel cell. *Biosensors Bioelectronics* 18: 327–338

Greene, N., F. E. Celik, B. Dale, M. Jackson, K. Jayawardhana, H. Jin, E. D. Larson, M. Laser, L. Lynd, D. MacKenzie, J. Mark, J. McBride, S. McLaughlin, and D. Saccardi. 2004. How biofuels can help end America's oil dependence. In *Growing Energy,* E. Cousins (ed.). New York: Natural Resources Defense Council, pp. 1–86

Gregoire, K. P. and J. G. Becker. 2012. Design and characterization of a microbial fuel cell for the conversion of a lignocellulosic crop residue to electricity. *Bioresource Technology* 119: 208–215

Guo, X. M., E. Table, E. Latrille, H. Carrère, and J. P. Steyer. 2010. Hydrogen production from agricultural waste by dark fermentation. *Journal of Hydrogen Energy* 35:10660–10673

Hawkes, F. R., R. Dinsdale, D. L. Hawkes, and I. Hussy. 2002. Sustainable fermentative hydrogen production: challenges for process optimization. *International Journal of Hydrogen Energy* 27: 1339–1347

Heijs, S. K., R. R. Haese, P. V. Wielen, L. J. Forney, and J. D. Elsas. 2007. Use of 16S rRNA gene-based clone libraries to assess microbial communities potentially involved in anaerobic methane oxidation in a Mediterranean cold seep. *Microbial Ecology* 53: 384–398

Hiibel, S. R., L. P. Pereyra, L. Y. Inman, A. Tischer, D. J. Reisman, K. F. Reardon, and A. Pruden. 2008. Microbial community analysis of two fieldscale sulfate - reducing bioreactors treating mine drainage. *Environmental Microbiology* 10: 2087–2097

Homles, D. E., D. R. Bond, and D. R. Lovely. 2004. Electron transfer by *Desulfobulbuspropionicus* to Fe (III) and graphite electrodes. *Applied Environmental Microbiology* 70: 1234–1237

Hon, D. N. S. 1992. New developments in cellulosic derivatives and copolymers. *Emerging Technologies for Materials and Chemicals from Biomass. ACS Symp.* 476: 176–196

Ibekwe, A. M., C. M. Grieve, and S. R. Lyon. 2003. Characterization of microbial communities and composition in constructed dairy wetland wastewater effluent. *Applied Environmental Microbiology* 69: 5060–5069

Ibrahim, M. M., W. K. El-Zawawy, Y. Jüttke, A. Koschella, and T. Heinze. 2013. Cellulose and microcrystalline cellulose from rice straw and banana plant waste - preparation and characterization. *Cellulose 20*: 2403–2416

IEA (International Energy Agency) 2009. CO_2 emission from fuel combustion.

Ieropoulos, I. A., J. Greenman, C. Melhuish, and J. Hart. 2005. Comparative study of three types of microbial fuel cell. *Enzyme and Microbial Technology* 37: 238–245

IPCC (Intergovernmental Panel on Climate Change). 2005. *IPCC Special Report on Carbon Dioxide Capture and Storage.* Cambridge, UK, and New York, USA: Cambridge University Press

Iqbal, H. M. N., G. Kyazze, and T. Keshavarz. 2013. Advances in the valorization of lignocellulosic materials by bio-technology: an overview. *Bioresources* 8(2): 3157–3176

Irwin, D. C., M. Spezio, L. P. Walker, and D. B. Wilson. 1993. Activity studies of eight purified cellulases: specificity, synergism, and binding domain effects. *Biotechnology and Bioengineering* 42: 1002–1013

Kalia, V. C., and H. J. Purohit. 2008. Microbial diversity and genomics in aid of bioenergy.*Journal of Industrial Microbiology and Biotechnology* 35: 403–419

Kim, H. J., H. S. Park, M. S. Hyun, I. S. Chang, M. Kim, B. H. Kim. 2002. A mediator-less microbial fuel cell using a metal reducing bacterium, *Shewenellaputrefaciens*. *Enzyme and Microbial Technology* 30: 145–152

Kim, M. S., C. Jaehwan, and D. H. Kim. 2012. Enhancing Factors of Electricity Generation in a Microbial Fuel Cell Using *Geobactersulfurreducens*. *Journal of Microbiology and Biotechnology* 22(10): 1395–1400

Lee, S. M. and Y. M. Koo. 2001.Pilot-scale production of cellulose using *Trichodermareesei* Rut C-30 in fed-batch mode. *Journal of Microbiology and Biotechnology* 11: 229–233

Leschine, S. B. 1995.Cellulose degradation in anaerobic environments. *Annual Review Microbiology* 49: 399–426

Logan, B. E. 2008. *Microbial Fuel Cells*. New York: Wiley

Logan, B. E. and K. Rabaey. 2012.Conversion of wastes into bioelectricity and chemicals using microbial electrochemical technologies. *Science* 337: 686–690

Logan, B. E. and J. M. Regan. 2006. Microbial challenges and fuel cell applications. *Environmental Science and Technology* 40: 172–180

Logan, B. E., B. Hamelers, R. Rozendal, U. Schroder, J. Keller, S. Freguia, P. Aelterman, W. Verstraete, and K. Rabaey. 2006. Microbial fuel cells: methodology and technology. *Environmental Science and Technology* 40: 5181–5192

Lovley, D. R. 2006. Microbial fuel cells: novel microbial physiologies and engineering approaches. *Current Opinion in Biotechnology* 17:327–332

Lovley, Derek R. 2006. Bug juice: harvesting electricity with microorganisms. *Nature Review Microbiology* 4:497–508

Lovley, Derek R. 2008. The microbe electric: conversion of organic matter to electricity. *Current Opinion in Biotechnology* 19:1–8

Lynd, L. R., M. S. Laser, D. Bransby, B. E. Dale, B. Davison, R. Hamilton, M. M. Himmel, J. Keller, D. McMillan, J. Sheehan, and C. E. Wyman. 2008. How biotech can transform biofuels. *Nature Biotechnology* 26: 169–172

Madden, R. H. 1983. Isolation and characterization of Clostridium stercorarium sp. nov., cellulolytic thermophile. *Internation Journal Systematic Bacteriology* 33: 837–840

Methé, B. A., K. E. Nelson, J. A. Eisen, I. T. Paulsen, and W. Nelson. 2003. The genome of *Geobactersulfurreducens*: Insights into metal reduction in subsurface environments. *Science 302*: 1967–1969

Milala, M. A., A. Shugaba, A. Gidado, A. C. Ene, and J. A. Wafar. 2005. Studies on the use of agricultural wastes for cellulase enzyme production by *A. niger*. *Journal of Agriculture and Biological Science* 1: 325–328

Ministry of Statistics and Programme Implementation. 2013. *Energy Statistics.* New Delhi: MOSPI

Modi, V. 2005. *Improving Electricity Services in Rural India: An Initial Assessment of Recent Initiatives and Some Recommendations.* New York: The Earth Institute at Columbia University

Mostafa, R., M. Najafpour, R. Daud, and W. Ghoreysh. 2009. Low voltage power generation in a biofuel cell using anaerobic culture. *World Applied Sciences* 6 (11): 1585–1588

Nwogu, N. G. 2007. Microbial fuel cells and parameters affecting performance when generating electricity. *Basic Biotechnology* 73–79

Osman, M. H., A. A. Shah, and F. C. Walsh. 2010. Recent progress and continuing challenges in biofuel cells. Part II: Microbial. *Biosensors and Bioelectronics* 26: 953–963

Pang, S. C., S. F. Chin, and V. Yih. 2011 Conversion of cellulosic waste materials into nanostructured ceramics and nanocomposites. *Advance Materials Letter* 2(2): 118–124

Park, H. S., B. H. Kim, H. S. Kim, H. J. Kim, G. T. Kim, M. Kim, I. S. Chang, Y. K. Park, and H. I. Chang. 2001. A novel electrochemically active and Fe (III)-reducing bacterium phylogenetically related to *Clostridium butyricum* isolated from a microbial fuel cell. *Anaerobe 7*: 297–306

Patra, A. 2008. Low-cost, single-chambered microbial fuel cells for harvesting energy and cleansing wastewater. *Journal of the U.S. SJWP* 1: 72

Postier, B. L., R. J. Di Donato, J. Nevin, K. P. Liu, B. Frank, D. R. Lovley, and B. A. Methe. 2008. Benefits of electrochemically synthesized oligonucleotide microarrays for analysis of gene expression in under studied microorganisms. *Journal of Microbiological Method* 74: 26–32

Potter, M. C. 1911. Electrical effects accompanying the decomposition of organic compounds source. *Proceedings of the Royal Society of London. Series B* 84 (571): 260–276

Prasad, D., S. Arun, M. Murugesan, S. Padmanaban, R. S. Satyanarayanan, S. Berchmans, and V. Yegnaraman. 2007. Direct electron transfer with yeast cells and construction of a mediatorless microbial fuel cell. *Biosensor Bioelectronics* 22: 2604–2610

Prasad, D., T. Sivaram, K. S. Berchmans, and V. Yegnaraman. 2006. Microbial fuel cell constructed with a micro-organism isolated from sugar industry effluent. *Journal Power Sources* 60: 991–996

Qureshi, M. E., M. K. Wegener, and T. Mallawaarachchi. 2001. The economics of sugar mill waste management in the Australian Sugar Industry: Mill mud case study. *45th Annual Conference of the Australian Agricultural and Resource Economics Society*, Adelaide, South Australia, pp. 23–25

Rabaey, K. and W. Verstraete. 2005. Microbial fuel cells: Novel biotechnology for energy generation. *Trends Biotechnology* 23: 291–298

Reguera, G., K. P. Nevin, J. S. Nicoll, S. F. Covalla, T. L. Woodard, D. R. Lovley. 2006. Biofilm and nanowire production leads to increased current in *Geobactersulfurreducens* fuel cells. *Applied Environmental Microbiology* 72: 7345–7348

Ryu, E. Y., M. Kim, and S. J. Lee. 2011. Characterization of Microbial Fuel Cells enriched using Cr (VI)-containing sludge. *Journal Microbiology Biotechnology* 21 (2): 187–191

Salgado, C. A. 2009 Microbial fuel cells powered by *Geobacter sulfurreducens*. *Basic Biotechnology* (5): 5–11

Schwartz, K. 2007. Microbial fuel cells: design elements and application of a novel renewable energy source. *MMG 445 Basic Biotechnology eJournal* 3(1): 20–27

Schwarz, W. H. 2001. The cellulosome and cellulose degradation by anaerobic bacteria. *Applied Microbiology and Biotechnology* 56 (6): 634–649

Sharma, V. and P. P. Kundu. 2010. Biocatalysts in microbial fuel cells. *Enzyme and Microbial Technology* 47: 179–188

Shiratori, H., I. Hironori, A. Shohei, K. Naoaki, M. Akiko, H. Kuniaki, B. Teruhiko, and U. Kenji. 2006. Isolation and Characterization of a new *Clostridium* sp. that performs active cellulosic waste digestion in a Thermophilic / Methanogenic bioreactor. *Applied Environmental Microbiology* 72(5): 3702–3709

Shukla, A. K., P. Suresh, S. Berchmans, and A. Rajendran. 2004. Biological fuel cells, and their applications. *Current Science* 87: 455–468

Swin, J. L. and W. R. Moomaw. 2009. Renewable revolution: Low carbon energy by 2030. Worldwatch Report. Details available on http://www.worldwatch.org/files/pdf/Renewable%20Revolution.pdf

Szekely, A. J., R. Sipos, B. Berta, B. Vajna, C. Hajdu´, and K. Ma´rialigeti. 2009. DGGE and T-RFLP analysis of bacterial succession during mushroom compost production and sequence-aided T-RFLP profile of mature compost. *Microbial Ecology* 57:522–533

Tender, L. M., C. E. Reimers, H. A. Stecher, D. E. Holmes, D. R. Bond, D. A. Lowy, K. Pilobello, S. J. Fertig, and D. R. Lovley. 2002 Harnessing microbially generated power on the seafloor. *Nature Biotechnology* 20: 821–825

Thambirajah, J. J., M. D. Zulkafli, and M. A. Hashim. 2005 Microbiological and biochemical changes during the composting of oil palm empty fruit bunches: Effect of nitrogen supplementation on the substrate. *Bioresource Technology* 52: 133–134

UNEP (United Nations Environmental Program Division of Technology, Industry, and Economics). 2009. Converting Waste Agricultural Biomass into a Resource Compendium of Technologies. International Environmental Technology Centre Osaka/Shiga.

UNIDO (United National Industrial Development Organization). Details available at www.unido.org

Walker, G. M. 2011 Fuel alcohol: Current production and future challenges. *Journal of the Institute of Brewing* 117(1): 3–22

Walker, A. L. and J. C. W. Walker. 2006. Biological fuel cell and an application as a reserve power source. *Journal Power Source* 160: 123–129

Wang, C., G. Ji, and Y. Wu. 2012. Analysis of microbial characterization in an upflow anaerobic sludge bed/biological aerated filter system for treating microcrystalline cellulose wastewater. *Bioresource Technology* 120: 60–69

Warnecke, F., P. Luginbuhl, N. Ivanova, and M. Ghassemian. 2007. Metagenomic and functional analysis of hindgutmicrobiota of a wood-feeding higher termite. *Nature* 450: 560–565

Whitford, M. F., R. J. Forster, C. E. Beard, J. H. Gong, and R. M. Teather. 1998. Phylogenetic analysis of rumen bacteria by comparative sequence analysis of cloned 16S rRNA genes. *Anaerobe* 4: 153–163

Wilson, D. B. 2004. Studies of Thermobifidafusca plant cell wall degrading enzymes. *Chemical Research* 4: 72–82

Wilson, D. B. 2008. Aerobic microbial cellulase systems. In *Biomass Recalcitrance: Deconstructing the Plant Cell Wall for Bioenergy,* pp. 374–392. Oxford UK: Blackwell

Yi, J. C., J. C. Sandra, A. B. John, *et al.* 2009. Production and distribution of endoglucanase, cellobiohydrolase, and β-glucosidase components of the cellulolytic system of *Volvariellavolvacea*, the edible straw mushroom. *Applied Environmental Microbiology* 65: 553–559

Yuan, X., Y. Cao, J. Li, B. Wen, W. Zhu, X. Wang, and Z. Cui. 2012. Effect of pretreatment by a microbial consortium on methane production of waste paper and cardboard. *Bioresource Technology* 118: 281–8

Zhang L. X., L. Zhuang, S. G. Zhou, J. R. Ni, M. Liu, and Y. Yuan. 2009. Bioelectricity generation by a Gram-positive *Corynebacterium sp.* strain MFC03 under an alkaline condition in microbial fuel cells. *Bioresource Technology* 101(6): 1807–1811

Zhang, T., C. Cui, S. Chen, H. Yang, and P. Shen. 2008. The direct electrocatalysis of *Escherichia coli* through electro-activated excretion in microbial fuel cell. *Electrochemical. Communication* 10: 293–297

17

Vermicomposting: A Potential Tool for Sustainable Management of Solid Waste

Sonal Dixit, Richa Kothari, and D.P. Singh[*]

Department of Environmental Science, Babasaheb Bhimrao Ambedkar University, Lucknow, Uttar Pradesh 226025

[*]E-mail:dpsingh_lko@yahoo.com

17.1 INTRODUCTION

Our world is facing several environmental challenges today, one of which is the rapid increase in the volume of waste. Waste may be defined as material produced from any source that has lost its economic value. According to a survey, India generates nearly 25 million tonnes of solid waste per annum (Aalok, Tripathi, and Soni 2008). Solid waste can be categorized into different categories, such as medical waste, municipal waste, radioactive waste, and hazardous waste (Table 17.1).

In search of better opportunities and lifestyle, people in developing countries migrate from rural areas to urban areas, and the consequent increase in population in these parts increases the volume of urban solid waste. India is trying to become an industrialized nation in the coming years and this too will significantly increase the amount of solid waste (Sharma and Shah 2005; CPCB 2004; Shekdar, Krishnaswamy, Tikekar, *et al.* 1992). The infrastructure of the country has not been able to keep up with the increase in urban waste and the municipalities are stretched to their limits trying to provide essential services (Kumari 2013). Currently, the majority of solid waste is disposed of unscientifically in landfills, which causes a hazardous impact on the environment. Management of solid waste has thus become the most vital environmental issue in India. The practice of using landfills for waste disposal should be avoided and new policies need to be initiated to save our environment. The government is trying to find ways to reduce the disposal of solid waste in open landfills.

A significant portion of solid waste is degradable and non-toxic, but this is dumped along with inorganic toxic waste in landfills. Existing methods for treatment and disposal of organic waste involve its collection, preservation, segregation, transportation, handling, and so on, and these are quite expensive. The main problem associated with the maintenance of organic waste is its high moisture content, which makes it very bulky and its collection, transport, and storage difficult. It also causes ground pollution by forming leachates. People are becoming aware of environmental problems and the high cost associated with conventional methods of waste disposal, and so they are gradually moving away from the practice of landfilling to reprocessing of organic waste on-site to produce useful end products. There is a need to investigate environment-friendly and sustainable alternatives to landfills. Vermicomposting happens to be one of these ecologically sound, economical, and commercially sustainable options for the management of solid wastes.

In Asian countries, more than 60% of the total waste is organic in nature, which can be further recycled and reused with the help of vermitechnology. This technology needs to be introduced worldwide because of the necessity to recover organic materials from waste. It is a new, organic technology that is a totally environmentally sound approach to solid waste management and has also been found to be useful to people in various other ways (Loehr, Martin, and Neuhauser 1998). The term "vermicomposting" originates from the Latin word "*vermis*", which means "worms". Aristotle described worms as the "intestines of the earth" and stated that there may not be any other creature that has played so important a role in the history of life on earth, as worms serve as versatile natural bioreactors to harness energy and destroy soil pathogens. In vermicomposting process, organic waste is converted into valuable manure by use of earthworms and certain microorganisms. Both earthworms and microorganisms work symbiotically to hasten the conversion of organic matter into fertile compost (Ronald and Donald 1977). This conversion of organic waste also makes use of livestock excreta. During the process of vermicomposting, essential nutrients such as N, P, K, and Ca present in organic waste gets converted into more soluble and readily available forms which aid in plant growth. Certain plant growth regulators are found in vermicompost as well. Besides plant growth, earthworms used for vermicomposting act as a protein source and can be used as animal feed. India is facing a pollution issue owing to a high load of organic waste, and the demand for organic manure is also growing day by day. Both these problems can be solved by using vermicomposting technology for solid waste management. There is an urgent need to apply the vermicomposting process to solve environmental and economic crises.

Table 17.1 Types of solid waste

Source	Source characteristics	Compositions
Residential	Single and multifamily dwelling	Food waste, paper, cardboard, plastics, textile, leather, yard waste, wood, glass, metals, ashes, special waste
Industrial	Light and heavy manufacturing, fabrication, construction sites, power and chemical plants	Housekeeping waste, packing and food waste, construction and demolition materials, hazardous waste, ashes, special waste
Commercial	Restaurants, markets, office buildings, and so on	Stones, bottles, paper, cardboard, plastics, wood, food waste, glass, metals, special waste
Institutional	Schools, hospitals, prisons, government institutes	Same as commercial
Construction and demolition	New construction sites, road repair, renovation site, demolition of buildings	Wood, steel, concrete, dirt, and so on
Municipal sites	Street cleaning, landscaping parts, beaches, other recreational areas, water and wastewater treatment plants	Street sweeping; landscape and tree trimming; general waste from parks, beaches, and other recreational areas; sewage sludge
Process	Heavy and light manufacturing, refineries, chemical plants, power plants, mineral extraction, and processing	Industrial process waste, scrap materials, off-specification products, slag tailing
Agricultural	Crops, farms, dairies	Spoiled food waste, agricultural residues, pesticides

Source World Bank (1999)

17.2   CLASSIFICATION AND COMPOSITION OF SOLID WASTE

The composition of solid waste differs to a great extent from one place to another, and the planning, designing, and operation of the technology needed for waste processing before its disposal at a landfill are based on the composition of the waste.

Solid waste can generally be divided into two groups: biodegradable or organic waste and non-biodegradable or inorganic waste. Organic waste primarily comprises kitchen garbage, straw, paper, and excreta, while non-biodegradable waste includes ash, stone, cinder, plastics, rubber, and metals. About 50%–70% of the total urban solid waste, also known as

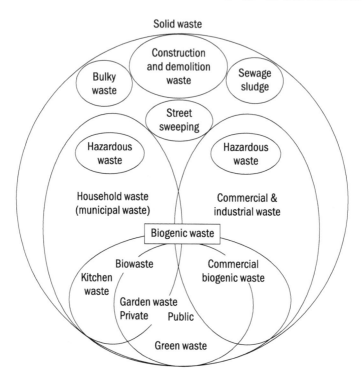

Fig. 17.1 *Classification of solid waste*
Source *Kothari, Kumar, Panwar, et al. (2014)*

municipal solid waste (MSW), comprises residential and commercial waste generated within a community. Figure 17.1 gives the classification of MSW. The distribution of the different components of solid waste depends on the following factors:

- Construction and demolition activities
- Service provided by municipalities
- Facilities used for wastewater treatment
- Standard of living of people

In India, the composition and toxicity of MSW vary greatly compared to the MSW of western nations (Gupta, Krishna, Prasad, *et al*. 1998; Shannigrahi, Chatterjee, and Olaniya 1997; Sharholy, Ahmad, Mahmood, *et al*. 2008). Compostable, recyclable, and inert are the three basic components of MSW in any given region (Table 17.2). The compostable fraction includes waste coming from kitchens, vegetables, and yards. The recyclable portion consists of paper, plastic, metal, and glass. The component of MSW that cannot be composted or recycled forms the inert portion, which includes stones, ash, and silt. Table 17.2 lists the region-wise compositions of MSW in India.

Table 17.2   MSW composition in different regions of India

Region	MSW (TPD)	Compostable (%)	Recyclable (%)	Inert (%)	Moisture (%)	Calorific value (kcal/kg)
Metros	51,402	50.89	16.28	32.82	46	1,523
Other cities	2,723	51.91	19.23	28.86	49	2,084
East India	380	50.41	21.44	28.15	46	2,341
North India	6,835	52.38	16.78	30.85	49	1,623
South India	2,343	53.41	17.02	29.57	51	1,827
West India	380	50.41	21.44	28.15	46	2,341
Total	130,000	51.3	17.48	31.21	47	1,751

TPD: tonnes per day
Source Annepu (2012)

17.3   CURRENT STRATEGIES USED FOR SOLID WASTE MANAGEMENT IN INDIA

The following actions need to be planned for designing a solid waste management system:

- Generation and storage of waste
- Primary collection of waste and its transfer to community bin
- On-site segregation of waste
- Management of community bin
- Secondary waste collection and transport to landfill
- Sanitary waste disposal in landfills
- Management of landfills
- Separation of recyclables at every step

The waste management system needs to be planned on the basis of the quantity and composition of waste generated in any country. Inappropriate management of solid waste systems causes environmental problems and climate change, hastens the degradation of natural resources, and degenerates the quality of life. The current solid waste management services in India are ineffective, costly, and substandard, thus posing a threat to the health of people and environment. The per capita waste generation rate in India has increased from 0.44 kg/day in 2001 to 0.5 kg/day in 2011, fuelled by changing lifestyles and the increased purchasing power of urban Indians. Within a decade, from 2001 to 2010, waste generation by Indian cities has increased by 50% due to population growth and urbanization. Such a sudden increase in waste generation has put a lot of stress on all available natural, infrastructural, and budgetary resources. Table 17.3 lists the state-wise generation, collection, and quantity of waste treated in India (CPCB 2012).

Table 17.3   State-wise data for MSW generation

State/union territory	Quantity generated (TPD)	Quantity collected (TPD)	Quantity treated (TPD)
Andaman and Nicobar	50	43	Nil
Andhra Pradesh	11,500	10,655	3,656
Arunachal Pradesh	94	NA	Nil
Assam	1,146	807	72.65
Bihar	1,670	1,670	Nil
Chandigarh	380	370	300
Chhattisgarh	1,167	1,069	250
Daman, Diu and Dadra Nagar Haveli	28+13=41	NA	Nil
Delhi	7,384	6,796	1,927
Goa	193	NA	NA
Gujarat	7,379	6,744	873
Haryana	537	NA	Nil
Himachal Pradesh	304	275	153
Jammu and Kashmir	1,792	1,322	320
Jharkhand	1,710	869	50
Karnataka	6,500	2,100	2,100
Kerala	8,338	1,739	1,739
Lakshadweep	21	21	4.2
Madhya Pradesh	4,500	2,700	975
Maharashtra	19,204	19,204	2,080
Manipur	113	93	2.5
Meghalaya	285	238	100
Mizoram	4,742	3,122	Nil
Nagaland	188	140	Nil
Odisha	2,239	1,837	33
Puducherry	380	NA	Nil
Punjab	2,794	NA	Nil
Rajasthan	5,037	NA	Nil
Sikkim	40	32	32
Tamil Nadu	12,504	11,626	603
Tripura	360	246	40
Uttar Pradesh	11,585	10,563	Nil

Contd...

Table 2.3 *contd...*

State/union territory	Quantity generated (TPD)	Quantity collected (TPD)	Quantity treated (TPD)
Uttarakhand	752	NA	Nil
West Bengal	12,557	5,054	606.5
Total	127,486	89,334	15,881

Source CPCB (2012)

Waste disposal techniques can be broadly divided into two categories: (i) non-eco-friendly techniques, which include landfilling, incineration, gasification, refuse-derived fuel (RDF), biomethanation, and (ii) eco-friendly techniques, which include mainly conventional composting and vermicomposting. Incineration, gasification, RDF, and biomethanation are all expensive processes and need to be handled by skilled professionals. On the contrary, composting is a cost-effective and easy method for organic solid waste management.

17.3.1   Non-eco-friendly Techniques

Landfilling

In many developing countries including India, the most common method of waste disposal is open dumping or landfilling (UNEP 2001; Kansal 2002; Sharholy, Ahmad, Mahmood, *et al.* 2008). In both cities and towns, more than 90% of total solid waste is dumped in landfills without prior treatment. Such a procedure is inherently unplanned and unacceptable. This sort of substandard and inadequately managed dumping gives rise to severe environmental issues. Even in waste disposal centres, the required standards for landfilling are not followed by the personnel. The waste is not even segregated at the source of waste collection and storage, and as a result toxic waste, including hospital waste and industrial waste, also makes its way to the landfill. At most waste disposal sites, important practices such as levelling and compression of waste, followed by covering the waste with a layer of soil, are rarely carried out, and neither are landfill gas monitoring systems and leachate collection systems in place, and this increases the threat of hazards to people (Bhide and Shekdar 1998; Gupta, Krishna, Prasad, *et al.* 1998). Sanitary land filling by following all set standards is recommended and necessary for proper MSW disposal.

Incineration

Incineration is the combustion process at high temperatures in which solid waste is burned completely. The temperature needed for incineration ranges from 980°C to 2000°C. This process is mainly applied to reduce the volume of waste and treat toxic wastes, such as hospital waste. It can reduce the original volume of waste by almost 90%, and some advanced incinerators currently in use can, at very high temperatures, produce a molten material and reduce the volume of waste to about 5% or less (Jha, Sondhi, and Pansare 2003; Ahsan 1999; Peavey, Donald, and Gorge 1985). The ashes produced by incineration are deep buried in a landfill. The incineration process for waste management is not much used in India because of the typical composition of solid waste in the country that interferes with the combustion process. Waste in India is rich in organic substances (40%–60%), moisture (40%–60%), and inert content (30%–50%), and it has a very low calorific value (800–1100 kcal/kg) (Kansal 2002; Joardar 2000; Bhide and Shekdar 1998; Sudhire, Muraleedharan, and Srinivasan 1996; Jalan and Srivastava 1995; Chakrabarty, Srivastava, and Chakrabarti 1995). However, for the treatment of hospital wastes, small incinerators are used in many cities of India (Sharholy, Ahmad, Mahmood, *et al.* 2005; Lal 1996; Chakrabarty, Srivastava, and Chakrabarti 1995; Dayal 1994).

Gasification

The thermal breakdown of solid waste under controlled oxygen at temperatures usually greater than 800°C is known as gasification. The end product of this incomplete oxidation of waste is a syngas, which mainly contains CO, CH_4 and H_2, and this can be further used as a fuel or for power generation. In India, only a few gasifiers are currently in operation, and they are mainly used for burning biomass such as agro-waste, forest waste, and sawmill dust. Gasification can be used for treating MSW, but it is carried out only after pre-treatment steps like drying and following the removal of inert from waste. For proper functioning of a gasifier, about one-fourth of the syngas formed should be recycled back into the system, and the remaining can be used for power generation and other purposes (CPCB 2004; Ahsan 1999).

Refused-derived fuel plants

The RDF method is employed to generate a better solid fuel from waste. RDF can be defined as a product obtained from waste that can be used as a high-calorific substitute fuel in industries (Gendebien, Leavens, Godley, *et al.* 2003). RDF can be used in place of coal in many industrial processes, such as power production, steel manufacturing, and cement kilns. This can enhance economic growth and also reduce environmental pollution caused owing to the use of coal (Gendebien, Leavens, Godley, *et al.* 2003). The

RDF method can be used to treat both organic and inorganic solid waste. At present, RDF plants are successfully running in Hyderabad, Guntur, and Vijayawada in Andhra Pradesh and Deonar in Mumbai (Yelda and Kansal 2003; Reddy and Galab 1998). The RDF method is a technology with a great potential, which has the ability to reduce the load on both landfills and the environment.

Anaerobic digestion (biomethanation)

Under anaerobic (without oxygen) conditions, the organic part of solid waste is neutralized by anaerobes, which results in the production of methane and carbon dioxide as by-products. This process reduces the amount of organic waste and helps to produce energy through biogas generation. Biogas is made of about 60% methane and can be used for power production or can be directly used as a fuel. The residue left after the digestion of organic waste can be used as manure for plants. However, the non-organic part of solid waste cannot be digested by anaerobes. It can be treated only by means of incineration or gasification.

Anaerobic digestion takes place naturally in landfills, but compared to aerobic composting, it is a slow process. It has been estimated that under controlled conditions, anaerobic digestion produces two to four times as much amount of methane in three weeks that a landfill will produce in 6–7 years from 1 tonne of waste (Ahsan 1999; Khan 1994; Saxena, Adhikari, and Goyal 2009). The government is trying to explore the commercial value of biomethanation process to generate energy by using solid wastes. In various cities of India, such as Delhi, Bengaluru, and Lucknow,

Fig. 17.2 *Non-eco-friendly techniques of solid waste management used in India*
Source *Annepu (2012)*

biomethanation technology has been used to generate a secondary source of energy by using industrial, municipal, and agricultural waste.

17.3.2   Eco-friendly Techniques

Aerobic composting

The bioconversion of organic matter present in solid waste in the presence of air by heterotrophic microorganisms in a warm and moist environment is called composting, and the end product that results from this process is called compost. Compost is used as an organic fertilizer, and it has all the nutrients required by plants for their growth. The compost formed by this method is also free from odour and pathogens (Ahsan 1999; Khan 1994). The process occurs naturally in the environment if moisture and the feed material exist for the growth of organic waste degrading microorganisms. By controlling all these factors, the rate of composting can be optimized. For waste discharged from any place, that is, rural, urban, or city, composting is the best technique if sufficient space is available because it reduces the load of solid waste on municipalities by about 50%–85% and also solves the problem of waste collection and transport. On the one hand, it reduces the load on landfills; on the other hand, it provides a valuable fertilizer for agriculture. Composting was used as a waste management technique for many years till 1980; however, compost formed from solid waste through this process was discontinued to be used as fertilizer owing to various reasons, and now only 9% of total waste is treated by this method (Gupta, Jha, Koul, *et al.* 2007; Sharholy, Ahmad, Mahmood, *et al.* 2006; Srivastava, Kushreshtha, Mohanty, *et al.* 2005; Malviya, Chaudhary, and Buddhi 2002; CPCB 2000).

Vermicomposting

Vermicomposting is the process of neutralizing the organic part of solid waste by the combined efforts of earthworms and aerobic microorganisms. Earthworms are also known as "ecosystem engineers" because they create favourable conditions for microbes and microorganisms to flourish. First, degradable organic waste is decomposed by extracellular enzymes secreted by microbes, and then earthworms ingest this partially degraded organic matter, which is further digested and broken down in its gut. Earthworms consume organic waste up to five times their body weight daily. The excreta or cast of earthworms is the end product of the organic matter decomposition, which is odourless and used as an agricultural fertilizer to enhance plant growth (Ghosh 2004; Bezboruah and Bhargava 2003; Jha, Sondhi, and Pansare 2003; Sannigrahi and Chakrabortty 2002). This process reduces the time required for composting.

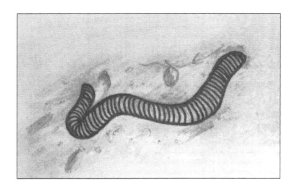

Fig. 17.3   *Earthworm*

17.4   EARTHWORMS

17.4.1   Anatomy of Earthworm

An earthworm has a long, rounded, and segmented body with a pointed head (Figure 17.3). Some species have a slightly flattened posterior, while in others the whole body is rounded. A fluid, known as coelomic fluid, fills the space between its body wall and alimentary canal, and provides shape to its body. The worm does not have a backbone; it can twist and turn with the help of the rings which surround its entire body. A broad band called clitellum is located in the upper part of its body. Owing to the lack of true legs, the earthworm moves with the help of setae present on its body. It breathes through its moist skin.

The alimentary canal of an earthworm is shown in Figure 17.4. The earthworm ingests food through the mouth, which passes through pharynx and oesophagus, and moves to a bag-like structure called crop. From the crop, the food passes to gizzard, where it is grinded with the help of ingested stones. After the grinding, digestion takes place in the intestine and the undigested food is eliminated in the form of cast from the body. This leftover cast is used as vermicompost.

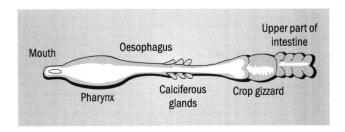

Fig. 17.4   *Alimentary canal of an earthworm (Lumbricus)*

Earthworms are hermaphroditic organisms. It means they possess both male and female sex organs, though they need another partner for mating. Clitellum secretes mucus after mating. This mucus slides over the worm's body and forms a sheath around sperms from one partner and eggs from the other. The sperms and eggs encased in the mucus fall from the worm's body and form a lemon-shaped cocoon approximately 1/8 inch long, depending upon the species. Hatching takes place in approximately 20 days, and two or more baby worms come out of the cocoon. A baby worm attains maturity and reproductive capability within 7–8 weeks after hatching (Figure 17.5).

17.4.2   Classification of Earthworms

Kingdom: Animalia

Phylum: Annelida

Class: Oligochaeta

Order: Opisthopora

Family: Lumbricidae

Genus: A large number of genera

Species: A large number of species

The following are the three types of earthworms that are ecologically strategic:

1. **Epigeic (litter dweller):** Epigeic worms live 10–30 cm deep beneath the soil and feed on humic substances and mineral matter. They are oesophageal, detritivorous, live in litter, and also consume litter. They have small size and are uniformly pigmented.

2. **Endogeic (subsoil dweller):** They are microphagous, geophagous (epiendogeic or hypoendogeic), live in horizontal and branching burrows in organo-mineral layers, consume soil, small to large in size, and weakly pigmented.

3. **Anecic (topsoil dweller):** Anecic earthworms can move very deep into the soil, up to 60–90 cm, and form complicated burrows for their movements. External abiotic parameters and poor soil nutrients appear to be the controlling factors for the increase in

Fig. 17.5 *Mating earthworms and their cocoon*

the population of these earthworms. They are macrophagous, detritivorous, and live in deep vertical burrows, but cast on the surface. They emerge at night to draw down organic matter (plant residue), are large as adults (200–1100 mm), and have pigmented skin that is brown.

17.4.3 Properties Required for Vermicomposting

Litter dwellers or epigeic earthworms, such as *Eudrilus eugeniae*, *Eisenia fetida*, and *Perionyx excavates*, are the most suitable species for vermicomposting. The properties required for worms used in vermicomposting are as follows:

- Earthworms used for vermicomposting should consume high biomass and also be able to convert organic biomass at a high rate.
- They should be tolerant to changing environmental conditions.
- They should be able to feed on different types of organic feed materials.
- Population growth rate should be high in species chosen for vermicomposting.
- Process of vermicomposting is more feasible when a mixture of different species of earthworms is used rather than a single species.
- Species of earthworm selected for vermicomposting should be disease resistant.

17.4.4 Species of Earthworms Found in India

There are five types of earthworms that are widely distributed in India. They show a definite pattern of distribution, soil salinity, and temperature tolerance range, which determine their habitat.

1. *Lampoon mauritii*: They have greater polytrophic adaptation and tolerance to changing micro-habitats and are found all over India.
2. *Polypheretima elongata*: They are found at a depth of 4–30 cm in low grasslands. They require a rich supply of water, high salinity, and a high level of organic matter for survival.
3. *Perionyx excavatus*: They are found near cowsheds where compost heaps, and they are highly adaptable to any kind of habitat.
4. *Pontoscolex corethrurus*: They are found in subsurface soils which contain very low salinity. They are also found in garden soil of shaded regions rich in organic matter.

5. *Dichogaster bolaui*: They belong to the class Oligochaeta and are a part of the major macrofauna of soil. *Dichogaster bolaui* is small in length, ranging from 25 mm to 40 mm.

17.5 VERMICOMPOSTING PROCESS

The vermicomposting process can be carried out on a small scale (either in a kitchen bin or in boxes, containers, and pits) and on a large scale by use of machinery which can accommodate tonnes of organic materials on a continuous basis. Vermicomposting should be done in a shaded area and at a high level in order to avoid waterlogging during heavy rains. First, a layer of bedding is made of broken bricks or stones (approximately 3–4 cm), followed by another layer of coarse sand (3–4 cm). Then a moist layer of loamy soil up to 15 cm, inoculated with desired earthworms, is spread over the bed. Loamy soil is used to ensure proper drainage. Cattle dung is then spread over the loamy soil and is covered with a layer of hay up to about 10 cm. Now the whole set-up is moistened with adequate water. Water should be added very carefully because an inadequate quantity of water or an excess can be lethal for worms. After this preparation, organic waste is spread over the bed. Organic waste and water can be added twice a week or as and when required. After applying the waste a few times,

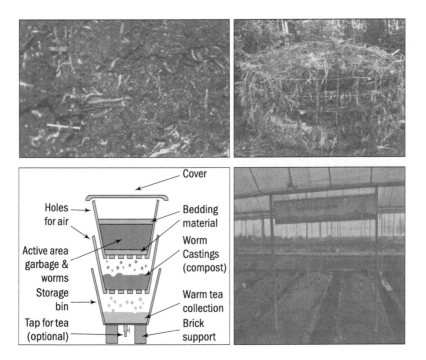

Fig. 17.6 *Vermicomposting process*

it should be turned upside down, ensuring the bedding material is not affected. The waste will change into a soft, spongy, sweet smelling, dark brown compost after some days and will be ready for harvesting (Ismail 1997) (Figure 17.6).

17.5.1   Vermicomposting Materials

The basic things needed for the process of vermicomposting are listed below.

Bedding

Any material which provides a suitable habitat and is the food source to earthworms forms the bedding for the vermicomposting process. The bedding material should be high in carbon and made to mimic decaying dried leaves on the forest floor—the natural habitat of worms. The bedding should be moist (approximately 75% moisture) and loose to enable worms to breathe and facilitate aerobic decomposition of the food that is buried in it; however, it should not be fully saturated, otherwise the worms will die owing to insufficient oxygen. A wide variety of bedding materials can be used, including shredded newspaper, sawdust, hay, cardboard, coir, burlap coffee sacks, peat moss, pre-composted manure, and dried leaves. Newspaper and phone books printed on regular, non-glossy paper with non-toxic soy ink are safe as well and make for good bedding material, particularly when pooled with straw and hay which provide high moisture retention and decompose relatively quickly. Most vermicomposters avoid using glossy paper from newspapers and magazines, junk mail, and shredded paper from offices because these may contain toxins which could disrupt the system. Cat litter and pet or human waste should not be used because they may carry disease. Coated cardboard that contains wax or plastic, such as milk boxes, should not be used either.

Food source

During the process, a regular supply of food material is necessary for the survival of earthworms. A variety of organic materials, such as livestock excreta, vegetable and fruit leftovers, tea bags, garden leaves, and eggshells, are commonly used as worm feedstock. Ideally, worms and other composting organisms should have a carbon to nitrogen (C:N) ratio of approximately 30:1. As some waste is richer in carbon and others in nitrogen, the waste must be mixed to approximate the ideal ratio. Brown matter or wood products such as shredded papers are rich in carbon. Green matter, such as food scraps, has more nitrogen, which is related to the amount of protein in the waste. If the waste is mostly vegetable and fruit scraps, and does not regularly include animal products or high-protein vegetable foods like beans, the resulting vermicompost will be low in nitrogen. About 90%–95% of waste ingested by worms as feed is excreted out in the form of vermicast and only 5%–10% is utilized within their body (CAPART 1998).

Moisture

For proper growth and survival of earthworms and other microbes, the bin content should be kept adequately moist but not soaked. A straw covering may be needed in exposed sites to prevent the bin from drying out during hot summer weather. If the amount of available moisture falls too low, the earthworm will begin to lose its internal water content, and a series of biological events begin to occur which, if unchecked, will eventually result in the death of the earthworm. Too high moisture content may also create anaerobic conditions which may be lethal to earthworms. Enough water should be added to completely moisten the pile because a dry compost pile will not decompose efficiently, but overwatering should also be avoided.

Aeration

Earthworms need to breathe, just like other living creatures. In the vermicomposting set-up, the available oxygen is used up and replaced with carbon dioxide and other miscellaneous waste gases. This can create a deadly condition for worms as on the one side oxygen will be reduced and on the other side toxic waste gases such as ammonia will be formed. The decreasing amount of fresh oxygen can also result in an increase in heat, which will further result in a similar increase in the oxygen requirement of worm. However, the whole situation is easily rectified and requires very infrequent attention. Factors such as high levels of fatty/oily substances in the feedstock or excessive moisture combined with poor aeration may render anaerobic conditions in the vermicomposting system.

Temperature

Optimum vermicomposting occurs at a temperature between 20°C and 30°C. However, the tolerance limit of different species of earthworms varies from 10°C to as high as 48°C. All the metabolic activities, growth, and reproduction of earthworms are significantly influenced by temperature, and the worms can bear cold and moist environments far better than a hot and dry environment (Slocum 2000). At 45–50°C temperature, adult worms may stop producing cocoons, and the growth rate of younger worms will diminish. When the temperature becomes unfavourable, the worms reorganize themselves within the bedding in order to have a favourable temperature for their growth.

pH

Earthworms are pH sensitive and generally show the optimum activity at the pH of 7.5–8.0, but they can also tolerate a pH value of 5–9 (Chan and Mead 2003). During the initial stage of decomposition, organic acids are produced, which decrease the pH of the bed. So, alkaline food materials should be used in vermicomposting which during initial stages will drop to neutral to slightly acidic and do not affect the whole process. However, if acidic food such as coffee grounds and peat mosses are used, they will cause a negative effect on vermicomposting. Acidic conditions in the bedding also attract pests. The pH can be increased by adding calcium carbonate. The end of composting is usually alkaline (pH 7.1–7.5).

Other important parameters

The following parameters are also important for the process of vermicomposting:

1. **Pre-composting of organic waste:** There are reports which suggest that pre-composting of waste is necessary because anaerobic conditions can develop in fresh organic waste, which then cause the death of earthworms even if all the other factors are appropriate for their proper growth.

2. **Salt content:** Earthworms do not prefer salt in food sources, and so if seafood is used as a food source in vermicomposting, it should be thoroughly washed to decrease its salt content. Certain types of manure also contain soluble salt, and if they are pre-composted before use, the salt will be leached out and not cause any problem.

3. **Urine content:** If the waste used as a feed source has a high urine content, it will be lethal to earthworms as it will produce toxic gases like ammonia in the bedding. This condition can easily be avoided if the waste is leached prior to use in the bedding.

4. **Other toxic components:** Various toxic components, such as detergent, industrial chemicals, pesticides, de-worming medicines, and tannins, can also be present in feed materials. Pre-composting of the feed material can reduce or sometimes even completely remove these toxic compounds, but pre-composting also decreases the nutrients present in the feed.

5. **Pests:** Several pests such as birds, moles, red mites, rat, and centipedes prey on earthworms, thus posing a threat to the vermicomposting system. The problem can be solved up to some extent by putting up some type of fence or cover on the bedding, such as a wire mesh, paving, or a layer of clay. One more pest

control solution is to pour water heavily on the bedding material. This draws the pests towards the surface where they can be easily identified and killed. Ants and mites are also an impediment to vermicomposting because they compete for the feed material with the worms. If the pH of the bedding is maintained at neutral to slightly alkaline, the problem of mites and ants can be checked.

17.6   VERMICOMPOST AND EARTHWORM HARVESTING

As earthworms grow very fast, their density becomes a problem in many vermicompost systems, especially the small ones. They should be harvested for proper functioning of the vermicompost system and can be used in another system as inoculant. Earthworm population density should always be estimated before the harvest. There are several methods which may be adopted for the harvesting of compost and earthworms (Figure 17.7).

17.6.1   Migration Method

This method is based on the ability of worms to move towards a more palatable source. The ripened old bedding is moved towards one side of the vermicompost bin and the new fresh bedding is kept in the old one's place. The feed material is then added into the new bedding. Slowly the worms will move from the old used bedding towards the fresh bedding. The old bedding is left as it is for about a month to allow the hatching of the new capsules of worms. After a month when all the worms have moved towards the new bedding, the compost is ready to be harvested.

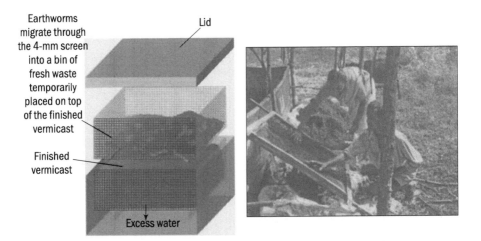

Earthworms migrate through the 4-mm screen into a bin of fresh waste temporarily placed on top of the finished vermicast

Lid

Finished vermicast

Excess water

Fig. 17.7 *Methods of harvesting earthworms and vermicompost*

17.6.2   Light Sensitivity Method

In this method, the old vermicomposting bin is emptied onto a plastic sheet or some other material, and is grouped into several small heaps, or a large heap is prepared according to the amount of the material. The heaps are then exposed to sunlight, or a bright light is placed on the top of the heap. As worms are sensitive to bright light, they will travel towards the bottom of the heap quickly, making the top of the heap free from worms. This worm-free compost is then scraped off after about 10–15 min of light exposure. This process is repeated several times till all the worms cluster themselves at the bottom of the pile. The worms at the bottom with very little bedding material are then shifted immediately into the new bedding.

17.6.3   Screening or Sieving Method

In the screening method, the bedding material to be separated is sieved through a coarse mesh. Castings fall down through the mesh and can be collected as compost, while the worms stay behind in the upper part of the mesh. The worms are then shifted into the new bedding with fresh, appetizing feed materials. This shifting is done very quickly and with moderate shaking to prevent the worms from escaping through the mesh wire. This technique is applied mainly in big worm farms, where large amounts of worms have to be collected rapidly.

17.6.4   Worm Swag Method

The worm swag used in this method has a small opening at the base from where the vermicast is squeezed out. This is a type of continuous flow system in which the feeding material is poured from the top, and the end product, that is, the vermicast, is collected from the bottom. After setting up the swag, the vermicastor vermicompost can be collected weekly. It is an easy and clean method of harvesting as there is no need to sort out worms or the decaying food material.

17.6.5   Vertical Separation

In the vertical separation method, three trays are placed one above the other. When the lowest tray is fully occupied by worms, a new tray with the feed material is kept on top of the exhausted tray. In search of food, the worms move upwards to the new tray and the used tray below is removed after sometime when all the worms have migrated upwards.

17.6.6   Manual Method

This technique is employed when just a few days old after the bedding is inoculated with worms, is needed to be harvested. The vermicast is simply

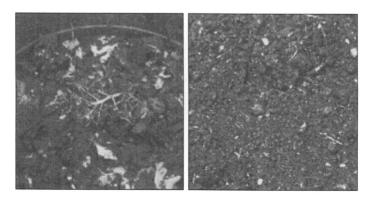

Fig. 17.8 *Vermicompost*

collected by hand and moved into a container. Earthworms, if present in the vermicast, are pulled out individually by hand and placed back into the compost pile. There is basically no preparation or technique required for this method, but only patience, as it is a slow and less organized process. Soon after removal, the worms should be placed into a bin with fresh bedding and feeding material.

17.7 COMPOSITION OF VERMICOMPOST

Nutrients present in matured vermicompost are given in Table 17.4. The concentrations of nitrogen, calcium, magnesium, sodium, potassium, molybdenum, and phosphates are higher in soil that contains earthworms. Earthworms increase the production of nitrate in soil by promoting the bacterial activity and also through their own decay. The movement of earthworms through the soil loosens it up and makes it valuable to all gardeners and anyone else who wants to utilize the soil (Figure 17.8).

Table 17.4 Nutrient status of matured vermicompost

Parameter	Expected value (%)
pH	6.8–7.5
Organic carbon	25.4–27.5
Nitrogen	1.2–1.6
C/N ratio	15–18
Phosphorus	0.3–0.5
Potassium	0.6–0.7
Calcium	4.2–6.7
Magnesium	0.2–0.3
Sulphur	0.4–0.5

17.8   ADVANTAGES AND DISADVANTAGES OF VERMICOMPOST

17.8.1   Advantages

Vermicompost has several advantages over the conventionally produced compost:

(i) Vermicompost is used as an inoculant in the manufacture of traditional compost.

(ii) Earthworms can also be used as feed for animals and for destruction of pathogens.

(iii) It does not produce any odour.

(iv) Vermitechnology generates a source of income for organic farmers.

(v) Vermitechnology could be used as a pre-treatment process for MSW, prior to landfilling.

Vermicompost also has several advantages over chemical fertilizers:

(i) The vermicomposting process re-establishes several beneficial microorganisms such as nitrogen-fixing microbes, phosphate-solubilizing microbes, among others.

(ii) It provides macronutrients and micronutrients for the growth of plants.

(iii) It boosts the quality of crops by increasing the sugar content.

(iv) It also improves the physical characteristics of soil such as texture, porosity, and water retention.

(v) It prevents soil erosion by improving the structural stability of the soil.

(vi) It decreases the use of toxic pesticides in farms.

(vii) It is a cost-effective, pollution-free, and valuable end product.

(viii) It is responsible for very low-greenhouse gas emissions and is, thus, an environmentally sustainable process.

17.8.2   Disadvantages

Although vermicomposting is a useful and eco-friendly process, it also suffers from the following disadvantages:

(i) Vermicomposting requires various start-up resources, including earthworms, and more time and labour as well.

(ii) The process requires large space because earthworms are surface feeders and able to degrade waste only at a depth of 1 m.

(iii) The process can also be disrupted by changes in environmental conditions, such as temperature, rain, and drought.

17.9 APPLICATIONS OF VERMICOMPOSTING

17.9.1 Medical Application of Vermicomposting / Vermiculture Biotechnology

The medical value of earthworms was discovered as far back as 2600 B.C. and it has been used for a range of diseases, from pyorrhea to post-partial weakness, jaundice to increase in sperm count, and as excellent aphrodisiacs.

17.9.2 Earthworms and Pollution Control

As solid wastes are generated continuously, they are undesirable pollutants for the environment and a menace to the health of the community. The sources of toxic substances reaching the soil surface are mainly the solid wastes containing heavy metals released by industries and pesticides used in agriculture. The accumulation of toxic chemicals in earthworm's tissue is significant ecologically very significant because these animals are important components in the food chain of several species of birds and mammals. This is very important for the minimization of soil pollution. Using earthworms in vermicomposting also creates jobs for people and are, thus, economically valuable as well.

17.9.3 Role of Vermicompost in Sustainable Agriculture

In tropical countries, due to high temperatures and moisture, the use of inorganic fertilizers is not economical as there is only a two-fold increase in yield on a ten-fold increased use of inorganic fertilizers. Soil in tropical countries has low organic carbon, which is a very important factor for soil organisms that improve the physical properties of soil such as porosity, bulk density, aggregate stability, and water holding capacity. They also contribute towards immobilization, as well as solubilization and mobilization, of nutrients as and when required. Vermicompost, which can be generated with a small investment (Figure 17.9), can be used as an organic manure to enhance biological processes in the soil. Thus, vermicompost can improve soil health and also supply nutrients to crops.

17.10 CONCLUSION

In developing countries like India, there is a great need to find an alternative way for the sustainable management of solid waste. The conversion of organic

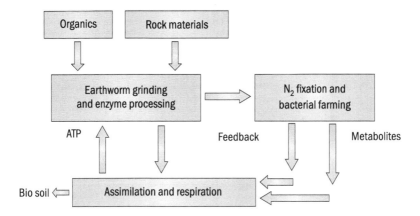

Fig. 17.9 *Vermiculture ecosystem*
Source *Singh (2004)*

wastes into useful products facilitated by earthworms is an ecologically safe and sustainable technology for the utilization of organic solid waste. The ill effects of using chemical fertilizers is usually overcome by the use of organic manure, green manure, composts, bioinoculants, biofertilizers, and so on. Besides providing nutrients to plants, the additional advantage associated with the use of vermicompost is that it acts as a soil conditioner and improves the quality of soil. The pathogenic microbial flora present in organic waste is also removed by earthworms. Vermicomposting process is also helpful in solving the domestic waste problem as it converts waste into harmless useful products which are non-toxic in nature.

With a tremendous increase in environmental pollutants like chemical fertilizers in most developing countries including India, vermicompost technology is indeed the key to overcome pollution load, maintaining the environment, and improving the economy. Vermicomposting is one of the fastest growing fields in the management of solid waste for recycling and recovery of resources. Vermicomposting is a green and viable technology for organic solid waste management that has the potential to generate income to the rural poor as well. There is a need to encourage developmental programmes and research for vermicomposting and solid waste management by earthworms, and raise civic awareness about its application at a commercial level.

REFERENCES

Aalok, A., A. K. Tripathi, and P. Soni. 2008. Vermicomposting: A better option for organic solid waste management. *Journal of Human Ecology* 24(1): 59–64

Ahsan, N. 1999. Solid waste management plan for Indian megacities. *Indian Journal of Environmental Protection* 19(2): 90–95

Annepu, R. K. 2012. Sustainable Solid Waste Management in India. [Masters thesis submitted to the Department of Earth and Environmental Engineering, Columbia University, New York]

Bezboruah, A. N. and D. S. Bhargava. 2003. Vermicomposting of municipal solid waste from a campus. *Indian Journal of Environmental Protection* 23(10): 1120–1136

Bhide, A. D. and A. V. Shekdar. 1998. Solid waste management in Indian urban centers. *International Solid Waste Association Times* 1: 26–28

CAPART. 1998. Vermicompost. Council for Advancement of People's Action and Rural Technology, Centre for Technology Development, New Delhi

Central Pollution Control Board (CPCB). 2000. *Status of Solid Waste Generation, Collection, Treatment and Disposal in Metrocities*, Series CUPS/46/1999–2000. New Delhi

Central Pollution Control Board (CPCB). 2004. Management of Municipal Solid Waste. New Delhi: Ministry of Environment and Forests

Central Pollution Control Board (CPCB). 2012. Status Report on Solid Waste Management. New Delhi: Ministry of Environment and Forests

Chakrabarty, P., V. K. Srivastava, and S. N. Chakrabarti. 1995. Solid waste disposal and the environment—a review. *Indian Journal of Environmental Protection* 15(1): 39–43

Chan, K. Y. and J. A. Mead. 2003. Soil acidity limits colonization by *Aporrectodeatrapezoides*, an exotic earthworm. *Pedobiologia* 47: 225–229

Dayal, G. 1994. Solid wastes: Sources, implications and management. *Indian Journal of Environmental Protection* 14(9): 669–677

Gendebien, A., K. Leavens, A. Godley, K. Lewin, K. J. Whiting, R. Davis, J. Giegrich, H. Fehrenbach, U. Gromke, N. del Bufalo, and D. Hogg. 2003. Refuse-derived fuel, current practice and perspectives. European Commission, Directorate of General Environment, Brussels

Ghosh, C. 2004. Integrated vermi-pisciculture—an alternative option for recycling of municipal solid waste in rural India. *Bioresource Technology* 93(1): 71–75

Gupta, P. K., A. K. Jha, S. Koul, P. Sharma, V. Pradhan, V. Gupta, C. Sharma, and N. Singh. 2007. Methane and nitrous oxide emission from bovine manure management practices in India. *Environmental Pollution* 146(1): 219–224

Gupta, S., M. Krishna, R. K. Prasad, S. Gupta, and A. Kansal. 1998. Solid waste management in India: Options and opportunities. *Resource, Conservation and Recycling* 24: 137–154

Ismail, S. A. 1997. *Vermicology: The Biology of Earthworms*. New Delhi: Orient Longman

Jalan, R. K. and V. K. Srivastava. 1995. Incineration, land pollution control alternative— design considerations and its relevance for India. *Indian Journal of Environmental Protection* 15(12): 909–913

Jha, M. K., O. A. K. Sondhi, and M. Pansare. 2003. Solid waste management: A case study. *Indian Journal of Environmental Protection* 23(10): 1153–1160

Joardar, S. D. 2000. Urban residential solid waste management in India: Issues related to institutional arrangements. *Public Works Management and Policy* 4(4): 319–330

Kansal, A. 2002. Solid waste management strategies for India. *Indian Journal of Environmental Protection* 22(4): 444–448

Khan, R. R. 1994. Environmental management of municipal solid wastes. *Indian Journal of Environmental Protection* 14(1): 26–30

Kothari, R., V. Kumar, N. Panwar, and V. Tyagi. 2014. Municipal solid-waste management strategies for renewable energy options. *Sustainable Bioenergy Production*, pp. 263–282

Kumari, S. 2013. Solid waste management by vermicomposting. *International Journal of Scientific and Engineering Research* 4(2): 1-5.

Lal, A. K. 1996. Environmental status of Delhi. *Indian Journal of Environmental Protection* 16(1): 1–11

Loehr, R. C., J. H. Martin, and E. F. Neuhauser. 1998. Stabilization of liquid municipal sludge using earthworms. In *Earthworms in Waste and Environmental Management*, C. A. Edwards and E. F. Neuhauser (eds). The Netherlands: SPB Academic Publishing

Malviya, R., R. Chaudhary, and D. Buddhi. 2002. Study on solid waste assessment and management: Indore city. *Indian Journal of Environmental Protection* 22(8): 841–846

Peavey, H. S., R. R. Donald, and G. Gorge. 1985. *Environmental Engineering.* Singapore: McGraw-Hill

Reddy, S. and S. Galab. 1998. An Integrated Economic and Environmental Assessment of Solid Waste Management in India—the Case of Hyderabad, India.

Ronald, E. G. and E. D. Donald. 1977. *Earthworms for Ecology and Profit, Vol. 1.Scientific Earthworm Farming.* California: Bookworm Publishing Company

Sannigrahi, A. K. and S. Chakrabortty. 2002. Beneficial management of organic waste by vermicomposting. *Indian Journal of Environmental Protection* 22(4): 405–408

Saxena, R. C., D. K. Adhikari, and H. B. Goyal. 2009. Biomass-based energy fuel through biochemical routes: A review. *Renewable and Sustainable Energy Reviews* 13: 167–178

Shannigrahi, A. S., N. Chatterjee, and M. S. Olaniya. 1997. Physico-chemical characteristics of municipal solid wastes in mega city. *Indian Journal of Environmental Protection* 17(7): 527–529

Sharholy, M., K. Ahmad, G. Mahmood, and R. C. Trivedi. 2006. Development of prediction models for municipal solid waste generation for Delhi city. In *Proceedings of National Conference of Advanced in Mechanical Engineering (AIME-2006)*, Jamia Millia Islamia, New Delhi, pp. 1176–1186

Sharholy, M., K. Ahmad, G. Mahmood, and R. C. Trivedi. 2005. Analysis of municipal solid waste management systems in Delhi—a review. In *Book of Proceedings for the Second International Congress of Chemistry and Environment*, Indore, India, pp. 773–777

Sharholy, M., K. Ahmad, G. Mahmood, and R. C. Trivedi. 2008. Municipal solid waste management in Indian cities—a review. *Waste Management* 28: 459–467

Sharma, S. and K. W. Shah. 2005. Generation and disposal of solid waste in Hoshangabad. In *Book of Proceedings of the Second International Congress of Chemistry and Environment*, Indore, India, pp. 749–751

Shekdar, A. V., K. N. Krishnaswamy, V. G. Tikekar, and A. D. Bhide. 1992. Indian urban solid waste management system—jaded systems in need of resource augmentation. *Waste Management* 12(4): 379–387

Singh, D. P. 2004. Vermiculture biotechnology and biocomposting. In *Environmental Microbiology and Biotechnology*, D. P. Singh and S. K. Dwivedi (eds), pp. 97–106. New Delhi: New Age International Ltd

Slocum, K. 2000. Maintaining the Flow in Continuous Flow Systems. Details available at http://www.wormdigest.org/articles/index.cgi

Srivastava, P. K., K. Kushreshtha, C. S. Mohanty, P. Pushpangadan, and A. Singh. 2005. Stakeholder-based SWOT analysis for successful municipal solid waste management in Lucknow, India. *Waste Management* 25(5): 531–537

Sudhire, V., V. R. Muraleedharan, and G. Srinivasan. 1996. Integrated solid waste management in urban India: A critical operational research framework. *Socio-Economic Planning Science* 30(3): 163–181

UNEP (United Nations Environment Programme). 2001. State of the Environment: South Asia. United Nations Environment Programme

V. K. Garg, Y. K. Yadav, and A. Sheoran. 2006. Livestock excreta management through vermicomposting using an epigeic earthworm *Eiseniafoetida*. *The Environmentalist* 26: 269–276

World Bank.1999. What a waste: Solid waste management in Asia. Urban Development Sector Unit, East Asia and Pacific Region. Details available at http://www.worldbank.org/html/fpd/urban/publicat/whatawaste.pdf

Yelda, S. and S. Kansal. 2003. Economic insight into MSWM in Mumbai: A critical analysis. *International Journal of Environmental Pollution* 19(5): 516–527

18

Biogas Potential from Sewage Treatment Plant and Palm Oil Mill Effluent for Electricity Generation in Malaysia

Kumaran Palanisamy[*], Hephzibah David, and
Sivasankari Ranganathan

Centre for Renewable Energy, Universiti Tenaga Nasional, Malaysia
*E-mail: Kumaran@uniten.edu.my

18.1 INTRODUCTION

Malaysia, a country with a population of 30 million, is a peninsula located in Southeast Asia and includes a part of Borneo. During British colonialism, the country relied heavily on tin mining and agriculture, especially rubber plantation. However, it transitioned to an industrial-based economy after independence and has witnessed a rapid economic growth, specifically in the manufacturing industry. According to Jomo and Edwards (1993), the total gross domestic product (GDP) increased from 10% in the late 1950s to 26% 30 years later. Meanwhile, the population

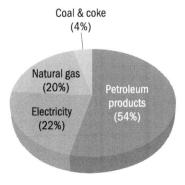

Fig. 18.1 *Energy demand in Malaysia*
Source *Economic Planning Unit (2013)*

and GDP grew by 48.67% and 92.47%, respectively, from 1984 to 2014 (Energy Commission 2014; Economic Planning Unit 2013).

According to Ang (2008), economic development is closely associated with energy consumption. The higher the population with more income, the higher the production and consumption of energy. Currently, electricity has the second highest final energy demand in Malaysia after petroleum products, which are approximately 22% and 54%, respectively (Figure 18.1).

Indeed, Malaysia's rapid economy progression has brought an increasing demand for energy, with the electricity demand in industrial sector the highest in Peninsular Malaysia, as shown in Figure 18.2. However, it is expected to drop by 6% from 2011 to 2020 owing to the growth of the commercial sector.

Correspondingly, the economic growth will directly impact the carbon dioxide (CO_2) emission as illustrated in Figure 18.3. The CO_2 emissions from 1990 to 2012 in Malaysia increased with respect to the GDP. The GDP increased by 65.6% from 1992 to 2012, while CO_2 emissions increased by 68.2% within the same period (Energy Commission 2013; The World Bank 2014). According to the reports of National Economic Advisory Council (NEAC) (2010) and Suruhanjaya Tenaga (Energy Commission) (2013), the Malaysian GDP is predicted to grow at the rate of 4.5%–5.8% during 2013–15, and the projected growth for 2016–20 is estimated to be 5.9%. A study on the Malaysian situation states that if the CO_2 emission is not reduced in Malaysia, 285.73 Mtoe of CO_2 will be emitted to the atmosphere by 2020 (Safaai, Noor, Hashim, *et al.* 2011).

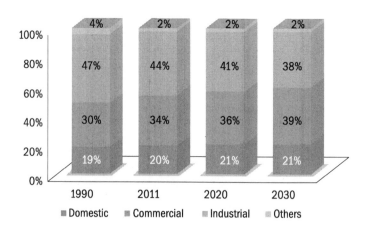

Fig. 18.2   *Sectoral electricity demand for Peninsular Malaysia*
Source *Energy Commission (2013)*

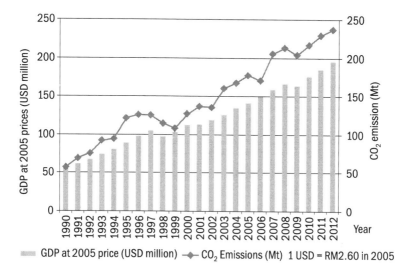

Fig. 18.3 *GDP and CO$_2$ emission in Malaysia during 1990–2012*
Sources *Energy Commission (2013); The World Bank (2014)*

According to a recent study, 74.2 Mtoe out of 76.3 Mtoe of primary energy consumed in Malaysia in 2012 was supplied by fossil fuels (British Petroleum 2013). However, fossil fuels are non-renewable resources and their reserves are depleting. A recent report states that Malaysia's primary energy resource, the petroleum reserves, will deplete by 2035, which will surge the fossil fuel prices (Rahim and Liwan 2012). Thus, efforts are being made to explore new energy resources that are renewable, readily available, and eco-friendly to minimize the environmental impact and supplement the depleting non-renewable fossil fuel reserves.

18.2 RENEWABLE ENERGY STATUS IN MALAYSIA AND MALAYSIAN RENEWABLE ENERGY TARGET

The Malaysian Government has adopted various strategies to achieve a 40% reduction in terms of emission intensity of GDP by the year 2020 compared to 2005 level, which includes a voluntary commitment by the honourable Prime Minister Datuk Seri Najib Tun Razak at the Copenhagen Summit in 2009 to reduce the dependency on fossil fuels. Indeed, renewable energy was introduced as the fifth fuel in the Five Fuel Strategy in the energy supply mix in addition to the existing fuels such as oil, gas, coal, and hydro in the Eighth Malaysia Plan in 1999. In the Tenth Malaysia Plan (2005–10), the National Renewable Energy Policy and Action Plan was introduced to increase the use of renewable energy and integrate it to the national electricity supply for a sustainable socio-economic development (Hashim and Ho 2011).

Apart from these measures, various incentives have been introduced by the government to promote the adoption of renewable resources to meet Malaysia's ambitious goal by 2015, as illustrated in Figure 18.4, to tap the available potential. On the basis of the National Renewable Energy Policy and Action Plan, it is anticipated that renewable energy will contribute 5.5% of Malaysia's total electricity generated by 2015, an increase from less than 1% in 2009. Among the potential capacities is biogas, which accounts for approximately 100 MW.

18.3   FEED-IN TARIFF IN MALAYSIA

The Government of Malaysia has introduced several financing and funding schemes as renewable energy requires high investment. The Feed-in Tariff (FiT) is Malaysia's new mechanism under the renewable energy Policy and Action Plan to stimulate the generation of renewable energy, which is up to 30 MW. FiT system was introduced by Sustainable Energy Development Authority (SEDA) of Malaysia in December 2011 under the Renewable Energy Act 2010 to encourage individuals and companies to invest in renewable energy projects by offering competitive buy-back rates. In comparison to other renewable energy policies, the FiT mechanism has been adopted by a large number of countries. Considering the advantages of FiT, at least 50 countries and 25 states/provinces adopted the FiT policy by early 2010 (Table 18.1).

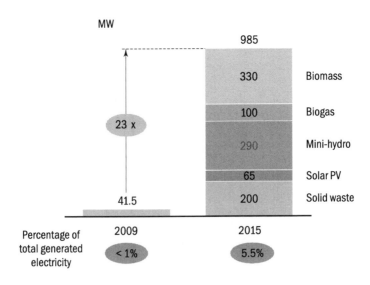

Fig. 18.4 *Planned increase in renewable energy capacity*
Source *Ministry of Energy, Green Technology and Water (2010)*

The main objective of FiT system is to increase the renewable energy contributions in the national energy mix, thereby facilitating the renewable energy industry growth by ensuring reasonable renewable energy generation costs so that the environment is conserved for the future generation. This mechanism allows electricity produced from indigenous renewable energy resources to be sold to power utilities at a fixed premium price. The system also obliges distribution licensees (DLs) to buy from Feed-in Approval Holders (FiAHs) the electricity produced from renewable energy and sets the FiT rate. The DLs will pay for RE supplied to the electricity grid for a specific period (Biogas-Renewable-Energy.Info. 2014). Hence, by guaranteeing access to the grid and setting a favourable price per unit of renewable electricity, the FiT mechanism would ensure that renewable energy becomes a viable and long-term investment for companies, industries, and individuals and at the same time enables to reduce the CO_2 emission. This provides conducive and secure investment conditions that will instil confidence in financial institutions to provide loans with a longer payback period (more than 15 years). As of April 2014, the installed capacity of electricity generated from biogas was 11.2 MW (Yaacob 2014).

However, as of 31 May 2014, a renewable energy capacity of only 73.99 MW was approved for biogas renewable energy resources. Out of the total capacity of 752.17 MW from other renewable resources from 2012 to 2016, only 10% of the total approved capacity is from biogas (Figure 18.5). Table 18.2 summarizes the released and remaining capacity quota for biogas from 2014 to 2017. Nevertheless, it can be observed that the utilization of biogas is gradually increasing and almost 62% of the available quota till 2016 for biogas has been taken up (Chen 2014).

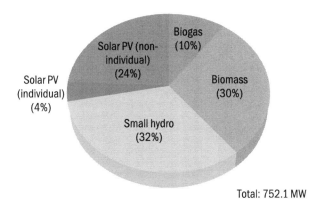

Total: 752.1 MW

Fig. 18.5 *Approved capacities of renewable energy (2012–16)*
Source *Chen (2014)*

Table 18.1   Advantages of FiT

Economic	Creates green jobs
	Creates foreign direct investment (FDI) and domestic direct investment (DDI) for manufacturing and export
	Creates stable conditions for market growth
	Provides hedge against the conventional fuel price volatility
Environmental	Reduces CO_2 emission and pollution
	Reduces fossil fuel dependencies and encourages energy efficiency measures
Political	Demonstrates the country's commitment to renewable energy deployment and creates routes for achieving renewable energy and emission reduction targets
	Promotes a more decentralized and democratized form of electricity organization
Social	More equitable wealth distribution and empowerment of citizens and communities
	Encourages community engagement in environment and climate protection activities
	Establishes renewable energy as a familiar component of the cityscape and landscape

Source Mendonca, Jacobs, and Sovacool (2010)

18.4   MALAYSIAN PALM OIL INDUSTRY

Oil palm history can be traced back to the days of Egyptian pharaohs (5000 B.C.). At the start of the 20th century, palm oil was introduced to Malaysia and has been commercially produced since 1917. Palm oil milling is basically the processing of oil palm fresh fruit bunches (FFBs) into crude palm oil (CPO) and palm kernel (Figure 18.6).

The unique composition of palm oil makes it versatile in its application in food manufacturing, chemical, cosmetic, and pharmaceutical industries. Its semi-solid physical properties are often sought in the food preparation industry. The low cholesterol quality and easy digestibility of the palm oil

Table 18.2   Released and remaining capacity quota for biogas (2014–17)

Year	Released capacity (MW)	Remaining capacity (MW)
2014	10	0.86
2015	15	1.79
2016	15	3.17
2017	15	15.00
Total	55	20.82

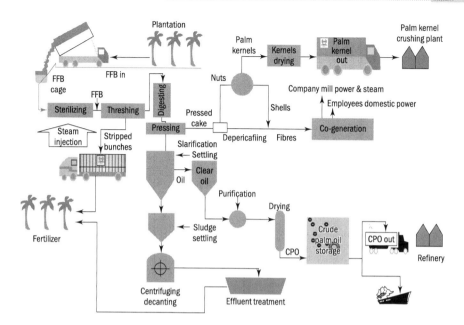

Fig. 18.6 *CPO milling process*
Source *Oil Refinery Plant (2014)*

make it popular as the source of energy, while its economic and technical superiority makes it preferable as a base material in the manufacturing of various non-edible products.

Palm oil has been recognized as the most traded oil in the world with its total export in 2011 reaching almost 40 million tonnes, of which Malaysia's share was 46%. The oil palm industry is a significant contributor to the Malaysian economy. The total planted area of oil palm in Malaysia for 2013 reached 5.23 million hectares, which is an increase of 6.12% compared to 2011 which recorded 5.00 million hectares in order to sustain the world's demand for CPO (Malaysian Palm Oil Board 2014). Figure 18.7 depicts the oil palm estates location around Malaysia.

Fig. 18.7 *Oil palm estates locations around Malaysia*

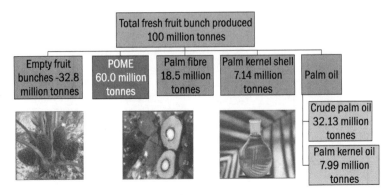

Fig. 18.8 *Oil palm biomass output (2012)*

As shown in Figure 18.8, out of the 100 million tonnes of FFB processed in 2012, production of CPO amounted to 32.13 million tonnes and 7.99 million tonnes of palm kernel oil was produced. In 2012, Malaysia's export of palm oil rose to 19.3 million tonnes and a similar trend was observed in the case of palm kernel oil, with the export volume reaching 1.08 million tonnes (Malaysian Palm Oil Board 2014).

Malaysia employs extensive and efficient refining and fractionation facilities to add economic value to CPO. This proved to be extremely successful during 1974–1999 as the export of processed palm oil grew drastically from 0.9 million tonnes to 8.9 million tonnes. In Malaysia, oil palm plantations make up 77% of the agricultural land or about 15% of the total land area (Sime Darby Plantation 2014). One of the key economic drivers of the agricultural sector in the developing countries such as Malaysia and Indonesia is the palm oil industry. Its economic prowess is greatest in the oil palm growing belt. In Malaysia, the industry provides direct employment to around 570,000 people (Malaysian Palm Oil Council 2014). The industry also offers a long term and stable source of income for its smallholders.

The process to extract CPO and palm kernel from the FFB results in the production of considerable amounts of by-products, such as fibre, shell, empty fruit bunch (EFB), and POME (palm oil mill effluent). The current uses of the by-products are given in the following table.

Empty fruit bunch	Palm oil mill effluent	Palm fibre
• Soil mulching, reduce input of inorganic nutrients, and increase soil organic content • Material for wood-based products (particle and fibreboards)	• Biologically treated and discharged • Compost, EFB+POME • No commercial applications as of now	• Raw material for pulp and paper manufacture • Animal feed supplement for cattle, goats, and sheep • Rich source of organic fertilizer for farming

However, these residues are the main energy resources that can be processed and converted to useful forms of energy, such as electricity, steam, and heat.

Carbon from palm biomass	Renewable materials	Environmental
• Palm biomass: EFB, shell, and fibre • Polymer and composite materials	• Palm biomass: EFB and POME • Oleochemical industries • High valued commodities	• Certified emission reduction through methane mitigation • Clean development mechanism

18.5 BIOGAS POTENTIAL IN PALM OIL MILLS

The extraction of CPO from the FFB generates large amounts of waste, such as oil palm trunks (OPT), oil palm fronds (OPF), EFB, and POME. POME is a thick brownish liquid and approximately 28% of weight of FFB (Sulaiman, Abdullah, Gerhauser, *et al.* 2011). It contains a mixture of water, oil, high chemical oxygen demand (COD), biochemical oxygen demand (BOD), total solids, and suspended solids. The characteristics of raw POME do not meet the Malaysian environmental standard requirements for discharge of effluent, and thus it has to be treated in series of open ponds for approximately 210 days prior to the discharge to waterbodies. POME is discharged during the last stage of the palm oil production in the mill. The common treatment system for POME in the palm oil mills are anaerobic and aerobic ponds. Thus, owing to high cost of treatment systems, the potential for the biodegradation remains untapped and during the degradation of POME, odour is released into the air which reduces the quality of surrounding areas. The important characteristics of POME for anaerobic digestion are given in Table 18.3.

Figure 18.9 shows the potential of biogas in a typical 45 t/h palm oil mill. Around 26.16 MWh/day electricity can be generated from 450 m³/day POME through anaerobic digestion and converted with a 40% efficiency gas engine. The EFB, kernel shell, and palm fibre can be processed through a combustor and the resulting heat is converted using steam

Table 18.3 Typical Malaysian POME characteristics

Characteristics	Palm oil mill effluent
Volatile solids/total solids (%)	0.70–0.85
Chemical oxygen demand (mg/L)	44,000–100,000
Carbon:nitrogen ratio	10:50
pH	4–5

Source Yacob, Hassan, Shirai, *et al.* (2005)

turbine generator at 30% efficiency to produce 240 MWh/day electricity. This total of 266.26 MWh/day electricity can be exported to grid. The remaining mesocarp fibre is then burned in a boiler and steam turbine is used to efficiently convert heat into electricity, and this resultant electricity can be used to meet the electricity requirement of the plant. Thus, in a typical 45 t/h FFB processing plant, the potential biogas revenue that can be generated is USD 733,347.88/year at USD 0.09/kWh for 300 days based on the FiT rates.

18.6   MALAYSIAN SEWERAGE TREATMENT

The Malaysia's sewerage industry has made advancement in the past half-century. Figure 18.10 summarizes the evolution of Malaysia's sewerage industry. Prior to the 1950s, the sewage industry mainly employed primitive methods such as pit and bucket latrines, overhanging latrines, and direct discharge to rivers and seas. In the early 1950s, pour flush latrines and septic tanks, which only happened to be the primary treatment, were introduced owing to the need of proper sanitation and better hygiene practices as a result of urbanization and industrialization. The sewerage system developed in the 1970s to oxidation pond system, in which priority was given to the biological treatment process. The fully mechanized plant was introduced in the late 1980s when biological filters and activated sludge systems were established. The mechanized plant was further upgraded in the late 2000s with the introduction of anaerobic digesters as pilot plants to tap the unexploited potential of energy from sewage sludge (SS). Prior to this, the main objective of the sewerage operator was to treat wastewater to produce a high-quality effluent that was within the stipulated DOE standard and could be safely discharged to the environment (Hamid and Mohd Baki 2005).

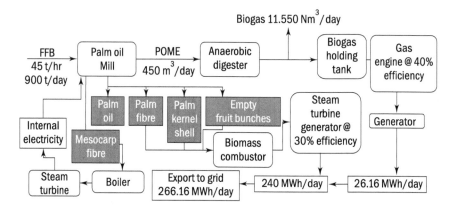

Fig. 18.9 *Biogas potential in a typical 45 t/h palm oil mill*

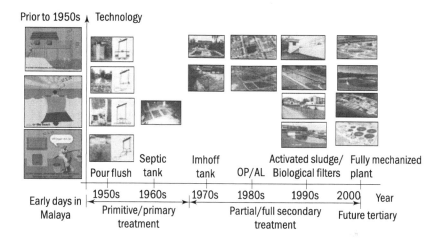

Fig. 18.10 *Evolution of Malaysia's sewerage industry*

Source *Wan Abdullah (2011)*

18.7   TYPICAL SEWAGE TREATMENT PLANT WITH ANAEROBIC DIGESTER SCENARIO IN MALAYSIA

The waste streams from primary and commercial sources are streamed to Sewage Treatment Plants (STPs) through the drainage system for processing. Figure 18.11 shows the process of a typical Malaysian STP. The raw wastewater is pumped to the STP, where the processing starts at coarse screens and fine screens. Here, large and small debris particles are removed to minimize the damage to the pump and equipment of the plant. Then, the raw wastewater is flowed into the grit chamber to eliminate grits and sands. Later, the wastewater is streamed to the primary clarifier to reduce the solid content of the wastewater through sedimentation. The settled solids, also known as primary sludge, are scrapped and then pumped into the gravity thickener, where solid processing takes place. Meanwhile, the clarified liquid which contains dissolved materials is flowed into the aeration tank. In the aeration tank, bacteria break down the organic materials in the liquid and clump together to form a microbial floc, which is known as activated sludge. Then, the activated sludge is pumped into the secondary clarifier, where it is allowed to settle. Later, the activated sludge will be pumped to the mechanical thickener, where flocculants will be added to concentrate the sludge to the desired concentration. The clarified liquid from the secondary clarifier will be then flowed to the measuring tank and the microorganisms present disinfected. It is ensured that the effluent is as per the stipulated standards before being discharged to rivers.

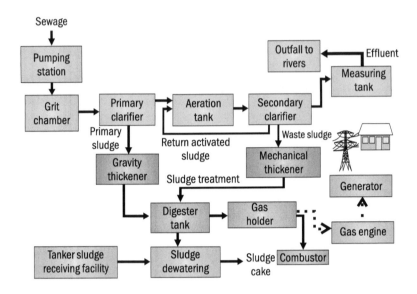

Fig. 18.11 *Process of a typical Malaysian STP*

Meanwhile, the thickened sludge from the gravity thickener and the mechanical thickener, which is rich in organic solids, are pumped into an anaerobic digester. In most of the anaerobic digesters in Malaysia, the temperature is maintained around 32–37°C and the contents are mixed continuously to enhance the rate of digestion. Through anaerobic digestion of the sludge, the organic solids are degraded and methane gas is produced. The generated methane gas can be utilized for electricity generation by employing gas engine generator.

Figure 18.12 shows the average electricity consumption and average influent flow rate of a typical Malaysian STP. It can be observed from the figure that the pattern of electricity consumption is influenced by the pattern of influent flow rate into the STP. As the influent flow rate increases, the electricity consumed also increases owing to the amount of wastewater needed to be treated by the plant. The average influent flow rate for a STP that treats wastewater from a population of approximately 250,000 people is approximately 54,554.86 m³/day. Meanwhile, the average daily electricity consumption is about 43,508.67 kWh. Therefore, about 0.50 kWh to 0.85 kWh of electricity is consumed to treat 1 m³ of wastewater (Cao 2011).

The average biogas production and the average energy potential of the biogas generated are illustrated in Figure 18.13. On the basis of the data, the daily average biogas production is around 1102 m³/day, which has the potential to generate 2930 kWh/day. The energy potential was calculated

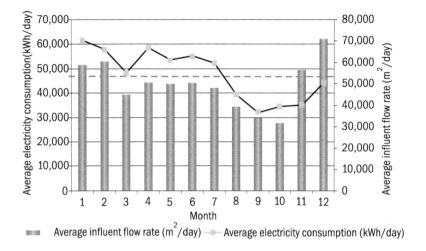

Fig. 18.12 *Average electricity consumption and average influent flow rate of a typical Malaysian STP*

based on the energy content of biogas, which is about 22,700 kJ/m^3 and at 60% methane content, while the gas engine efficiency is 40%. The biogas production pattern is not influenced by the influent flow rate. However, it is affected by the characteristics of the wastewater and the operational process of the STP. Therefore, attention should be given to the operational process of a STP as it affects the characteristics of the feed sludge into the anaerobic digesters that directly affects the biogas production from the anaerobic digestion.

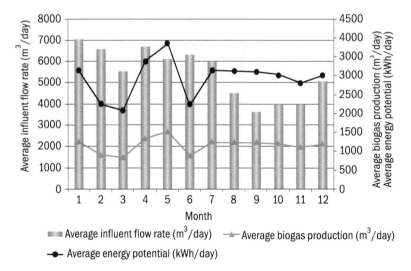

Fig. 18.13 *Average biogas production and average energy potential of a typical Malaysian STP*

Table 18.4 Typical Malaysian SS characteristics

Characteristics	Sewage sludge
Volatile solids/total solids (%)	0.70–0.80
Chemical oxygen demand (mg/L)	10,800–22,800
Carbon:nitrogen ratio	2.07:10
pH	6–7

18.8   BIOGAS POTENTIAL AT STP

The characteristics of SS prior to anaerobic digestion are crucial for biogas production. The important characteristics of SS for anaerobic digestion are given in Table 18.4.

Although biogas production can be utilized for various purposes, its most common use in Malaysia is for the generation of electricity. The electricity generated from biogas can be used within the plant or sold to the utilities through the FiT system. However, the current process in the STP can be further optimized and improved. Table 18.5 summarizes the simulation of electricity generation from the average biogas generated via anaerobic digestion of SS before and after process optimization and improvement. At the current fixed operational process of STP, 1102 m^3 of biogas can be generated daily, which has the potential to produce 2780 kWh of electricity per day using gas engine with the efficiency of 40%. Electricity generated from biogas can account for 6.4% of the total electricity consumption of the STP and reduce the CO_2 emission of the electricity generated by fossil fuels by 1865 kg CO_2/day. Besides, if the STP operator plans to sell the generated electricity to the utilities through FiT, the revenue of USD 119,262 can be generated per year.

The existing operational processes can be optimized and improvised according to the characteristics of the influent flowing into the plant daily. Upon further refining the anaerobic sludge and anaerobic digestion processes, it is expected that 2836 m^3/day of biogas can be generated from the same amount of influent flowing into the STP. The biogas generated has the potential to produce 7155 kWh of electricity after a gas engine conversion. It is plausible to replace 16.45% of the total electricity needed to operate the STP by the electricity generated from biogas. It can also reduce 4801 kg of CO_2 emission produced by fossil fuel combustion. Moreover, the electricity generated from the biogas can be sold to the utility company through FiT, and the revenue earned can be USD 306,949 per year.

Table 18.5 Simulation of electricity generation from the biogas generated via anaerobic digestion of SS before and after process optimization and improvement

Parameter	Unit	Value Before process optimization and improvement	Value After process optimization and improvement
Population	Person	250,000	250,000
Wastewater generated per person	m^3/person	0.22	0.22
Influent flow rate per day	m^3/day	54,554.86	54,554.86
SS generated per day	m^3/day	172	172
Biogas generated per day	m^3/day	1,102	2,836
Biosolids generated per day	kg/day	4,800	2,736
Energy content of biogas from STP	kJ/m^3	22,700	22,700
Efficiency of gas engine	%	40	40
Electricity needed	kWh/day	43,503.61*	43,503.61*
Electricity generation potential	kWh/day	2,780	7,155
Electricity recovered from biogas	%	6.4	16.45
Electricity from utility	kWh/day	40,723.61	36,348.61
Reduction of CO_2 emission from fossil fuel combustion	kg CO_2/day	1,865.38	4,801
FiT rate per kWh	USD/kWh	0.13	0.13
Potential revenue from biogas per year	USD/year	119,262	306,949

*Based on 0.80 kWh per m^3 of specific energy consumption

18.9   CO-DIGESTION OF POME WITH SEWAGE SLUDGE

As discussed previously, the characteristics of raw POME do not fulfil the Malaysian environmental standard requirements for discharge of effluent. Thus, it has to be treated in a series of open ponds for approximately 210 days prior to discharging to waterbodies. This primitive technology is a relatively inferior treatment method, but it has been widely practised owing to lower cost compared to modern treatment systems such as anaerobic digestion. However, in the sewage treatment system, as mentioned earlier, the process of treatment is continued in a state-of-the-art anaerobic digester till the specified regulatory environmental standards are met. Although most of the treatment plants are built to handle 250–300 mg/L of incoming SS solids, it has been found that only 50% of solids are obtained from incoming influent, thus affecting the optimum operation of an anaerobic digester. Low concentration of SS after treatment also affects the hydraulic retention time (HRT) in the anaerobic digester (Beszedes, Laszlo, Szabo, et al. 2008), thus resulting in reduced biogas productivity and delay in the complete breakdown of organic materials. Biogas produced currently is simply flared to reduce the greenhouse gas (GHG) emission since the quantity and quality are not economically favourable for investment to harness the biogas potential as an energy source.

An alternative to this problem is to increase the anaerobic digestion loading by introducing other available organic waste materials such as POME sludge. Table 18.6 shows the characteristics of POME and SS combined at the optimum ratio of 70:30.

Characterization studies performed on POME and SS have shown that when these are combined, the optimum C:N ratio is achieved at 20 (Table 18.6). COD was recorded to be at 50,600 mg/L, which falls within the optimal value of 50,000–55,000 mg/L for the anaerobic digestion process.

Co-digestion is a process in which a combination of two or more substrates is digested in a single digester to improve the digestion process. Usually, co-digestion enhances the biogas yields from anaerobic digester

Table 18.6   Characteristics of POME and SS at an optimum ratio

Characteristics	70% POME + 30% SS
MLVSS (mg/L)	23,600
Total solids (mg/L)	45,400
COD (mg/L)	50,600
Total kjeldahl nitrogen (mg/L)	1,230
C:N ratio (mg/L)	20.0

owing to positive synergism established in the digestion medium and the supply of possible missing nutrients by the co-substrates (Mata-Alvarez, Mace, and Llabres 2000). The improved digestion is usually achieved when the micro- and macronutrients in the sludge get more balanced, optimal pH and C:N ratio are reached, and inhibitors are diluted to lower levels (Mata-Alvarez 2003). Better nitrogen:phosphate:potassium (NPK) ratio can also be obtained through co-digestion, which will aid in higher methane production (Fabien 2003). The advantage of co-digestion concept is that it has an economic edge because it generates surplus energy which provides additional income to plant owners. Co-digestion of substrates is also cost efficient as one plant can be used to treat more than one kind of waste.

Figure 18.14 shows the total amount of energy that can be harnessed by co-digesting SS with POME. The potential green energy that can be generated from co-digestion of SS and POME is 60,621 kWh/day. The total electricity required for operating the STP is 43,503 kWh/day, calculated at USD 0.13/kWh for 365 working days. This energy demand is lesser than the amount that can be generated from the plant, which provides an excess of 17,118 kWh/day that can be sold to the grid through the FiT programme at USD 0.13/kWh. The potential biogas revenue from co-digesting SS with POME is 2.54 million USD per year, taking in account the potential savings from using in-house generated electricity and sale of excess electricity to the grid. Thus, if a 2.5 MW capacity biogas engine is installed with USD 7.58 million investment cost, the payback period for this project will be 3 years from its start date.

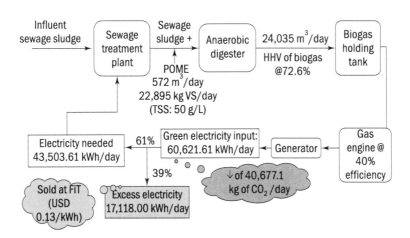

Fig. 18.14 *Total energy generation after co-digestion of POME and SS*

Table 18.7   Energy generation potential and CO_2 reduction potential of SS:POME Co-digestion estimated for 2015

Substrate	Methane production (2015) (m^3/day)	Electricity potential (2015) (TWh/year)	Economic potential (USD/year)	CO_2 emission reduction potential (2015) (Mt CO_2 eq/year)
POME[a,b,c]	1,071,000	4.49	583,700,000	3.01
SS[a,b,d]	1,150,874	1.59	206,700,000	1.07
Co-digestion of POME and SS	2,221,874	6.08	790,400,000	4.08

[a]FiT rate up to and including 4 MW is USD 0.13 (Yaacob 2014).
[b]Average value of CO_2 avoidance per kWh is 0.671 kg.
[c]Estimate based on CPO generated in 2015 and the case when the POME generated is anaerobically digested.
[d]Estimate based on the wastewater generated by the population in 2015 and the case when all the STPs are upgraded to modern mechanized plants with anaerobic dige.

Co-digesting SS with POME can be economically and environmentally beneficial (Table 18.7), in which it is assumed that all the 532 palm oil mills have used anaerobic digester for POME treatment and domestic wastewater of total 30 million population of Malaysia is treated in a modern mechanized STP. The total CO_2 emission reduction achievable by merely combining these substrates is 4 million tonnes annually. This is in line with the mission of SEDA Malaysia to ensure that sustainable energy plays an important role in the nation's economic development and environment conservation.

18.10   IMPACT OF BIOGAS RECOVERY ON SUSTAINABILITY

There are three potential advantages to installing and utilizing biogas recovery systems among others and these are given in Table 18.8.

Table 18.8   Impact of biogas recovery on sustainability

Advantages	Opportunities
Environmental	• Reduces methane and CO_2 emissions from STPs • Serves as a platform to reduce nutrient run-off • Reduces dependency on fossil fuels • Reduces foul odour emission
Society	• Increases job opportunities due to opening of a new industry • Increases energy access and energy security
Economic	• Creates temporary and permanent jobs • Turns cost item/waste products into revenue generating opportunity • Can operate in conjunction with composting operations • Improves rural infrastructure and diversifies rural income streams

Source US Agency for International Development (2014)

18.11 CONCLUSION

Anaerobic digestion has proven to be feasible and economically viable for renewable energy generation from waste. Malaysia has a wide range of industries which generate wastewater stream high in organic content, such as palm oil mills and STPs. These wastes can be utilized to generate methane, which subsequently can be used to produce energy. Biogas produced from wastewater treatment has the potential to be an economical alternative renewable fuel for electricity generation. This industry has a great potential in Malaysia to become one of the renewable energy sources. However, the quantity of biogas generated in STP is not economically viable to be harnessed and converted to electricity owing to the less enriched solid content. Moreover, anaerobic digesters are not usually installed in palm oil mills owing to high investment cost. Thus, SS and POME can be co-digested to enhance the biogas generation, leading to plants being energy-wise self-sustaining and also at the same time able to produce profitable excess energy. It reduces the fossil fuel dependency to meet the energy demand and also results in CO_2 emission reduction. Co-digestion of SS with POME results in the reduction of 4.08 Mt CO_2 eq/year, which contributes to 6% of the 40% emission reduction pledged by the Malaysian Prime Minister at the United Nations Climate Change Conference 2009. There are other areas of renewable energy that can be explored further to realize the emission reduction pledge. The Government of Malaysia has provided financial support in the form of FiT to increase the uptake of renewable energy projects and mitigate CO_2 emission. These efforts are essential to stretch the available fossil fuel resources and reduce the dependency on fossil fuels. In conclusion, this is the best method available, which is economically beneficial, technically practical, and environmentally beneficial. Furthermore, biogas energy recovery and utilization in STP and palm oil mills will also contribute to sustainability. However, implementation issues such as logistics and inter-organizational negotiation exist and must be tackled to exploit the above-mentioned benefits.

ACKNOWLEDGEMENTS

The authors would like to extend their gratitude to Akaun Amanah Industri Bekalan Industri Elecktrik (AAIBE) and Kementerian Tenaga, Teknologi Hijau dan Air (KeTTHA) for funding this project. In addition, the authors wish to thank Indah Water Konsortium (IWK) management for providing their support in completing this research. Furthermore, they would like to express their sincere gratitude to all the research assistants and laboratory staff members of the Department of Civil Engineering,

UNITEN, Centre for Renewable Energy, UNITEN, and Department of Mechanical Engineering, UNITEN and the management of UNITEN for their support and direct involvement in this study.

REFERENCES

Ang, J. 2008. Economic development, pollutant emissions and energy consumption in Malaysia. *Journal of Policy Modeling* 30(2): 271–278

Beszedes, S., Z. Laszlo, G. Szabo, and C. Hodur. 2008. Enhancing of biodegradability of sewage sludge by microwave irradiation. *Hungarian Journal of Industrial Chemistry* 36: 11–16

Biogas-Renewable-Energy.Info. 2014. Biogas composition. Details available at http://www.biogas-renewable-energy.info/biogas_composition.html, last accessed on 11 January 2014

British Petroleum. 2013. *BP Statistical Review of World Energy 2013*. London: British Petroleum

Cao, Y. 2011. *Mass Flow and Energy Efficiency of Municipal Wastewater Treatment Plants*. London: Internation Water Association

Chen, W. 2014. Update on feed-in tariff in Malaysia. *Malaysia Biomass Industry Networking Seminar*

Economic Planning Unit. 2013. *The Malaysian Economy in Figures 2013*. Putrajaya: Economic Planning Unit, Prime Minister's Department

Energy Commission. 2013. *Statistics—Malaysia Energy Information Hub*. Details available at http://meih.st.gov.my/statistics, last accessed on September 2014

Energy Commission. 2014. *Malaysian Energy Statistics Handbook 2014*. Putrajaya: Suruhanjaya Tenaga (Energy Commission)

Fabien, M. 2003. An introduction to anaerobic digestion of organic wastes. *Final Report*, Remade, Scotland

Hamid, H. and A. Mohd Baki. 2005. Sewage treatment trends in Malaysia. *The Ingenieur* 3: 46–53

Hashim, H. and W. Ho. 2011. Renewable energy policies and initiatives for a sustainable energy future in Malaysia. *Renewable and Sustainable Energy Reviews* 15(9): 4780–4787

Jomo, K. and C. Edwards. 1993. Malaysian industrialisation in historical perspective. In *Industrialising Malaysia: Policy, Performance, Prospects*, edited by K. S. Jomo, 1st edn, pp. 14–15. London: Routledge

Malaysian Palm Oil Board. 2014. Export of Palm Kernel Oil by Destination 2012. Details available at http://bepi.mpob.gov.my/index.php/statistics/export/104-export-2012/545-export-of-palm-kernel-oil-by-destination-2012.html, last accessed on 22 September 2014

Malaysian Palm Oil Council. 2014. Malaysian Palm Oil Council (MPOC). Details available at http://www.mpoc.org.my/, last accessed on 11 February 2014

Mata-Alvarez, J. (ed.). 2003. *Biomethanization of the Organic Fraction of Municipal Solid Wastes*. London: IWA Publishing

Mata-Alvarez, J., S. Mace, and P. Llabres. 2000. Anaerobic digestion of organic solid wastes: An overview of research achievements and perspectives. *Bioresource Technology* 74: 3–16

Mendonca, M., D. Jacobs, and B. Sovacool. 2010. *Powering the Green Economy: The Feed-in Tariff Handbook*. London: Earthscan

Ministry of Energy, Green Technology and Water (KeTTHa). 2010. Details available at http://www.kettha.gov.my/content/sejarah-kementerian

National Economic Advisory Council (NEAC). 2010. *New Economic Model for Malaysia*. Putrajaya, Malaysia: National Economic Advisory Council, Federal Government Administrative Centre

Oil Refinery Plant. 2014. Palm Oil Processing | Oil Refinery Plant. Details available at http://www.oilrefineryplant.com/palm-oil-processing/, last accessed on 22 September 2014

Rahim, K. and A. Liwan. 2012. Oil and gas trends and implications in Malaysia. *Energy Policy* 50: 262–271

Safaai, N., Z. Noor, H. Hashim, Z. Ujang, and J. Talib. 2011. Projection of CO_2 emissions in Malaysia. *Environmental Progress & Sustainable Energy* 30(4): 658–665

Sime Darby Plantation. 2014. Sustainability Report 2014. Details available at http://www.simedarby.com/upload/Sime_Darby_Plantation_Sustainability_Report_2014.pdf, last accessed on 28 September 2014

Sulaiman, F., N. Abdullah, H. Gerhauser, and A. Shariff. 2011. An outlook of Malaysian energy, oil palm industry and its utilization of wastes as useful resources. *Biomass and Bioenergy* 35(9): 3775–3786

Suruhanjaya Tenaga (Energy Commission). 2013. Peninsular Malaysia Electricity Supply Industry Outlook. Details available at http://www.st.gov.my, last accessed on 22 September 2014

The World Bank. 2014. Malaysia Data. Details available at http://data.worldbank.org/country/malaysia, last accessed on 25 September 2014

US Agency for International Development. 2014. U.S. Agency for International Development. Details available at http://www.usaid.gov/, last accessed on 8 February 2014

Wan Abdullah, W. 2011. An Overview of Malaysia's Sewerage Management. *Water Environment Partnership in Asia (WEPA), 3rd International Workshop.*

Yaacob, A. 2014. SEDA PORTAL. Details available at http://seda.gov.my/, last accessed on 29 September 2014

Yacob, S., M. Hassan, Y. Shirai, M. Wakisaka, and S. Subash. 2005. Baseline study of methane emission from open digesting tanks of palm oil mill effluent treatment. *Chemosphere* 59(11): 1575–1581

19

Passive Energy in Residential Buildings

Uraimindi Venkata Kiran

School for Home Sciences, Babasaheb Bhimrao Ambedkar University,
Lucknow, Uttar Pradesh 226025
E-mail: druvkiran@gmail.com

19.1 INTRODUCTION

Ancient buildings were pro-environment. The constructions were carried out considering the existing environmental and climatic conditions, ensuring comfort inside buildings in a natural way. In contemporary constructions, the conventional energy sources are predominantly used to achieve comfort levels in buildings. Complete dependence on artificial sources of energy is alarming. Conserving energy and environment has been a matter of concern in the recent past; hence, an urgent need was felt to enhance natural comfort in the residential architecture. A better understanding of energy flow in residential buildings will definitely aid in achieving this endeavour. Increasing population, urbanization, and modernization is enhancing and predominantly contributing towards global energy demand. A significant impact on the environment is due to the increased needs and levels of comfort, specifically in the built-in environment. Energy use in buildings has escalated and it has become a major contributing factor to growing demand for energy in the global scenario (Pérez-Lombard, Ortiz, González, et al. 2009).

The total electricity consumption in India has drastically increased, and the domestic energy consumption contributed to 22% of the total energy till 2012, with an increase from 80 TWh to 186 TWh. By 2030, the aggregate floor area of buildings will see an increase of 400%, and the floor area of new buildings will be 20 billion m^2 (Kumar 2011). Increase in the purchasing power of consumers leads to a greater use of home appliances and can be counted as one of the attributing factors for rise in energy demand. The requirements for enhanced electricity production owing to

Indian GDP, escalation of residential floor space, and high expectations of domestic comfort have proved to be the major reasons for a significant rise in the release of dangerous emissions.

As predicted, by 2050, energy consumption from residential buildings will increase to more than 8%, and, hence, it is high time that India focuses on energy conservation strategies and tries to adopt energy-efficient methods, especially to control energy consumption in the residential sector.

Sustained economic growth in various sectors, namely, residential, transportation, and industrial, leads to escalation in aggregate primary energy demand, which is expected to grow by 2.3 times in the next two decades (Chaturvedi, Eom, Clarke, et al. 2011). The total power consumption is rising at 8% annually, accounting to 30% use from the residential and commercial sectors (22% residential and 8% commercial) (Kumar 2011).

A doubling floor space is expected by 2030 owing to the rising demand to accommodate upcoming industries and migration (Kumar, Kapoor, Deshmukh, et al. 2010). The intensity of energy per unit floor area, along with an increased floor area, has put pressure and heightened the gravity of energy demands in the residential sector.

The primary energy utilized during the entire span of a building is greatly influenced by environmental conditions, building materials used, and the operational costs of the building accounted to various additional features. Scheuer, Keoleian, and Reppe (2003) conducted a primary energy analysis on a newly constructed engineering building located on the campus of the University of Michigan. They reported in their study that 97.7% of the total primary energy consumed was accounted for the primary energy of the operations phase and a lifespan of 75 years was desired. The carbon footprint of the building can be reduced through energy-efficient building designs. One of the most cost-effective sectors where energy consumption can be reduced is the residential sector, and, hence, this sector has to be concentrated to overcome the energy crisis (IEA 2010).

The spending capacity of Indian consumers has increased owing to the rising GDP and greater affordability of consumer goods, and this has resulted in changed patterns of energy consumption. The increased energy consumption is due to heavy installation of artificial fixtures in residential and commercial buildings invoking a major percentage of energy use, leading to higher energy performance index in buildings. Even though the appliances are consuming a major part of energy, the characteristics of a building envelope play a very vital role. Buildings can be made thermally and acoustically comfortable through proper planning and placement of architectural features, inclusive of walls, windows, roof, and floor. Artificial lighting and comfort systems have to be relied upon if the building does not meet the comfort criteria. Buildings have to be constructed considering the local climatic conditions to enhance the natural energy flow.

Passive solar design refers to the use of sun's energy for heating and cooling of living spaces. Solar energy is utilized by the building as a whole, or some element of the building takes advantage of natural energy characteristics through building materials or the building orientation. Passive systems are simple, have very few moving parts, and require minimal maintenance without any mechanical system.

The passive solar concept encompasses a wide range of strategies in the design and construction of a building in order to take the maximum advantage of sunlight and the surrounding environment. The building itself collects, stores, and uses solar energy. The buildings are designed to take advantage of site-specific features for natural heating, cooling, and lighting. Orientation of the building, size, location of windows, shading, ventilation devices, and colour of outer surfaces help in building efficient thermal capacities of building elements and the maximum usage of passive technologies through proper planning of all the design elements may enhance the energy efficiency of the building. Passive technologies aim at maximizing the effect of this approach by choosing proper orientation of the building, size and location of windows, shading and ventilation devices, colours of outer surfaces, thermal resistance, and absorption capacities of building elements.

A building separates the inside spaces from outside environment for creating stable and comfortable thermal conditions for occupants. The building has to be designed in such a manner that it opens itself to the favourable conditions of sun, wind, humidity, ambient temperature, and sky and closes itself to unfavourable conditions. The term passive refers to techniques of utilizing the natural modes of energy transfer, that is, radiation, conduction, convection, and evaporation, including the use of simple devices such as fans and pumps for moving air.

Ancient architecture had many passive features, supported for creating comfortable thermal environment in the building. The shape of the building and placement of different parts of the building, that is, doors, windows, indoor spaces, and balconies, used to take the maximum advantage of the environment. Greenery around buildings can also be considered as a favourable factor in determining the thermal comfort.

Modern-day construction ignores these passive concepts and more mechanical systems are introduced for heating and cooling of buildings. Construction of towns and cities without any greenery also hampers usage of passive solar energy. Every building should have an interface

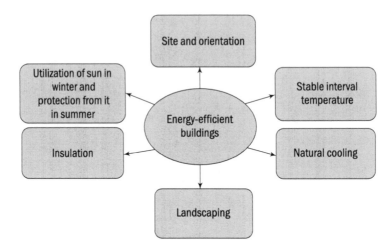

Fig. 19.1 *Parameters considered in energy-efficient buildings*

with its occupants. Visually, colours rendered by natural and artificial illumination are produced by combinations of doors and windows. Heat transfer between the human body and its surroundings affects the thermal comfort of a house—conduction of heat from point of contact with the structure, heat removed by natural convection currents in the room, and radiated heat transfers between the body and its surroundings.

The architectural features of a building, inclusive of windows, walls, and floors, in a passive solar design are planned and located in such a manner that the building is able to use the required amount of solar energy during winters and reject solar heat during summers. Unlike the active solar system, in which mechanical and electrical systems are used to utilize solar energy, passive solar buildings themselves collect, store, and distribute solar energy. Passive solar buildings act based on local climatic conditions (Figure 19.1).

19.2   PASSIVE SOLAR COOLING

Passive cooling aims at improving indoor thermal comfort by focusing on controlling and distribution of heat gain in a building and consuming low or nil energy (Santamouris and Asimakopoulos 1996; Leo Samuel, Shiva Nagendra, and Maiya 2013). Prevention of heat from entering the interiors (heat gain prevention) or elimination of heat for providing natural cooling in the building is termed as passive cooling. The onsite energy owing to local climatic conditions and the location of architectural elements in a building rather than mechanical systems improve the scope of natural cooling (Niles and Haggard 1980).

19.2.1   Operable Windows and Ventilation

The vital parameter, especially in hot humid climates, to be considered is employing natural ventilation. Avoiding mechanical devices for cooling of buildings is the primary strategy adopted in passive cooling. Placement of fans in the building and landscaping plays a prominent role in providing proper ventilation to inmates. South glazing poses a major problem, making the building very hot, and the following strategies may be adopted to improve indoor climate:

(i) Operable windows may be fixed towards the south exposure.

(ii) Best airflow can be offered through casement windows. Fully opened awning or hopper windows will direct air to the ceiling. They also offer best rain protection and perform much better than double-hung windows.

(iii) When windows can be placed only on one wall of the room, two widely spaced windows can be used for better airflow.

19.2.2   Wing Walls

Vertical solid panels at the side of the windows perpendicular to the direction of the wind are called wing walls. The pressure created by them will accelerate the speed of the natural airflow.

19.2.3   Thermal Chimney

Exterior exhaust outlet is provided for creating a warm or hot zone which will draw hot air from the building through convection currents. Ventilators are provided for drawing cool air into the house. A thermal chimney is constructed in a narrow configuration with a black metal absorber inside, which can be easily heated and, hence, the house may be insulated. The chimney must be placed in such a manner that it is terminated just above the roof level. A rotating metal scoop at the top which opens opposite the wind will allow heated air to exhaust without being overcome by the prevailing wind. Thermal chimney effects can be integrated into the house with open stairwells and atria, the approach which can add an aesthetic value too.

19.2.4   Sunrooms

Rooms located towards south can be designed as sunrooms. Upper vents can be provided to vent the excessive heat generated. Lower vents in the living space, along with windows on the north side, will allow air through the living space to be exhausted through the sunroom upper vents. The upper vents from the sunroom to the living space and any side operable windows must be closed and the thermal mass wall in the sunroom must be shaded for effective functioning.

19.2.5   Ventilation Strategies

(i) Inlets close to a wall result in air "washing" along the wall. Ensure centrally located inlets for air movement in the central areas of the room.

(ii) The insect screens on windows decrease the velocity of slow breezes more than strong breezes (60% decrease at 1.5 mph and 28% decrease at 6 mph). The air speed cannot be reduced by screen porches or insect screens. The ventilation rate of 30 air changes per hour or greater should be provided for better night ventilation. Minimization of fabric furnishings will aid in better cooling of high mass buildings.

(iii) Living comfort in buildings is of vital importance. The main factors governing this comfort are thermal, ventilation, lighting, and acoustical properties of building materials and components with respect to the outside climate. A basic feature for achieving an acceptable building environment is the consideration of the external environment (sun, wind, rain) and its effects on the internal environment (comfort). Between the two are the barriers, that is, the fabric and the building envelope, which refer to walls, roofs, opening, and so on of the building.

(iv) The house should be oriented to take advantage of natural elements vital to a particular climate. 'Orientation' is the placement of features of a building so that it may obtain the best advantages in relation to its physical location. The major considerations in orientation of a building and dwelling unit are sunlight, prevailing breezes, and view.

Sunlight

The objective of orientation for the sun is to obtain sunlight when it is desired and to block out sunlight when it is not desired. Sunshine in every room is desirable, but it cannot be achieved if the building is placed with a due north exposure. The best position for the house is to face the sun at about 30° south to east. In winter, the sun hits such a house on all four sides.

Orientation

The orientation of the building and rooms plays a vital role in gaining and utilizing the maximum natural energy from the environment. The various orientations are as follows:

(i) **South orientation:** It is the best position to obtain the maximum sunlight during the day. Sunlight will occur from late morning to early afternoon.

(ii) **East orientation:** Sunlight will occur only in the morning during the winter months, sun will rise more towards the south-east, thus providing a shorter period of sunlight.

(iii) **South-east orientation:** Sunlight will occur from early morning to late morning and possibly to noon.

(iv) **South-west orientation:** Sunlight will occur from early to late afternoon.

(v) **West orientation:** Sunlight will be present from mid- to late afternoon. During the summer months, the west will be very intense.

(vi) **North orientation:** No sunlight will be obtained from a direct north orientation.

There have been attempts to derive the benefits of sunlight in the house by solar orientation and to achieve the maximum thermal comfort. The passive solar design proves to be a vital concept for heating and cooling of living spaces. Solar energy is utilized by the building as a whole, or some element of the building takes advantage of natural energy characteristics through building materials or the building orientation. The passive solar concept encompasses a wide range of strategies in the design and construction of a building in order to take the maximum advantage of site-specific features. The physical and mental health of an individual is greatly influenced by the built-in environment, as more than 90% of an average person's life is spent in buildings (Evans 2003).

In summer, over 50% of the heat comes from the top and most of our roofs—the concrete/stone/tile roof starts radiating heat into the room. In winter, the slab conducts the heat from the room to the external space. In his study of 'low cost roof designs to keep the heat away', Rajan (2000) concluded that to stop heat from coming in, one could take the following measures:

(i) Insulate the roof from top using earthen pots.

(ii) Insulate the roof from below by building a false ceiling made of jute fabric and timber framework.

(iii) Shade the roof using a shade net or by growing creepers.

(iv) Combination of any of the above three measures.

Ventilation

Good ventilation means adequate supply of sunlight and fresh air into the house to promote comfortable and healthy living. Effective ventilation is achieved by maintaining the desirable ratio between air measurements going inside and coming out, expressed as a percentage. For good cross ventilation, windows should be built on walls facing each other, and if windows are provided only on one wall, one wide window or two windows at the corners should be preferred.

Prevailing breezes

Wind is an important factor in the orientation of the house. It is desirable to avoid its disadvantages and take advantages of its benefits. Winds are vigorous cooling agents. Heat loss through the walls of a house will occur to a lesser degree. A study of the wind pattern will help to decide the placement of doors and windows in the house, location of sleeping rooms, exhaust fans, porches, and wind breaks. Proper recognition of the effect of wind when designing the house will reduce the cost of heating or cooling.

In order to carry out all the activities smoothly and efficiently, artificial lighting fixtures, heating appliances, and cooling appliances play an important role. Improper orientation or ill planning leads to an excessive use of artificial fixtures, and thus excessive flow of energy takes place, which is conventional in nature. In order to minimize the conventional energy flow in the house, one needs to have a better understanding about orientation and the various sources of non-conventional energy.

Every individual spends a maximum amount of time in buildings, that is, at offices or homes. Hence, energy used in these buildings is highly accountable, and it accounts for a significant percentage of a country's total energy consumption. The prevailing local climatic conditions and the amount of building area per capita highly influence the degree of electrification, and in turn will have an impact on the energy consumption. Local, national, and international policies for energy have to be made considering the energy conservation measures.

Energy consumption data in many countries have revealed that buildings consume more energy than transport and industry. Statistics, as estimated by International Energy Agency (IEA), showed that electricity consumption by the building sector accounts for 42% of energy used globally (IEA 2004a, 2004b, 2004c).

The vital parameters to be considered under natural energy gain flow are passive heating, passive cooling, natural airflow, and daylight. Utilization of natural energy in an intelligent manner can significantly reduce the energy load dependency on other sources for meeting a building's energy needs. Wastage of energy can also be avoided and each unit of energy can be used efficiently if buildings are made environment friendly.

Buildings can be made environment friendly by adopting the following strategies to improve natural energy gains:

(i) The location of the site and building plan for minimizing the use of artificial energy.

(ii) Shape of the building which increases the use of daylight and use of natural ventilation in order to reduce energy losses.

(iii) Glare and overheating of the building should be avoided, and benefits from solar gain should be maximized through proper building orientation.

(iv) Unwanted solar gain has to be avoided with proper utilization of natural daylight.

(v) Mechanical ventilation strategies should be minimized by maximizing the provision of natural ventilation.

(vi) Building envelope should be capable of providing good thermal insulation and preventing unwanted air infiltration.

(vii) Employing natural ventilation wherever practical and appropriate, and avoiding mechanical ventilation and air conditioning to the extent possible.

(viii) Good levels of thermal insulation and prevention of unwanted air infiltration through the building envelope.

(ix) Intrinsically efficient and well-controlled building services, well matched to the building fabric and to the expected use.

The maximum utilization of natural energy can be achieved through proper planning during the design phase of the building, and minor changes can be made through refurbishment to avoid maximum losses. Winston Churchill once said, "We shape our buildings and then they shape us." The degree of livability that a house design provides depends upon how well the architectural features of that house meet the needs of those who carry out the activities. Artificial lighting fixtures, as well as heating and cooling devices, also play an important role in controlling the solar or heat radiation in the house. Improper orientation or ill planning leads to excessive use of air conditioners, heaters, lighting fixtures, and so on, and, thus, excessive energy flow takes place, which is conventional in nature.

The complete advantage of prevailing breezes and protection against cold winds has to be considered as an important factor in placement of rooms in a building. Room orientation should take cognizance of different climatic conditions. Natural ventilation can be accomplished only when the exterior and interior openings are of an adequate size and so arranged that the natural flow of fresh air is not impeded or does not leave a major portion of the room unventilated.

Predicted energy requirements of a building for various purposes, namely, heating, cooling, lighting, ventilation, and air conditioning, accounting for gross energy demands are determined through local environmental conditions, indoor and outdoor requirements, the building properties, and architectural elements of a building.

Therefore, for achieving a proper building environment, an understanding of the external environment and an assessment of the functional barriers provided by walls, roofs, and openings are necessary, and this requires a considerable amount of data on physical properties of the building components. Therefore, the study is conducted with the following objectives:

(i) To study the energy flow in selected residential architectural designs through case analysis.

(ii) To estimate the energy costs of buildings through payback charts and analyse the energy losses of dwellings.

For a particular type of building under defined climatic conditions, the energy consumption benchmarks can be calculated and its energy efficiency can be predicted (energy consumption per square metre of the floor area). Increasing generation capacity in developing countries cannot be ignored, but the peak demand constraints the balance between energy supply and demand. Quality of services provided in a building cannot be compromised by adopting energy efficiency measures. This study envisages to relate energy flow in the residential architecture to the orientation, location, building materials used, colour scheme, openings in the dwellings, cracks, joints, and so on present in buildings so as to analyse whether the energy flow in the residential dwelling is low, high, or optimum.

19.3   METHODOLOGY

The study is analysed in the form of case studies. Three different kinds of dwellings with regard to location, orientation, inhabitants, and materials used are taken. The information was collected through a checklist, where information pertaining to the various aspects of energy flow in residential architectural building was gathered.

Information gathered is further analysed and discussed as different "case studies". In order to have a better understanding, dwellings at four different locations and styles have been chosen for the study. To bring in more clarity in orientation and placement of openings such as doors and windows, house plans have been drawn along with the analysis, which will help in examining or visualizing the dependencies on the plan and orientation of building and openings, thus revealing the energy flow in the dwelling. The energy cost of the building was also calculated through payback charts to estimate energy losses.

19.3.1   Analysis of Case Study I

This residential building is a two-storey building. The total height of the building is 10 m, and the total plinth area of the first floor which is considered for the study is 120 m². The dwelling is inhabited by two persons. The plan for the floor consists of one bedroom, one drawing room, one kitchen, and one bathroom. It has a long front corridor and a side balcony. The building is not surrounded by any trees or any other green plantations.

The kitchen faces the east direction and the bedroom the west. The drawing room is towards north and the bathroom is also towards the north. A balcony is placed in the eastern aspect and a long corridor in the northern direction.

The whole floor is plastered with mortar, and even the partitions in cupboards and almirahs are made up of concrete slabs. The terrace above is fully exposed to sunlight, and the overhead tank is just above the bathroom.

The colour of the floor in the entire house is grey with walls in white colour. There is white ceiling, and the height of the walls is 3 m. The thickness of the walls is 0.2 m in the outer and 0.1 m in the inner. The thickness of the roof is 15 cm, and there is no sheathing material used. The construction is done by bricks and cement, and the floor is tiled throughout.

The entrance is through the corridor towards north, leading to drawing room. The front faces north and one of the sides faces east. The northern side has one opening, one main door, and one window on the east. There are two other doors, one leading to kitchen and the other to bedroom. The drawing room that faces the northern direction does not have any opening into the exterior environment, but it has an opening (window) towards east, and so the room gets sufficient sunlight in the morning. However, as the day passes, artificial lights have to be resorted to. The kitchen is on to the eastern aspect, and hence it has an advantage of morning sunlight, but as the day passes and the direction of the sun changes, the kitchen becomes dark and artificial lighting has to be resorted to. The kitchen is not only fitted with an exhaust fan and has a window towards east, but it has also cupboards with mortar partitions, which conduct and radiate heat when the exhaust proves to be ineffective.

The bedroom is in the southern direction, has an opening (window) towards south, and so it obtains maximum sunlight during the day, especially in summers. Therefore, the room is exposed to maximum solar radiation, which makes it very hot.

The bathroom is placed on to the west with a door connecting to the bedroom. It is tiled throughout the floor, and three-fourth of the walls is also covered with tiles. Overhead tank is placed exactly above the bathroom, and, hence, the bathroom remains very cool but dark during the daytime.

As the sun moves, the drawing room and kitchen get dark, and hence artificial lighting has to be resorted to. No cracks or holes are found in the house. Placement of some potted plants and creepers around the house, especially towards west and south, will make the interiors cool and pleasant. To avoid conduction and radiation of heat owing to mortar partitions and cupboards, the same may be replaced with wood. Provision of sunshades for windows on west and south aspect will help avoid excess sunlight.

19.3.2   Analysis of Case Study II

This residential building is an independent house. The total height of the building is 8 m, the total plinth area is 75.78 m², and the total plot area is 167 m². The staircase area is 6.86 m² and the portico area is 12.52 m². The dwelling is inhabited by four persons. The plan for the floor comprises one drawing room, one dining room, one kitchen, one *pooja* room, and two bedrooms with attached bathrooms.

The kitchen faces the south-east direction. The drawing room is towards east and dining room is towards south. One bedroom is in the south-west direction and the other on the north-west. The *pooja* room is towards east. The dwelling is exposed to sunlight on all the sides during the daytime. There is no dwelling adjacent to this building.

The whole floor is plastered with mortar. The entire construction is done with cement plastering of the partitions in the built-in cupboards, and almirahs are made up of wood. The terrace above is fully exposed to sunlight. A low height compound wall is built around the building.

The colour of the floor in the house is white, and the colour of walls and ceiling is also white. The height of the wall is 3.5 m, the thickness of the outer wall is 0.2 m and that of the inner wall is 0.1 m, the thickness of the roof is 15 cm, no sheathing material is used, the construction is done by bricks and cement, and there is marble flooring throughout.

The entrance is through the portico and sits out towards eastern aspect. It leads to drawing-cum-dining room, which is a huge space and in which the front faces east, one side faces north, and the other side faces south. The eastern and northern sides have one opening each, that is, window, and one main door, which is the entrance, and there are three other doors, one leading to kitchen and the other two to bedrooms. As the drawing room faces east, there is good sunlight in the morning, and as the day passes, the room becomes cool owing to less intense heat, and the room requires no artificial lighting as it has openings on three sides.

The kitchen is east facing and, hence, exposed to morning sunlight, but with the change in the direction of the sun, the room is exposed to maximum solar radiation as it is having openings towards south. The kitchen is fitted with an exhaust fan and the partitions in the cupboards are made up of wood, and, hence, there is no complaint of conduction and radiation of heat through them.

Both the bedrooms are towards the west, with two windows each. The bedroom that is in the south-west has a window on both south and west, and the other bedroom is towards north-west, with windows on north and west. Two bedrooms have attached bathrooms with fixed ventilators towards west. There is western aspect of the bedrooms, low levelled compound wall, and no greenery around. The sunlight falls directly into the room, and, hence, the walls of the room get very hot and the heat radiates even during night, making the room hot and uncomfortable for inmates.

The bathrooms have small ventilators and tiled up to three-fourth of the walls. The ceiling is built of mortar, and the height of the wall is low and provided with an attic space, above which there is overhead tank, and, hence, the bathrooms remain very cool but dark during the daytime.

There is good cross ventilation in the house except the bedrooms as they are in west aspect. No cracks or holes are found in the house, and the doors and windows are varnished with wooden colour. Plantation of trees around the house will help maintain the building temperature low and comfortable. Provision of balcony on the western aspect will act as buffer to avoid sunlight during noon time and radiation during nights.

19.3.3 Analysis of Case Study III

This residential building has a total plinth area of 160 m^2. The dwelling is inhabited by six persons. The plan comprises two bedrooms, one verandah, one living room, one kitchen, one storeroom, and two bathrooms.

The kitchen is situated towards north. One bedroom is in the north and the other in south. The living room is in the south-east direction. One bathroom is in west and the other in north. Kitchen and storeroom are adjacent to each other.

The whole floor is plastered with mortar. The partitions in the cupboards and almirahs except in kitchen are made up of wood. The partitions in kitchen are made up of cement slabs.

The colour of the floor in the whole house is grey, with walls in pink and ceilings white. The height of the wall is 3 m. The thickness of the walls is 0.2 m and the thickness of the roof is 15 cm. No sheathing material is used, and the construction is done by bricks and cement. The floor is tiled throughout.

The entrance is through the main door facing south and leading to verandah. There is one door from the verandah leading to living room, with one double window towards east. It has five doors, one leading to kitchen, one to bedroom, one to bathroom, and one to the other bedroom. Since the verandah faces southern direction, it obtains maximum sunlight during the day, especially in summers, but as there is a portico on eastern side, it hinders some of the radiation. The living room is located almost in the middle of the house with one opening towards east. Hence, in the morning, there is sufficient sunlight, but as the day passes, the direction of the sun changes, the room becomes dark, and artificial lighting has to be resorted to.

The kitchen is north facing (north-east) and windows are towards east. There is sufficient sunlight during daytime. Adjacent to the kitchen, there is a storeroom with no openings, and, hence, the room remains dark throughout, which necessitates a need for artificial lighting. The kitchen is fitted with an exhaust fan, but it has a cupboard with mortar partitions, and, hence, they conduct and radiate heat.

The bedroom is in the south-west. It obtains maximum sunlight during the day, especially in summers, and, hence, the room is exposed to maximum solar radiation and as one of the openings is towards west, the room becomes very hot and uncomfortable.

The other bedroom is in the northern direction with west facing and north facing windows. As one of the openings is towards west, the room becomes hot as the day progresses, but there is provision of good cross ventilation in the room. Therefore, there is no need for artificial lighting. It has an attached bathroom, which is very small, with an opening towards north (ventilator). It is tiled throughout the floor and three-fourth of the walls. Sufficient lighting is not available in the bathroom, and, hence, artificial lighting has to be resorted to. One more bathroom is towards west between the two bedrooms, with entry from the living room. It has a window-cum-ventilator on the west, and so it receives good daylighting, with no need of artificial lighting.

No cracks or holes are found in the house. The doors and windows are painted in cream colour. In summer, the house becomes hot. Plantation of trees around the house will help in making the internal environment of the dwelling cool and comfortable. The openings towards west may be half painted to avoid direct sunlight in the house. Provision of sunshades and covering the western openings with awnings will also help in avoiding solar radiation in the rooms. Changing of partitions in the kitchen and replacing the slabs in the kitchen cupboards from mortar to wood will make the interiors cool.

19.4   ESTIMATION OF ENERGY COST

Total energy consumption of any country is contributed to maximum by the building sector. Maximum (42%) usage of electricity is evident in the building sector in comparison to any other sector. The generation capacity of developing countries is still an issue, and, hence, the usage of energy in developed countries is more compared to developing countries. The energy cost of a dwelling can be estimated through various methods, such as payback chart and energy efficiency housing options evaluation (ENEHOPE) model.

19.4.1   Payback Chart

The chart shows step-by-step calculation in months or years of the payback time for investments in a typical group of heat/energy saving improvements. Most of the data for the chart are acquired by visiting the house and measuring or estimating dimensions, insulation, and turnover. The rest of the chart is filled by simple arithmetic based on the data.

The chart will be most useful if it is used to compare savings from different improvements, particularly via the payback figures. When planning such steps as solar heating, upgrading the efficiency, or turning down the thermostat, insert the estimate of fuel bill savings, and they will provide the money saved annually. Also, deduct these savings from the prior fuel cost estimate. Then enter the cost of the new improvement and re-evaluate payback for all the improvements.

The payback charts for the residential plans are given as energy cost calculations:

Case Study I

House elements	Area (m²)	R value	Heat loss	Total heat loss
Walls	8.98	1.37	6.56	
Windows	3.43	1.22	2.81	
Doors	5.88	3.50	1.68	
Floors	105	1.24	0.85	194.05
Ceiling	105	0.63	98.0	

$$R \, \alpha \, \frac{1}{\text{Thermal conductivity}} \, \text{(resistivity)}$$

$$\text{Heat loss} = \frac{\text{Area}}{R}$$

Case Study II

House elements	Area (m²)	R value	Heat loss	Total heat loss
Walls	12.09	1.37	8.8	
Windows	9.27	1.22	7.5	
Doors	6.38	3.54	1.8	
Floors	75.00	1.24	62.09	134.89
Ceiling	75.00	1.37	54.70	

Case Study III

House elements	Area (m²)	R value	Heat loss	Total heat loss
Walls	9.28	1.37	6.7	
Windows	10.26	1.22	8.4	
Doors	8.76	3.50	2.5	
Floors	148.50	1.24	111.0	522.00
Ceiling	148.50	0.63	235.0	

The building sector, inclusive of residential and commercial, is a major energy consumer. At a relatively low cost, the maximum usage of natural energy is possible if it is adopted at planning and designing phase of a building.

Adopting suitable measures in placement of rooms may reduce the cooling demand of a building. The energy for ventilation can be minimized through proper design and placement of windows along with a mixed mode of ventilation. The energy efficiency of buildings can be maximized by adopting the following parameters:

- Buildings should not be more than 12–15 m in depth.
- Central atria in the buildings may be used to achieve natural ventilation in the building.
- Stack effect in the building will aid in drawing air from the outer perimeter and up through the centre of the building.
- Natural light into the building can be maximized and energy can be saved through proper placement of openings in the building, and, thus, attractive environment may be provided to inmates.

- Use of high ceilings and clerestory windows can be effective in providing good daylight.
- Solar control glass can be used. It will enhance the provision of daylight while reducing the solar gains. Glazing may be selected with the highest light transmittance and lowest solar gain factors. The properties of the glass may be altered through a range of selective coatings.

Built-in environment has to be made comfortable and congenial for inmates to improve their quality of life. Natural energy resources in coordination with local climatic conditions will maximize the energy flow in buildings and, thus, reduce the need for resorting to mechanical devices and artificial energy sources. The demand of building sectors for heating, cooling, ventilation, lighting, water heating, electricity consumption due to appliances, and so on has also to be reduced through planning and efficient energy utilization.

19.5 CONCLUSION

Systematic energy conscious design and management is necessary for the successful and economic operation of built facilities and services, with the minimum consumption of energy and materials. Energy-efficient buildings and environment can be achieved through knowledge of climate control and awareness of latest technologies. Passive system is the one that collects or uses solar energy without direct resource or any sources of the conventional power, such as electricity, to aid to the collection. Passive design seeks to reduce the energy budget of a house by paying close attention to design features of the building and energy transfer properties of building materials. Energy cost of the building may be reduced by adopting solar passive principles in building construction. Enforcing energy audit and encouraging energy-efficient technologies by offering financial assistance, incentives, and so on will promote energy savings and consequently reduce the cost.

REFERENCES

Álvarez Bel, C., M. A. Ortega, G. E. Escrivá, and A. Gabaldón Marín. 2009. Technical and economical tools to assess customer demand response in the commercial sector. *Energy Conversion and Management* 50(10): 2605–2612

Arendse, G. 2011. *Stellenbosch New Engineering Building Electrical System Design.* Stellenbosch: De Villiers & Moore Consulting Engineers

Ashpole, M. 2009. Improving energy efficiency in buildings: SANS 204 published. *Civil Engineering* 17(2): 58–59

ASHRAE. 2000. *Advanced Energy Design Guide for Small Office Buildings.* Atlanta: American Society of Heating, Refrigerating, and Air-conditioning Engineers, Inc.

ASHRAE. 2007. *ASHRAE Standard 90.1: Energy Standard for Buildings Except Low-rise Residential Buildings.* Atlanta: American Society of Heating, Refrigerating, and Air-conditioning Engineers, Inc.

Auliciems, A. and S. V. Szokolay. 2007. Thermal Comfort. PLA in association with Department of Architecture, University of Queensland, Brisbane

Balaras, C. 1995. The role of thermal mass on the cooling load of buildings. An overview of computational methods. *Energy and Buildings* 24: 1–10

Begemann, S., G. Van den Beld, and A. Tenner. 1996. Daylight, artificial light and people in an office environment, overview of visual and biological responses. *International Journal of Industrial Ergonomics* 20: 231–239

Cengel, Y. A. 2006. *Heat and Mass Transfer: A Practical Approach*, 3rd edn. Reno, Nevada: McGraw Hill

Central Electricity Authority. 2013. Growth of electricity section in India from 1947–2013. Technical Report. New Delhi: Ministry of Power

Chapman, A. J. 1984. *Heat Transfer.* New York: Macmillan

Chaturvedi, V., J. Eom, L. E. Clarke, and P. Shukla. 2011. Energy Scenario. Details available at http:// WWW.global changes umd. Edu/wp-content/uploads/2011/12/CMM_India Building_Chaturvedi.pdf.

Cheng-wen, Y. and J. Yao. 2010. Application of ANN for the prediction of building energy consumption at different climate zones with HDD and CDD. In Proceedings of the 2nd International Conference on Future Computer and Communication, Vol. 3, pp. 286–289

Corgnati, S. P., E. Fabrizio, and M. Filippi. 2008. The impact of indoor thermal conditions, system controls and building types on the building energy demand. *Energy and Buildings* 40(4): 627–636

Cornell University. 2011. College of Architecture. Details available at http://aap.cornell.edu/milstein/design/sustainability.cfm

Corobrick. 2009. *Towards a balanced and better understanding of the thermal performance and sustainability of walling envelopes as marketed in South Africa,* Johannesburg: Energy and buildings 35(2): 214-232.

Crawley, D. B., L. K. Lawrie, C. O. Pedersen, F. C. Winkelmann, M. J. Witte, R. K. Strand, R. J. Liesen, W. F. Buhl, Y. J. Huang, R. H. Henninger, J. Glazer, D. E. Fisher, and D. Shirey. 2002. EnergyPlus: New, Capable and Linked. Green Building International Conference and Expo, U.S. Green Building Council, Austin, Texas, 13–15 November

Crawley, D. B., L. K. Lawrie, F. C. Winkelmann, W. F. Buhl, Y. J. Huang, C. O. Pedersen. 2001. EnergyPlus: Creating a new-generation building energy simulation program. *Energy and Buildings* 33(4): 319–331

Crawley, D. B., L. K. Lawrie, F. C. Winkelmann, W. F. Buhl, Y. J. Huang, C. O. Pedersen, *et al.* 2004. EnergyPlus: An Update. SimBuild 2004, IBPSA-USA National Conference, Boulder, CO, USA, 4–6 August

Djongyang, N., R. Tchinda, and D. Njomo. 2010. Thermal comfort: A review paper. *Renewable and Sustainable Energy Reviews* 14: 2626–2640

Ekici, B. B. and U. T. Aksoy. 2007. Prediction of building energy consumption by using artificial neural networks. *Advances in Engineering Software* 40: 356–362

EnergyPlus. 2010. EnergyPlus Engineering Reference: The Reference to EnergyPlus Calculations. U.S. Department of Energy

Eskom. 2011. *ESKOM Retail Tariff Adjustment for 2011/12*. Johannesburg: Eskom

Evans, G. W. 2003. The built environment and mental health. *Journal of Urban Health: Bulletin of the New York Academy of Medicine* 80(4): 536–555

Ascione, F., L. Bellia, P. Mazzei, and F. Minichiello. 2010. 'Solar gain and building envelope: The surface factor. *Building Research and Information* 38(2): 187–205

Fanger, P. 1970. *Thermal Comfort: analysis and application in environmental engineering.* Copenhagen: Danish Technical Press

Flager, F., J. Basbagill, M. Lepechand, and M. Fischer. Multi-objective building envelope optimization for life cycle cost and global warming potential. Ninth European Conference on Product and Process Modeling, Reykjavik, Iceland, G. Gudnason and R. Scherer (eds), pp. 193–200. London: Taylor & Francis Group

GBCSA (Green Building Council of South Africa). 2008. *Green Star SA Office Design & Office Built: technical manual.* South Africa: Green Building Council of South Africa

Gulma, M. A., S. L. Lorenzo, and J. O. Falaiye. 1989. Passive solar houses in Northern Nigeria: The west African sub-region. *Solar and Wind Technology* 6(4): 427–431

Gunnell, K. 2009. *Green buildings in South Africa: emerging trends.* South Africa: Department of Environmental Affairs and Tourism

Harris, R. 2010. *Specifying the minimal thermal performances for external walling.* WSP & Clay Brick publishers.

Horvat, M. and P. Fazio. 2005. Comparative review of existing certification programs and performance assessment tools for residential buildings. *Architectural Science Review* 48: 69–80

Hydeman, M. and K. L. Gillespie 1999. Tools and techniques to calibrate electric chiller component models. *ASHRAE Transactions* AC-02-9-1

IEA. 2004a. *Energy Balances for OECD Countries.* Paris: International Energy Agency

IEA. 2004b. *Energy Balances for Non-OECD Countries.* Paris: International Energy Agency

IEA. 2004c. *Energy Statistics for Non-OECD Countries.* Paris: International Energy Agency

IEA. 2010. *Energy Technology Perspectives 2010: Strategies and Scenarios to 2050.* Paris: International Energy Agency

Incropera, F. P. and D. P. DeWitt. 2002. *Fundamentals of Heat and Mass Transfer.* New York: Wiley

Kanagaraj, G. and A. Mahalingam. 2011. Designing energy efficient commercial buildings—A systems framework. *Energy and Buildings* 43: 2329–2343

Kneifel, J. 2010. Life-cycle carbon and cost analysis of energy efficiency measures in new commercial buildings. *Energy and Buildings* 42(3): 333–340

Kumar, S. USAID ECO-III Project, 2011. Energy use in commercial buildings-key findings from the national bench marking study- USAID-India

Kumar, S., R. Kapoor, A. Deshmukh, M. Kamath, and S. Manu. 2010. Total commercial floor space estimates, Energy conservation & Commercialisation–ECO 3.

Leo Samuel, D. G., S. M. Shiva Nagendra, and M. P. Maiya. 2013. Passive alternatives to mechanical air conditioning of building: A review. *Building and Environment* 66: 54–64

Levy, S. M. 2010. *Green and sustainable buildings: Construction process planning and management,* Boston: Butterworth-Heinemann

Li, K., S. Hongye, and C. Jian. 2011. Forecasting building energy consumption using neural networks and hybrid neuro-fuzzy system: A comparative study. *Energy and Buildings* 43(10): 2893–2899

Lützkendorf, T. and D. Lorenz. 2005. Sustainable property investment: valuing sustainable buildings through property performance assessment. *Building Research & Information* 33(3): 212–234

Magnier, L. and F. Haghighat. 2009. Multiobjective optimization of building design using TRNSYS simulations, genetic algorithm, and artificial neural network. *Building and Environment* 45(3): 739–746

Marble Institute of America. 2006. History of Green Building. Details available at http://www.marbleinstitute.com/industryresources/historystoneingreenbuilding.pdf

Masoso, O. and L. Grobler. 2008. A new and innovative look at anti-insulation behaviour in building energy consumption. *Energy and Buildings* 40(10): 1889–1894

Moult, R. 1999. VAV systems for office buildings—Part 1. *Air Conditioning and Refrigeration Journal,* issue January – March, pp. 1–7

Muldavin, S. R. 2010. *Value Beyond Cost Savings: how to underwrite sustainable properties.* San Rafael, CA: Green Building Finance Consortium

Neto, A. H. and F. A. Sanzovo. 2008. Comparison between detailed model simulation and artificial neural network for forecasting building energy consumption. *Energy and Buildings* 40(12): 2169–2176

Niles, P. L. B., and K. L. Haggard. 1980. *Passive Solar Handbook*. California Energy Resources Conservation

Pérez-Lombard, L., J. Ortiz, and C. Pout. 2008. A review on buildings energy consumption information. *Energy and Buildings* 40(3): 394–398

Pérez-Lombard, L., J. Ortiz, R. González, and I. R. Maestre. 2009. A review of benchmarking, rating and labelling concepts within the framework of building energy certification schemes. *Energy and Buildings* 41(3): 272–278

Rajan, S., V. 2000. Low cost roof designs to keep the heat away. *Invention Intelligence* 35(2):85–87

Rawal, R., Y. Shukla, S. Didwania, M. Singh, V. Mewada. 2014. Residential buildings in India: Energy Use Projection and Savings Potential. Technical Report, CEPT-GBPN

Rey, F., E. Velasco, and F. Varela. 2007. Building energy analysis: A methodology to assess building energy labelling. *Energy and Buildings* 39(6): 709–716

Reynolds. L. 2010. *Energy Efficiency in Buildings—Evolution of a Standard Series*. Pretoria: SANS 204 Working Group

SAM. 2011. Solar Advisor Model (SAM). National Renewable Energy Laboratory

M. Santamouris and D. Asimakopoulos 1996. *Passive Cooling of Buildings*. London, UK: James & James

Scheuer, C., G. A. Keoleian, and P. Reppe. 2003. Life cycle energy and environmental performance of a new university building: Modelling challenges and design implications. *Energy and Buildings* 35(10): 1049–1064

Shaw, M. R., K. W. Treadaway, and S. T. P. Willis. 1994. Effective use of building mass. *Renewable Energy* 5: 1028–1038

Singhaputtangkul, N., S. P. Low, A. L. Teo, and B.-G. Hwang. 2014. Criteria for architects and engineers to achieve sustainability and buildability in building envelope designs. *Journal of Management in Engineering* 30(2): 236–245

Strand, R., F. Winkelmann, F. Buhl, J. Huang, R. Liesen, C. Pedersen D. Fisher, R. Taylor, D. Crawley, and L. Lawrie. 1999. Enhancing and extending the capabilities of the building heat balance simulation technique for use in EnergyPlus. In *Building Simulation '99*, Vol. I, Kyoto, Japan, International Building Performance Simulation Association, pp. 81–88

Summers, J. 2010. 4 Green Building Trends 4U. Details available at http://www.santamonicapropertyblog.com/

Swan, L. G. and V. I. Ugursal. 2009. Modeling of end-use energy consumption in the residential sector: A review of modeling techniques. *Renewable and Sustainable Energy Reviews* 13(8): 1819–1835

Tu, J. V. 1996. Advantages and disadvantages of using artificial neural networks versus logistic regression for predicting medical outcomes. *Journal of Clinical Epidemiology* 49(11): 1225–1231.

U.S. Environmental Protection Agency. 2010. Green Building. Details available at http://www.epa.gov/greenbuilding/pubs/about.htm, last accessed on 08 July 2011

Ulrich, R., X. Quan, C. Zimring, A. Joseph, and R. Choudhary. 2004. *The Role of the Physical Environment in the Hospital of the 21st Century: a once-in-a-lifetime opportunity.* Concord, CA: The Centre for Health Design

U.S. Department of Energy, 2010. Getting Started with EnergyPlus: Basic Concepts Manual—Essential Information You Need about Running EnergyPlus. U.S. Department of Energy: ENERGYPLUS™

Veeraboina, P. and G. Yesuratnam. 2013. Significance of design for energy conservation in buildings: building envelope component. *International Journal of Energy Technology and Policy* 9: 34–52

Wong, S. L., K. K. Wan, and T. N. Lam. 2010. Artificial neural networks for energy analysis of office buildings with daylighting. *Applied Energy* 87(2): 551–557

Yang, J., H. Rivard, and R. Zmeureanu. 2005. Building energy prediction with adaptive artificial neural networks. *Building Simulation* August, 1401–1408

Yezioro, A., B. Dong, and F. Leite. 2007. An applied artificial intelligence approach towards assessing building performance simulation tools. *Energy and Buildings* 40(1): 612–620

Solar Photocatalytic Treatments of Wastewater and Factors Affecting Mechanism: A Feasible Low-cost Approach

Arya Pandey[a], Shamshad Ahmed[a], Virendra Kumar[a], Pratibha Singh[b], Richa Kothari[a,*]

[a]Bioenergy and Wastewater Treatment Laboratory, Department of Environmental Science, Babasaheb Bhimrao Ambedkar University, Lucknow, Uttar Pradesh 226025
[b]Department of Chemistry, Kamala Nehru Institute, Ram Manohar Lohia Awadh University, Faizabad, Uttar Pradesh, 224001
* E-mail: kothariricha21@gmail.com

20.1 INTRODUCTION

One of the most significant problems that affect people around the world is scarcity of clean water. Clean water is not only a basic need but also important for the functioning of a civil society. Water crisis is an issue the world over, and the clean water crisis is expected to become worse in the coming decades. Therefore, scientists are developing innovative technologies that can clean, detoxify, and decontaminate wastewater so that it can be reused later (Riaz, Mohamad, Azmi, *et al.* 2014). Waste that contaminates surface and groundwater comes from various sources; it can be industrial waste effluent, agricultural waste like pesticides and fertilizers running off to rivers and ponds, domestic waste, landfill, and so on. Non-degradable organic pollutants, known as bio-recalcitrant pollutants, are also present in wastewater. Earlier, wastewater was commonly treated through physical, chemical, and biological methods. Conventional methods of wastewater treatment have long been known to remove many contaminants, both chemical and microbial, which are a danger to public health and the environment. However, the effectiveness of these processes has been increasingly challenged with the identification of more and more contaminants that they cannot remove, rapid growth in population and industrial activities, and diminishing water resources. Now innovative technologies are being put in to use. Research is being conducted to develop new technologies to remove such pollutants from wastewater at a low cost and with less energy input. However, research should not be limited to the areas of reducing costs and energy input. It should also work towards finding ways to minimize the use of chemicals to avoid adverse effects on the environment as well as human health. It is known that the use of ultraviolet (UV) radiation-generated artificial lamps and ozone production can be expensive and are not economically feasible, hence scientists are working to improve wastewater treatment technologies by using solar photocatalysts and solar radiation. Out of these new technologies, photocatalytic-based wastewater treatment is a novel, economically feasible, and promising approach which has been in practice for a few decades. The main advantage of photocatalytic treatment of wastewater is that it is a low-cost and efficient approach that can bring about photodegradation of organic pollutants. It is a clean, eco-friendly technology which uses solar radiation for photocatalytic reaction through which one can minimize electricity use. The photocatalytic method involves a combination of catalyst and light, and has proven to be very effective.

Both homogeneous and heterogeneous catalytic reactions are important in wastewater treatment, but nowadays advanced oxidation processes are widely used due to their many advantages, including possibilities of different •OH radical productions. To enhance the efficiency of wastewater treatment, a large number of oxidants such as TiO_2/UV, H_2O_2/UV, photo-Fenton and ozone (O_3, O_3/UV and O_3/H_2O_2) are currently used in photocatalytic reactors. Over the years, a lot of innovative research has enhanced the rate of reaction and efficiency of photocatalytic reactors used for water disinfection.

20.2 SOLAR SPECTRUM

The energy that the earth receives from the sun is about 1.7×10^{14} kWh/day which amounts to 1.5×10^{18} kWh per year (Figure 20.1).

Radiation coming from the atmosphere has a wavelength between 0.2 m and 5 m, which is reduced between 0.3 m and 3 m when reaching the ground surface because of the absorption of some part of it by different atmospheric components (ozone, oxygen, carbon dioxide, aerosols, steam, and dust particle clouds). The earth receives a sufficient amount of solar UV radiation for almost ten months of a year. Solar radiation that reaches the ground without being absorbed or scattered to the atmosphere is called

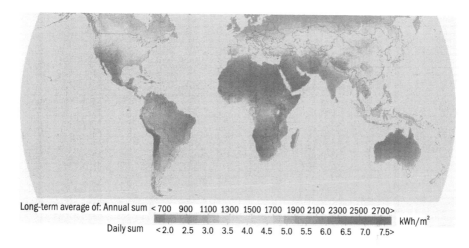

Long-term average of: Annual sum <700 900 1100 1300 1500 1700 1900 2100 2300 2500 2700> kWh/m²

Daily sum <2.0 2.5 3.0 3.5 4.0 4.5 5.0 5.5 6.0 6.5 7.0 7.5>

Fig. 20.1 *World map of global horizontal irradiation*
Source *Details available at http://solargis.info/doc/free-solar-radiation-maps-GHI*

direct radiation. Radiation that reaches the ground but gets dispersed is called diffuse radiation. The sum of both direct radiation and diffuse radiation is known as global solar radiation.

On cloudy days, the direct component of global radiation is less and the diffuse component is higher. The case is opposite on clear days. Figure 20.2 shows the standard solar radiation spectra (Hulstrom, Bird, and Riordon 1985) at ground level on a clear day. Table 20.1 lists the available solar spectrum.

The percentage of global UV radiation (direct and diffuse), worldwide, generally increases when the atmospheric transitivity, such as dust, cloud, aerosol decreases as shown in Figure 20.2.

The UV radiation required for photocatalysis may come from an artificial source or from the sun. Artificially generated UV radiation by lamps is an expensive process as they consume a lot of electricity and

Table 20.1 Available solar spectrum

Wave	Available electromagnetic radiation	Wavelength (nm)	Available radiation (%)
Short wave	UV-C	100–280	0
	UV-B	280–315	2–5
	UV-A	315–400	95–98
Visible	VIS	400–700	43
Long wave	NIR	700–1100	27.6
	FIR	1100–2500	25.4

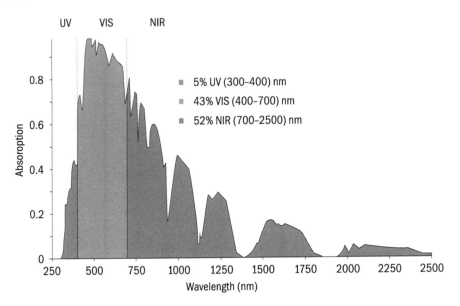

Fig 20.2 *Ultraviolet spectra at the Earth's surface*
Source *Biernat, Malinowski, and Gnat (2013)*

their maintenance and operation costs are high as well. By using natural available sunlight, these problems can easily be overcome. There are only a few examples of solar photocatalysis; others focus on artificial light as the UV source. It is important that we choose solar radiation as the UV source, since it is a naturally available energy source and can be conveniently harvested for the irradiation of semiconducting materials (Neppolian, Choi, Sakthivel, *et al.* 2002).

20.3   DESIGN OF SOLAR PHOTOCATALYTIC REACTORS

Photocatalysis requires a specific device known as the photoreactor, which can efficiently bring photons coming from the sun and chemical reagents into contact with the photocatalyst. Solar photoreactors significantly differ from classic chemical reactors as their geometry is of critical importance ensuring that solar radiation is efficiently collected. Different parts of the reactor play different important roles during the wastewater treatment process. In the case of photocatalytic detoxification of wastewater pollutants, the reactor should be made in such a way that the UV light can be efficiently transmitted. The material used to design the photocatalytic reactor must be impervious to UV light in order achieve high durability. This kind of reactor requires a reflective surface and a catalyst/sensitizer (Malato, Blanco, Vidal, *et al.* 2002). The absorber tube and the reflective surface are significant parts of this kind of reactor.

20.3.1 Absorber Tube

In a photocatalytic reactor, working fluid is an important part and its flow distribution inside the reactor must be assured. The material used in the absorber tube must be able to transmit UV light as well as resist UV light degradation. Usually fluoropolymers and acrylic polymers are used to make the absorber tube but it is not very economically viable due to its high cost. Therefore, low iron-content glass is the preferred material for manufacturing this tube.

20.3.2 Reflective Surface

In photocatalytic reactors, reflective surfaces play a significant role in achieving high optical quality and solar concentration. Aluminium is a better option than a traditional silver-coated mirror for the reflector/concentrator because of its high reflectivity and low cost. In these reactors, conventional operating parameters such as pressure, temperature, and mixing are less important for an optimal photocatalytic process. However, it should be noted that for the catalytic process the increase in temperature caused by the thermal reaction can have an effect on the performance of the reactor. When designing and scaling-up a solar photoreactor, one of the major problems faced is achieving a uniform distribution of sunlight inside the reactor. This is not always possible due to absorption, reflection, and scattering of the catalyst particles or the supporting material. It is worth going through some critical reviews that there is some difficulties in comparing the design and performance of the more common pilot and the commercial photoreactor (Zapata, Oller, Bizani, *et al.* 2009; Braham, and Harris 2009; Malato, Blanco, Maldonado, *et al.* 2004). Solar photoreactors can usually be divided into three different families: parabolic trough collectors (PTCs), non-concentrating collectors (NCCs), and compound parabolic collectors (CPCs).

Parabolic trough collectors

Parabolic trough collectors have a parabolic structure, with concentrated solar radiation on a treatment tube along the parabolic focal line through which the reactant fluid flows. The maximum efficiency of the PTC system is defined and calculated on the basis of the ratio between the collector aperture area and the absorbed area, which is technically known as concentrator factor (CF). The functioning of the PTC system is controlled by azimuth and elevation tracking system (Spasiano, Marotta, Malato, *et al.* 2015) (Figure 20.3). Thus, for maximum efficiency, the aperture plane of the collector should always be perpendicular to solar radiation that is reflected onto the reactor tube by a parabola. The area that intercepts solar

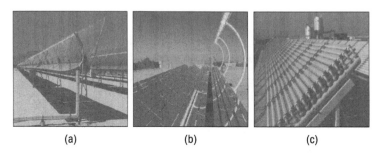

(a)   (b)   (c)

Fig. 20.3 *Types of solar collectors: (a) concentrating solar collector, (b) non-concentrating solar collector, and (c) compound parabolic collector*
Source *Galvez and Rodriguez (2003)*

radiation is the aperture area while the area that receives the sunlight is the absorber area. For photocatalytic applications with PCT reactors, the concentration factor ranges from 5 to 35 suns (Malato, Blanco, Maldonado, *et al.* 2004). The trough surfaces are usually made of aluminium to ensure highest reflectivity. The system supports turbulent flow with efficient homogenization. Several overviews on commercial PTCs designed for supplying thermal energy are available for reading.

The reactor tube is a closed system that prevents vaporization of volatile compounds. The photocatalyst is generally suspended in the fluid although research has been done on some designs for a supported photocatalyst. The major advantage of increased intensity of the incident radiation, which allows smaller photocatalyst loads than other photoreactors with the same solar light collecting area. This results in smaller receiver tube diameter requirements, which in turn means that operating pressures can be higher and construction material quantities lower (Spasiano, Marotta, Malato, *et al.* 2015).

One of the main disadvantages of PTCs is their geometry, due to which the collectors can use only direct beam radiation, making them practically useless on cloudy days. Their tracking system makes them expensive, and quantum efficiencies are low due to high electron/hole recombination (Malato, Blanco, Vidal, *et al.* 2002). Moreover, for intermediate and high solar zenith angles, the diffuse component in the UV solar spectrum may be equal to or greater than the direct component of UV solar radiation, even on cloudless days (Paulescu, Paulescu, Gravila, *et al.* 2012).

Non-concentrating collectors

The non-concentrating collector or inclined plate collector (IPC) is a flat or corrugated inclined plate over which fluid flows in a thin film. The photocatalyst is usually supported on the surface of the inclined plate, with the photons first travelling through the reactant fluid before reaching the photocatalyst. The back plate may be made of glass, metal, or stone. As the solar radiation is not concentrated, this reactor design is able to capture diffuse light. The top may also be left open to the atmosphere (Goslich, Dillert, and Bahnemann 1997), thus further increasing efficiency by excluding light absorption by the reactor cover and removing the potential for a suspended photocatalyst to form an opaque film on its inner surface. In the case of a NCC, its main drawback is high volatile chemical and water losses due to evaporation and interference from the atmosphere. It could also be dangerous to use this collector under windy conditions when wastewater to be treated is toxic. With higher flow rates, reactor residence times are shorter and film thickness increases, thus leading to increased mass-transfer constraints and a decrease in NCC efficiency. NCCs, due to their design simplicity and low capital costs (Enzweiler, Mowery, Wagg, *et al.* 1994) (no moving parts or tracking mechanisms), have become known for being effective for small-scale operations, particularly in less developed regions where other wastewater treatment plants are unfeasible. However, non-concentrating reactors require larger surfaces than the concentrating ones, and therefore must be designed to resist the high operating pressure necessary for pumping the fluid, which substantially increases their cost. NCCs are mainly used in solar pilot plants for photocatalytic treatment of various industrial or agricultural wastewaters containing bio-refractory organic pollutants and for agro-industrial water disinfection using a heterogeneous catalyst (Khan, Reed, and Rasul 2012).

Compound parabolic collectors

When it comes to treating wastewater by photocatalytic treatment, compound parabolic collectors are an interesting cross between the two aforementioned photoreactors. They remove pollutants from wastewater in an efficient manner (Malato, Blanco, Vidal, *et al.* 2002). CPCs consist of stationary collectors along with a parabolic reflective surface around a cylindrical reactor tube. The main advantage of a CPC is that the reflector geometry reflects indirect light onto the receiver tube, and can, therefore, capture both direct and diffuse sunlight (Malato, Blanco, Richter, *et al.* 1997; Kalogirou 2014). Demonstration and pilot-scale CPCs with collector areas ranging from 3 m^2 to 150 m^2 have been used for homogeneous and heterogeneous photocatalytic removal of very toxic compounds, such as chlorophenols, pathogenic organisms, dyes, pesticides, chlorinated solvents, bacteria, and biorecacitrant compounds from water. Other CPC applications include treatment of sanitary landfill leachate, olive mill waste, and urban wastewater.

20.4   PHOTOCATALYTIC (SOLAR) TREATMENT OF POLLUTANT

Photocatalysis is a process that accelerates photoreaction in the presence of a catalyst, while photolysis can be defined as a chemical reaction in which a chemical compound gets broken down by photons. In photocatalytic treatment of wastewater, solar radiation is absorbed by the adsorbed surface. Photocatalytic detoxification has been described as an alternative method for cleaning polluted water (Malato, Blanco, Alarcón, *et al.* 2007, Bahnemann 2004). Homogeneous and heterogeneous photocatalytic treatment of pollutants with solar technologies to achieve the mineralization of toxics present in wastewater is a significant achievement, but now advanced oxidation processes (AOPs) are being applied to remove refractory organic pollutants and xenobiotics as well (Glaze, Kang, and Chapin 1987). Figure 20.4 explains solar photocatalysis-based wastewater treatment.

20.4.1   Homogeneous Catalytic Treatment

During homogeneous catalytic treatment (HCT), the adsorbed contaminated compounds present on the surface of the semiconductor absorb solar radiation in the visible range (Fernandez-Ibanez, Planko, Maitato, *et al.* 2003; Ohno 2004; Alkhateeb, Hussein, and Asker 2005). Ozone and H_2O_2 are usually used in homogeneous catalytic treatment to detoxify the wastewater. There are three major catalysts used in HCT: (i) H_2O_2, (ii) ozone, and (iii) photo-Fenton (Fe and H_2O_2). The use of H_2O_2 is very common in wastewater detoxification and has the following

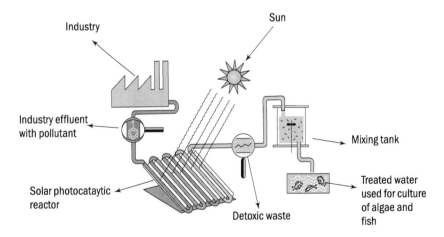

Fig. 20.4 *Schematic diagram of a photocatalysis-based wastewater treatment system*
Source *Galvez and Rodriguez (2003)*

advantages: (i) H_2O_2 is easily available and soluble in water to give a wide range of concentrations; (ii) no air emissions; and (iii) it gives a high-quantum yield of hydroxyl radicals; but the major drawback, however, is its molar extinction coefficient, which means that in water with high UV absorption the fraction of light absorbed by H_2O_2 may be low unless very large concentrations are used. The photocatalytic decolorization of dyes is also characterized by a free radical mechanism.

$$H_2O_2 + H_2O \longrightarrow HO_2^- + H_3O^+ \qquad ...(20.1)$$

$$H_2O^- + O_3 \longrightarrow O_3^- + HO_2^\circ \qquad ...(20.2)$$

$$O_3^- + H_2O \longrightarrow HO_3^\circ + OH^- \qquad ...(20.3)$$

$$HO_3^\circ \longrightarrow OH^\circ + O_2 \qquad ...(20.4)$$

The highly reactive radicals (such as $HO\bullet$ and $HO_2\bullet$) oxidize almost all organic substances to produce carbon dioxide, water, and inorganic salts. In a photo-Fenton reaction, Fe^{2+} ions are generally oxidized by H_2O_2 and produces one $\bullet OH$, and finally Fe^{3+} and other complexes are obtained which act as light-absorbing element and produce another radical while the initial Fe^{2+} is recovered.

$$Fe^{+2} + H_2O_2 \longrightarrow Fe^{+3} + OH^- + H^+ + OH^- \qquad ...(20.5)$$

$$Fe^{+3} + H_2O + hv \longrightarrow Fe^{+2} + H^+ + OH^- \qquad ...(20.6)$$

$$Fe(OOC\text{-}R)^{+2} + hv \longrightarrow Fe^{+2} + CO_2 + R^\circ \qquad ...(20.7)$$

Dissociation of some organic pollutants is restricted to UV radiation so the pollutant must keep in front of high intensity of radiation emitted by the artificial lamp or solar light to dissociate the organic pollutants. Usually, organic pollutants absorb solar radiation over a wide range of wavelengths, but strong absorption of light takes place at lower wavelengths, especially below 250 nm and the quantum yield of photodissociation increases at lower wavelengths because photon energy increases during the process and the free radicals generated can react with dissolved oxygen in water. Table 20.2 shows the impact of solar photocatalytic treatment (with the help of catalytic movements) on organic substances responsible for contamination in industrial wastewater.

20.4.2 Hetero Catalytic Treatment

Hetero photocatalytic treatment of wastewater is characterized on the basis of irradiation of the catalyst. In this process, the photocatalytic degradation takes place on simple, solid and stable semiconductors which irradiate and stimulate the rate of reaction. The process is called heterogeneous because it takes place in two phases—solid and liquid.

Table 20.2 Homogenous catalytic treatment of organic substances present in contaminated water sources

Substance	[Fe] (mM)	[H$_2$O$_2$] (mM)	pH	Solvent	Collector geometry	Removal (%)	References
Direct Black 38 (C$_{34}$H$_{25}$N$_9$Na$_2$O$_7$S$_2$)	0.1	5	2	SW	CPC	80	Bandala, Pelaez, Garcia-Lopez, et al. (1998)
Direct Black 38 (C$_{34}$H$_{25}$N$_9$Na$_2$O$_7$S$_2$)	1	50	2	SW	CPC	90	
Direct Black 38 (C$_{34}$H$_{25}$N$_9$Na$_2$O$_7$S$_2$)	1	50	2	RWW	CPC	60	
Phenols (from olive mill wastewaters)	1	588	2.8	RWW	CPC	97	Silva, Silva, Cristina Cunha-Queda, et al. (2013)
Phenols (from olive mill wastewaters)	5	147	2.8	RWW	FFR	88	
Mixture of 10 commercial pesticides	1	1	2.8	RWW	CPC	88	Bauer, Waldner, Fallmann, et al. (1999)
4-Chlorophenol (ClC$_6$H$_4$OH)	0.75	45	2.8	DW	Pool	100	Krutzler, Fallmann, Maletzky, et al. (2001)
4-Chlorophenol (ClC$_6$H$_4$OH)	0.75	45	2.8	DW	Pool	100	
Imidacloprid (C$_9$H$_{10}$ClN$_5$O$_2$)	0.05	15	2.8	DeW	CPC	100	Malato, Caceres, Aguera, et al. (2001)
Diethyl phthalate (C$_{12}$H$_{14}$O$_4$)	0.305	0		DeW	CPC	100	Mailhot, Sarakha, Lavedrine, et al. (2002)
Red wine (WV)	55	12	3	MQW	CPC	46	Lucas, Mosteo, Maldonado, et al. (2009)

SW = Synthetic water, RWW = Real wastewater, DW = Distilled water, DeW = Desalined water

When the semiconductor comes in contact with the solution containing liquid electrolyte along with redox couple, a charge transfer takes place across the interface to create a balance between the potentials of the two phases; simultaneously an electric field is formed on the surface of the semiconductor. Now the bending vibration of bands starts from the mass of the semiconductor towards the interface, as the electric field develops. During the photodetoxification process, the photon absorbs appropriate energy while the bending of bands provides an appropriate condition for carrier separation. In the case of the semiconductor or electrolyte, two charge carriers should react at its interface with the species present in the solution. In a steady state, the amount of charge transferred towards the electrolyte must be equal and opposite for two different types of charge carriers. A charge transfer takes place across the interface in a semiconductor-mediated redox process and the generated electron moves to the mass of the semiconductor and hole-migrate towards the surface. The oxidation or reduction of the pollutant, the redox reaction in other words, takes place when the charge carriers separate quickly from one another. For photocatalytic treatments, oxides of metal and sulfides are used as the semiconductor. Table 20.3 lists selected semiconductors, along with their band gap energy, which are used as catalysts in photocatalytic processes.

Table 20.3 List of selected semiconductor materials with specified band gap energy

Material	Band gap (eV)	Wavelength corresponding to band gap (nm)
PbS	0.286	-
SnO_2	3.9	318
ZnS	3.7	336
$SrTiO_3$	3.4	365
$BaTiO_3$	3.3	375
TiO_2	3	390
ZnO	3.2	390
ZrO_2	3.87	-
WO_3	2.8	443
CdS	2.5	497
GaP	2.3	540
Fe_2O_3	2.2	565
Cu_2O	2.17	-
CdO	2.1	590
CdSe	1.7	730
GaAs	1.4	887

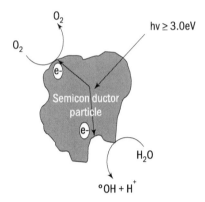

Fig. 20.5 *Effect of UV radiation on a TiO$_2$ particle dispersed in water*

Photoholes and photoelectrons produced during the photoreaction process migrate to the surface of adsorbed species. This is known as photocatalysis. It has been mentioned earlier that TiO$_2$ and ZnO are catalysts that are widely used in photocatalytic reactions because of their efficient absorption of long-wavelength radiation and their stability towards chemicals. There are some other semiconductors such as WO$_3$, CdS, GaP, CdSe, and GaAs that absorb a wide range of solar spectrum and form chemically activated surface-bound intermediates, but these photocatalysts usually degrade during the repeated catalytic cycles involved in heterogeneous photocatalysis. Table 20.4 points to a significant removal of different organic pollutants through the photocatalytic treatment of wastewater by using TiO$_2$ as a catalyst.

Table 20.4 Hetero photocataytic treatment of different organic pollutants by using TiO$_2$ as a catalyst, using synthetic water as solvent and compound parabolic concentrator as collector

Substance	Chemical composition	TiO2 (mg/l)	pH	Removal (%)
Progesterone	$C_{21}H_{30}O_2$	335	-	100
Triclosan	$C_{12}H_7Cl_3O_2$	335	-	100
Hydroxybiphenyl	$C_6H_5C_6H_4OH$	336	-	100
Diclofenac	$C_{14}H_{11}Cl_2NO_2$	337	-	100
Ibuprofen	$C_{13}H_{18}O_2$	338	-	100
Ofloxacin	$C_{18}H_{20}FN_3O_4$	339	-	100
Caffeine	$C_8H_{10}N_4O_2$	340	-	100
Acetaminophen	$C_8H_9NO_2$	341	-	100
Sulfamethoxazole	$C_{10}H_{11}N_3O_3S$	342	-	100

Contd...

Table 20.4 *Contd...*

Substance	Chemical composition	TiO2 (mg/l)	pH	Removal (%)
Antipyrine	$C_{11}H_{12}N_2O$	343	-	70
Flumequine	$C_{14}H_{12}FNO_3$	344	-	100
Isoproturon	$C_{12}H_{18}N_2O$	345		100
Ketorolac	$C_{15}H_{13}NO_3$	346		100
Carbamazepine	$C_{15}H_{12}N_2O$	347		100
Atrazine	$C_8H_{14}ClN_5$	335		80

20.4.3 Advanced Photocatalytic Treatment

Besides these two conventional homogeneous and heterogeneous catalytic treatments, there are some advanced photocatalytic treatments that are also in use. These techniques are executed by noble co-catalyst or platinum deposition on the catalyst and remove pollutants in a more efficient manner.

Advanced oxidation processes of wastewater treatment

The advanced oxidation process (AOP) for wastewater treatment is a process in which highly reactive free radicals such as hydroxyl radicals (OH) are produced in adequate quantity for wastewater detoxification. It is a simple and easy-to-handle technique, widely used for the dissociation of organic pollutants present in wastewater (Zhou and Daniel 2004). The AOP involves chemical oxidation or redox reaction of wastewater using different oxides such as hydrogen peroxide, ozone, combined ozone and hydrogen peroxide, hypochlorite, Fenton's reagent, ultraviolet-enhanced oxidation such as UV/O_3, UV/H_2O_2, UV/air, wet air oxidation, catalytic wet air, and others, out of which hydroxyl radicals are a highly reactive species that dissociate a wide range of organic pollutants present in wastewater. Table 20.5 indicates that hydroxyl radical is the second strongest oxidant used in wastewater treatment.

Table 20.5 Various oxidants used for wastewater treatment in AOPs

Oxidant	$E°$ (V)
Iodine (I_2)	1.36
Bromine (Br_2)	1.45
Chlorine (Cl_2)	1.49
Hypochlorous acid (HClO)	1.50
Chlorine dioxide (Cl_2O)	1.59
Potassium permanganate ($KMnO_4$)	1.67

Contd...

Table 20.5 Contd...

Oxidant	$E°$ (V)
Hyrdroperoxyl radical (H_2O+)	1.70
Hydrogen peroxide (H_2O_2)	1.78
Ozone (O_3)	2.07
Atomic oxygen (O_2)	2.42
Hydroxyl radical (OH)	2.8
Fluorine (F_2)	3.03

Incorporation of noble metal co-catalyst onto semiconductor

To enhance the photocatalytic activity under UV radiation, TiO_2 material is doped by noble metal to increase the potential of catalysts (Chawdhury, Gomma, and Ajay 2013). To enhance the photocatalytic degradation of chloroform, elements such as Fe^{3+}, Ru^{3+}, V^{4+}, Mo^{5+}, Os^{3+}, Re^{5+}, and Rh^{3+} are doped by TiO_2 (Choi, Park, and Hoffmann 2010) under UV radiation. It has also been observed that photocatalytic activity decreases in the case of Co^{3+} and Al^{3+} ion doped by TiO_2. It is, therefore, clear that a noble metal used for doping the catalyst plays an important role in enhancing the efficiency of the photocatalytic reaction.

Platinum-deposited TiO_2 methods

There are several methods available for the deposition of platinum on the catalyst surface, some of them being photo deposition of metal (Li and Li 2002), impregnation (Kryukova, Zenkovets, Shutilov, et al. 2007), chemical vapour deposition, and chemically reduced Pt salts (Mei, Sharma, Lu, et al. 2005). The use of this process allows a different degree of surface modification and subsequent catalytic efficiency.

Photodeposition method

In photodeposition, TiO_2 powder is usually suspended in a deaerated solution and dispersed by sonication. The solution contains a platinum precursor such as H_2PtCl_6, K_2PtCl_6, $H_2Pt(OH)_6$, $Pt(NH_3)_2 (NO_2)_2$, water, and a sacrificial organic reagent such as methanol, ethanol, propan-2-ol, 2-methylpropan-2-ol, oracetic acid. Ideally the pH of the solution should be 3.0, that is acidic before the reaction starts. The suspension is stirred and irradiated with either UV or solar light generated from an Hg vapour lamp or Xe arc lamp (after a 1-to-2-hour irradiation, the colour of the suspension changes from white to black owing to Pt deposition).

Impregnation

In the impregnation process, oxides of metal or metal complexes are dissolved in liquid solution in order to get into contact with TiO_2. In anaqueous solution, the noble metal adsorbs onto the high surface area of the porous oxide catalyst. The catalyst slurry is then filtered if an excess of solution has been employed, or just evaporates to dryness (Spieker and Regalbuto 2001). Then it is treated further to transform the metal from its precursor state into its active form.

20.5 FACTORS AFFECTING SOLAR PHOTOCATALYSIS

The rate of photodegradation of the pollutants present in wastewater depends on the type of solvent, that is the characteristics of the wastewater and its composition and physico-chemical parameters such as pH, mixing, temperature, catalyst.

20.5.1 Nature and Concentration of Pollutant (Organic and Inorganic)

The rate of photodegradation during photocatalytic treatment is directly related to the concentration and type of pollutants present in wastewater. An increase in the concentration of pollutants in wastewater saturates the surface area of the photocatalyst and reduces the radiation efficiency which in turn deactivates the photocatalyst. The chemical composition and nature of the target compound that has to be degraded also plays a significant role in the degradation performance in the photocatalytic reactor. Table 20.6 lists different kinds of photodegradation of organic pollutants in the presence of different homogeneous and heterogeneous photocatalysts.

Inorganic ions such as phosphate, nitrate, sulphate, magnesium, zinc, copper, aluminium, bicarbonate, chloride, inhibit the photocatalytic reaction because they adsorb onto the surface area of the photocatalyst. It has been noted that photocatalytic deactivation occurs while using the photocatalyst in a slurry that possesses inorganic ions, which leads to the deactivation process by attaching itself (slurry) onto the surface area of the TiO_2. Inorganic ions, such as nitrate, chloride, carbonate and sulphates play a significant role in deactivating the photocatalytic activity of its surface area. Phosphate, iron and copper, if present in adequate amounts, decrease the photocatalytic degradation of wastewater, while magnesium, calcium, and zinc have little effect on the photodegradation process. Table 20.7 lists the photodegradation of inorganic compounds in the presence of homogeneous and heterogeneous photocatalysts.

Table 20.6 Degradation of organic pollutants by homogeneous and heterogeneous photocatalytic treatments

Compound degraded	Photocatalyst used	Source of light	Intermediate detected	References
		Aliphatic compounds		
$CHCl_3$	Degussa P-25 TiO_2	1000 W Xe arc lamp	C_2Cl_6, C_2Cl_4	Choi and Hoffman (1996)
$CHBr_3$	Degussa P-25 TiO_2	1000 W Xe arc lamp	C_2Br_4	
CCl_4	Degussa P-25 TiO_2	1000 W Xe arc lamp	CO, CO_2	
Dichloro Methane	Betonite clay pillared by TiO_2 and titanium wx changed clays	1000 W Xe arc lamp	Not analysed	Tanguay, Suib, Buddiman, et al. (1989)
2-Propanol	Degussa P-25 attached to glass spiral	20 WNEC black light fluorescent tube,	Not analysed	Abdullah, Fgary, and Matthews (1996)
Trichlororthylene	TiO_2	Sunlight	Not analysed	Mehos and Turchi (1998)
Methyl-tert-Butyl ether	TiO_2 Degussa P-25	450 W medium-pressure mercury lamp	Tert-butyl formate, tert-butyl alcohol, acetone	Barreto, Gray, and Anders (1989)
Glycolic acid	TiO_2 Degussa P-25 on stainless steel	40 W Hg lamp	Not analysed	Mazzarino and Piccinini (1999)
Citric acid	TiO_2 Degussa P-25 on stainless steel	40W Hg lamp	Not analysed	Mazzarino and Piccinini (1999)
Monocrotophos	Degussa P-25 TiO_2	20 W blacklight fluorescent tube	A few detected and identified	Hua, Manping, Zongfeng, et al. (1995)

Contd...

Table 20.6 Contd...

Compound degraded	Photocatalyst used	Source of light	Intermediate detected	References
		Aromatic compounds		
Malic acid	Degussa P-25 TiO$_2$ on quartz	Philips	Not analysed	Herrmann, Tahiri, Air-lochou, et al. (1997)
Benzene	Degussa P-25	Seven 15W black light blue fluorescent bulbs	Phenol and 1,4-benzoquinone as major intermediates	Truchi and Ollis (1989)
Chlorobenzene	Anatase TiO$_2$	UV lamp	Not analysed	Butler and Davis (2000)
Nitrobenzene	Degussa P-25	Sunlight	p-nitrophenol, o-nitrophenol, 2-nitroresorcinol, 4-nitrocatechol	Bhatkhande, Pangarker, and Beenackers (2001)
Phenol	TiO2 Degussa P-25 immobilized on glass pearls	125 W	Hydroquinone, p-benzoquinone, and hydrohydroquinone (neg)	Trailas, Peral, and Doneneck (2000)
Phenol	Degussa P-25	Sunlight	Benzoquinone, pyrocatechol, resorcinol, hydroquinone	Yawalkar, Bhatkahnde, Pangarkar, et al. (2001)
Toluene	Anatase TiO$_2$	UV lamp	Not analysed	Butler and Davis (2000)

Table 20.7 Degradation of inorganic components by homogeneous and heterogeneous photocatalytic treatments

Compounds degraded	Photocatalyst used	Source of light	Intermediate detected	References
$AgNO_3$	TiO_2 anatase and rutile	400W high-pressure mercury lamp	None	Ohantani, Okugawa, Nishimoto, et al. (1997)
$HgCl_2$	Degussa P-25 TiO_2	Simulated solar light	None	Serpone, Ah-you, Tran, et al. (1997)
CH_3HgCl	Degussa P-25 TiO_2	Simulated solar light	None	Serpone Ah-you, Tran, et al. (1997)
Reduction of Cr(VI) to Cr(III)	Degussa P-25 TiO_2 Hombikat UV100, WO_3ZnO	Medium-pressure mercury lamp	None	Khalil, Mourad, and Rophael (1998)

20.5.2   Physico-chemical Parameters

Solar catalytic treatment of wastewater involves a physical process, chemical process, or combination of both the processes. These parameters, whether homogeneous, heterogeneous, or an AOP, strongly influence the treatment of wastewater.

pH

Radiation-induced treatment of wastewater depends on pH because it imparts a significant effect through the charge on catalyst particles, position of conductance, valence bond, and size of aggregates. Protonation and deprotonation of the surface area of the TiO_2 changes according to the acidic and alkaline conditions of the reaction. The reaction given below represents the condition of protonation and deprotonation (Galvez and Rodriguez 2003).

$$TiOH + H^- \longrightarrow TiOH_2 \qquad \qquad ...(20.7)$$

$$TiOH + OH^- \longrightarrow TiO + H_2O \qquad \qquad ...(20.8)$$

The pH value of wastewater is a highly significant parameter for treatment processes because it affects the adsorption of pollutants which takes place on the surface area of the photocatalyst during the process of photocatalytic degradation of wastewater. Discharge of wastewater from different industrial (dye, textile, pharmaceutical, dairy) as well as domestic and food waste sources have pH values in different ranges. In the case of photocatalytic treatment of wastewater, the generation of hydroxyl

Table 20.8 Total degradation of DDT mixture at different pH values

pH	Total degradation after 0.5 hours %	Total degradation after 1.5 hours	Total degradation after 2.5 hours %
2	31.6	69.1%	98.1
7	31.4	68.7%	98.0
10	31.2	69.2%	98.1

Source Alawi (2015)

radical is very important for the reactions, which in turn depend on the pH value of the solution. The pH plays a crucial role in the chemical nature of wastewater as well as hydroxyl radiation generation. The effect of pH on wastewater treatment using UV and solar radiation and the consequent degradation of organic and inorganic pollutants are processes that have been extensively studied. Table 20.8 documents an experimental study, involving the degradation of a DDT (pesticide) in percentage and the total degradation time. The results show that change in pH also affects the degradation but not to a great extent and the degradation process can complete in 2.5 hours.

Both heterogeneous and homogeneous solar photocatalysis are affected by the pH of the solution. Heterogeneous photocatalysts have a strong pH dependency on the band gap potential as well as on the surface charge.

Mixing

It has been documented that the mixing process or proper aeration of wastewaer does not affect the total percentage degradation, and, therefore, mixing is not necessary when using conventional methods of wastewater treatment. However, in treatments involving photocatalytic degradation, mixing plays a very important role in achieving an equal distribution and circulation of the solid catalyst. Table 20.9 shows the effect of mixing and not mixing waste with a solid catalyst and the consequent percentage degradation.

Table 20.9 Effect of mixing on the degradation of DDT (%) with flow rate of 3.5 L/min

Status	Duration	TiO_2 concentration (g/L)	Total degradation (%)
With mixing	5.5	2.0	96.8 ±0.35
Without mixing	2.5	2.0	98.0 ± 0.40

Source Alawi (2015)

Temperature

It has been observed that an increase in temperature also increases the rate of photocatalytic reaction during wastewater treatment. There are some studies indicating if the temperature rises above 80°C, it would lead to the recombination of charge carriers which would make the absorption of organic compound on the surface of the catalyst difficult. The most favourable temperature for radiation-induced wastewater treatment is 20–80°C. If it falls below 20°C, the photocatalytic reaction would not proceed any further. Temperature plays a significant role in increasing or decreasing the kinetic rate of homogeneous photocatalysis (Hussen and Abass 2010). The rate of reaction increases with increasing temperature up to a certain limit; if the temperature increases further, the reaction rate falls due to thermal decomposition. It has also been observed that at high temperatures, adsorption capacity of the solid substrate gets reduced.

Catalyst

In a photocatalysis-based treatment of wastewater, the catalyst plays a significant role in enhancing the rate of reaction. Surface morphology of the catalyst is a significant character in radiation-induced photocatalytic reaction during wastewater treatment because there is a correlation between the surface coverage of the photocatalyst and the organic compounds present in the wastewater.

Catalytic loading

The rate of photocatalysis reaction is directly affected by the concentration of the catalyst used in a heterogeneous catalytic regime. If the concentration of catalyst crosses its saturation point, the light photon adsorption coefficient decreases (Agustina, Ang, and Vareek 2015). In the presence of excess photocatalyst, a light screening effect may occur which reduces the surface area exposed to irradiation and, hence, reduces the photocatalytic efficiency of the process.

20.5.3   Light Intensity and Solar Irradiance

The rate of degradation of organic and inorganic pollutants present in wastewater can be magnified by increasing exposure towards solar radiation. When the intensity of solar light is high, electron hole formation is predominant and, hence, it promotes electron hole recombination which plays a significant role in achieving photodegradation of the organic compound. However, poor light intensity decreases the formation of free radicals, and there is subsequently a lower percentage of wastewater degradation (Gonzalez-Martin 2000). The rate of solar photocatalytic reaction increases with increasing solar irradiance which is indirectly related to the quantum yield of the overall process. In fact, the total number of photons received by a photoreactor can be calculated by measuring and integrating the solar irradiance around the reactor walls.

20.5.4 Optical Density, Photoreactor Diameter, and Specific Area

Optical density, photoreactor diameter, and specific area are interrelated. It is important to note that in heterogeneous slurries, the rate of reaction increases with increasing catalyst load but decreases up to a certain limit (Akpan and Hameed 2009). Therefore, in a slurry reactor, the load of photocatalyst should be optimized for best performance of photocatalytic activities. A photoreactor with a larger diameter has a lower optimum catalyst load and in the case of a small diameter the catalyst load is higher.

20.6 RECOMMENDED ANALYTICAL METHODS

There are some analytical methods we can use to analyse the photodecomposition of organic pollutants present in wastewater. Table 20.10 lists these analytical methods and their descriptions. Photocatalytic treatment of wastewater (containing organic and inorganic pollutants) follows a stoichiometry which can be identified by appropriate analytical methods. During the treatment process of wastewater, analysis of water toxicity is very significant as it helps determine if the toxicity level of the contaminated water is being decreased or not. By-products or intermediates generated during the photocatalytic reaction can also be identified using these recommended analytical methods.

Table 20.10 Some common methods used for the analysis of wastewater in solar photocatalytic treatments

S.No.	Analytical method	Description
1.	Total organic carbon (TOC)	During the photodecomposition process, identification of generated intermediate compound is quite difficult. Therefore, to find out the moment at which only water and CO_2 remains is difficult as well. For this reason, TOC analysis is recommended as it is a reliable, simple and a rapid way to close the mass balance at any moment to get an idea of the remaining amount of intermediate compounds generated.

Contd...

Table 20.10 *Contd...*

S.No.	Analytical method	Description
2.	Intermediates analysis (GC-MS/HPLC-MS)	This method is helpful in that it gives a large amount of structural and spectral information. By using this method of analysis, separation of even similar intermediate compounds can be done.
3.	Extraction method	It is a common handling method for analysis of samples containing trace organic compounds. Liquid–liquid extraction method and solid-phase extraction method are widely used in the treatment of wastewater containing organic pollutants.

Source Galvez and Rodriguez (2003)

20.7   ECONOMICAL ASSESSMENT

Cost always plays a significant role when it comes to developing innovative technologies. It indicates whether any new standard commercial procedures or technologies are providing a significant reduction in processing cost and whether they are economically viable or not. The cost of wastewater treatment using solar detoxification systems depends on several parameters, such as rate constant, reactor type, catalyst, and pre- and post-treatment costs. In the case of radiation-induced detoxification of pollutants present in wastewater, its main contribution is the use of solar energy to solve contamination problems present in wastewater. Photocatalytic processes are mostly preferred for hazardous non-biodegradable organic pollutants. Photocatalytic technologies are used for the treatment of biologically recalcitrant compounds to achieve biodegradability of organic compounds, after which they can be transferred to a conventional biological plant for further treatment (Galvez and Rodriguez 2003). This type of treatment based on the advanced oxidation process reduces treatment time and optimizes the overall economics. It has been estimated that operation costs and investment involved in photocatalytic treatment of wastewater are lower than other wastewater treatment technologies such as ozone treatment, chlorination of water, wet oxidation process of water, incineration, and membrane separation.

20.8   CONCLUSION

Polluted wastewater always poses a problem for society. But the search for wastewater treatment technologies that are a low-cost/cost-effective option may end with the use of solar photocatalytic treatment approach. Although a variety of methods are being explored for the decomposition of organic and inorganic constituents/contaminants, solar radiation-induced degradation of pollutants is now a chief method currently under investigation. A large number of photocatalytic reactors have been designed and fabricated for the photodegradation of wastewater by using solar as well as artificial light, but it has been reported that photocatalytic reactors using solar light are economically viable in terms of saving electricity as well as operation cost. Out of the various processes, AOP solar photocatalytic treatment provides a new paradigm for treating wastewater. It can, thus, safely be concluded that photocatalytic treatment of wastewater is a much more eco-friendly and cost-effective method when compared to conventional treatments.

REFERENCES

Abdullah, M., K. C. L. Fgary, and R. W. Matthews. 1990. Effect of common inorganic anion on rates of photocatalytic oxidation of hydrocarbon over illuminated TiO_2. *Journal of Physical Chemistry* 94: 6620–6825

Adina, E. S., O. Cristina, L. Carmen, S. Paula, V. Paulina, B. Cornelia, and G. Ioan. 2013. Wastewater treatment methods. Details available at http://dx.doi.org/10.5772/53755

Agustina, T. E., H. M. Ang, and V. K. Vareek. 2005. A review of synergistic effect of photocatalysis and ozonation on wastewater treatment. *Journal of Photochemistry and Photobiology C: Photochemistry Reviews* 6(4): 264–73

Akpan, U. G. and B. H. Hameed. 2009. Parameters affecting the photocatalytic degradation of dyes using TiO_2-based photocatalysts: A review. *Journal of Hazardous Materials* 170(2–3): 520–9

Alfano, O., D. Bahnemann, A. Cassano, R. Dillert, and R. Goslich. 2000. Photocatalysis in water environment using artificial and solar light. *Catalyst Today* 58: 199–230

Alkhateeb, A., F. Hussein, and K. Asker. 2005. Photocatalytic decolorization of industrial wastewater under natural weathering conditions. *Asian Journal of Chemistry* 17(2): 1155–1159

Alwai, M. A. 2015. Solar bath-like system for photocatalytic degradation of organic pollutants. *European International Journal of Science and Technology*, pp. 2304–9693

Andreozzi, R., M. Canterino, and R. Marotta. 2006. Advanced oxidation processes (AOP) for water purification and recovery. *Water Research* 40: 3785–3792

Bahnemann, D. 2004. Photocatalytic water treatment: solar energy applications. *Solar Energy* 77: 445–459

Bandala, E. R., M. A. Pelaez, A. J. Garcia-Lopez, M. J. Salgado, and G. Moeller. 2008. Photocatalytic decolourisation of synthetic and real textile wastewater containing benzidine-based azo dyes. *Chemical Engineering Process* 47: 169–176

Barreto, R. D., K. A. Gray, and K. Anders. 1995. Photocataltic degradation of methyl-tetrt-butyl ether TiO_2 slurries: A proposed reaction scheme. *Water Research* 29: 1243–1248

Bauer, R., G. Waldner, H. Fallmann, S. Hager, M. Klare, T. Krutzler, S. Malato, and P. Maletzky. 1999. The photo-Fenton reaction and the TiO_2/UV process for wastewater treatment: novel developments. *Catalysis Today* 53(1): 131–144

Bhatkhande, D. S., V. G. Pangarker, and A. A. C. M. Beenackers. 2003. Photocatalytic degradation of nitrobenzene using TiO_2 chemical effects. *Water Research* 37(6): 1223–1230

Biernat, K. , A. Malinowski, and M. Gnat 2013. The possibility of future biofuels production using waste carbon dioxide and solar energy. Biofuel "Biofuels - Economy, Environment and Sustainability. *DOI: 10.5772/53831*

Bossmann, S. H., E. Oliveros, S. Göb, S. Siegwart, E. P. Dahlen, L. M. Payawan, Jr, S. Matthias, M. Wörner, and A. M. Braun. 1998. New evidence against hydroxyl radicals as reactive intermediates in the thermal and photochemically enhanced Fenton reactions. *Journal of Physical Chemistry* A102(28): 5542–5550

Braham, R. J. and A. T. Harris. 2009. Review of major design and scale-up considerations for solar photocatalytic reactors. *Industrial & Engineering Chemistry Research* 48: 8890–8905

Butler, E. C. and A. P. Davis. 1993. Photocatalytic oxidation in aqueous titanium dioxide suspension: The influence of dissolved transition metal. *Journal of Photochemistry and Photobiology* 70(3): 273–283

Chandan, S., C. Rubina, and S. T. Rajendra. 2011. Performance of advanced photocatalytic detoxification of municipal wastewater under solar radiation: A mini review. *International Journal of Energy and Environment* 2(2): 337–350

Chawdhury, Punkaj, H. Gomma, and K. Ajay. 2013. Dye-sensitized Photocatalyst: A breakthrough in green energy and environmental detoxification. ACS symposium series

Choi, W. and M. R. Hoffmann. 1996. Novel photocatalytic mechanisms: $CHCl_3$, $CHBr_3$, CCl_3CO_2 degradation and the fate of photo generated trihalomethyl acids over TiO_2 in a flow system. *Journal of Chemical Technology and Biotechnology* 67: 237–242

Choi, J., H. Park, and M. R. Hoffmann.2010. Effects of single metal-ion doping on the visible-light photo reactivity of TiO_2. *Journal of Physical Chemistry* 114(2): 783–792

Enzweiler, R. J., D. L. Mowery, L. M. Wagg, and J. J. Dong. 1994. A pilot scale investigation of photocatalytic detoxification of BETX in water. In *Solar*

Engineering, D. E. Klett, R. E. Hogan, and E. Tanaka (eds), pp. 155–62. New York: Asme

Fernandez-Ibanez, P., J. Planko, S. Maitato, and F. de las Nieres. 2003. Application of colloidal stability of TiO_2 particles for recovery and reuse in solar photocatalysis. *Water Research* 37(13): 3180–3188

Galvez, J. B. and S. M. Rodriguez. 2003. Solar detoxification technology. In *Solar Detoxification*. Paris: UNESCO Publishing. Details available at http://unesdoc. unesco.org/images/0012/001287/128772e.pdf

Glaze, W. H., J. W. Kang, and D. H. Chapin. 1987. The chemistry of water treatment processes involving ozone, hydrogen peroxide and UV-radiation. *Ozone: Science Engineering* 9: 335–352

Gogate, P. R. and A. B. Pandit. 2004. A review of imperative technologies for wastewater treatment I: oxidation technologies at ambient conditions. *Advance Environmental Research* 8: 501–551

Gonzalez-Martin, A., O. J. Murphy, and C. Salinas. 2000. Photocatalytic oxidation of organics using a porous titanium dioxide membrane and an efficient oxidant. *US patent* 6, P. 186

Goslich, R., R. Dillert, and D. Bahnemann. 1997. Solar water treatment: principle and reactors. *Water Science and Technology* 36: 137–148

Herrmann, J. M., H. Tahiri, Y. Air-lochou, G. Lassaletta, A. R. GonzalerElipe, and A. Fernandez. 1997. Characterization and photocatalytic activity in aqueous medium a of TiO_2 and $Ag\text{-}TiO_2$ coating on Quartz. *Applied Catalysis B: Environmental* 13: 219–228

Hua, Z., Z. Manping, X. Zongfeng, and G. K. C. Low. 1995. Titanium dioxide mediated photocatalytic degradation of monocrotophos. *Water Research* 29: 2681–2688

Hussen, F. H. and T. A. Abass. 2010. Photocatalytic treatment of textile industrial wastewater. *International Journal of Chemical Science* 8(3): 1353–64

Hulstrom, R., R. Bird, and C. Riordon. 1985. Spectral solar irradiance data sets for selected terrestrial condition. *Solar cells* 15: 365–391

Kalogirou, S.A. 2014. *Solar Energy Engineering: Processes Systems*. UK: Academic Press

Khalil, L. B., W. E. Mourad, and M. W. Rophael. 1998. Photocatalytic reduction of environmental pollutant Cr(VI) over some catalysis and environmental. *Applied Catalysis B: Environmental* 70(3): 267–273

Khan, S. J., R. H. Reed, and M. G. Rasul. 2012. Thin-film fixed-bed reactor for solar photocatalytic inactivation of Aeromonashydrophila: influence of water quality. *BMC Microbiology* 12(285): 1–13

Krutzler, R., H. Fallmann, P. Maletzky, R. Bauer, and S. Malato. 1999. Solar-driven degradation of 4-Chlorophenol Blanco. *Journal of Catalysis Today* 54: 321–327

Kryukova, G. N., G. A. Zenkovets, A. A. Shutilov, M. Wilde, K. Gunther, D. Fassler, and K. Richter. 2007. Structural peculiarities of TiO$_2$ and Pt/TiO$_2$ catalysts for the photocatalytic oxidation of aqueous solution of Acid Orange 7 Dye upon ultraviolet light. *Applied Catalysis B: Environmental* 71(3): 169–176

Li, F. B. and X. Z. Li. 2002. The enhancement of photodegradation efficiency using Pt-TiO$_2$ catalyst. *Chemosphere* 48(10): 1103–1111

Lucas, M. S., R. Mosteo, M. I. Maldonado, S. Malato, and J. A. Peres. 2009. Solar photochemical treatment of winery wastewater in a CPC reactor. *Journal of Agricultural and Food Chemistry* 57: 11242–11248

Mailhot, G., M. Sarakha, B. Lavedrine, J. Caceres, and S .Malato. 2002. Fe(III)-solar light induced degradation of diethyl phthalate (DEP) inaqueous solutions. *Chemosphere* 49: 525–532

Malato, S., J. Blanco, D. C. Alarcón, M. I. Maldonado, P. Fernández, and W. Gernjak. 2007. Photocatalytic decontamination and disinfection of water with solar collectors. *Catalysis Today* 122:137–149

Malato, S., J. Blanco, M. I. Maldonado, P. Fernandez, D. Alarcon, M. Collares, J. Farinha, and J. Correia. 2004. Energy of solar photocatalytic collector. *Solar Energy* 77: 513–524

Malato, S., J. Blanco, A. Vidal, and C. Richter. 2002. Photocatalysis with solar energy at pilot plant scale: an overview. *Applied Catalyst B: Environmental* 37: 1–15

Malato, S., J. Caceres, A. Aguera, M. Mezcua, D. Hernando, J. Vial, and A. R. Fernandez-Anda. 2001. Degradation of imidacloprid in water by photo-Fenton and TiO$_2$ photocatalysis at a solar pilot plant: A comparative study. *Environmental Science and Technology* 35: 4359–4366

Malato, S., J. Blanco, C. Richter, D. Curco, and J.Gimenez. 1997. Low-concentrating CPC collectors for photocatalytic water detoxification—comparison with a medium-concentrating solar collector. *Water Science and Technology* 35(4): 157–164

Malato', Sixto, Manuel I. Maldonado, PilarFernández-Ibáñez, Isabel Oller, Inmaculada Polo, and Ricardo Sánchez-Moreno. 2015. Decontamination and disinfection of water by solar photocatalysis: The pilot plants of the Plataforma solar deAlmeria. *Material Science in Semiconductor Processing* 42(1): 15–23

Mazzarino, I. and P. Ppiccinini. 1999. Photocatalytic oxidation of organic acids in aqueous media by supported catalyst. *Chemical Engineering Science* 54: 3107–3011

Mehos, M. S. and C. S. Truchi. 1993. Field-testing solar photochemical detoxification of TCE contaminated groundwater. *Environmental Progress* 12: 194–199

Mei, Y., G. Sharma, Y. Lu, M. Ballauff, M. Drechsler, T. Irrgang, and R. Kempe. 2005. High catalytic activity of platinum nanoparticles immobilized on spherical polyelectrolyte brushes. *Langmuir* 21(26): 12229–12234

Neppolian, B., H. C. Choi, S. Sakthivel, B. Arabindoo, and V. Murugesan. Solar/ UV-induced photocatalytic degradation of three commercial textile dyes. *Journal of Hazardous Materials* 89(2-3): 303–317

Ohantani, B., Y. Okugawa, S. Nishimoto, and T. Kagiya. 1997. Photocatalytic activity of TiO_2 powder suspended in aqueous silver nitrate solution: correlation with pH dependent surface structures. *Journal of Physical Chemistry* 91: 355–3555

Ohno, T. 2004. Preparation of visible light active S-doped TiO_2 photocatalysts and their photocatalytic activities. *Water Science Technology* 49(4): 159–163

Paulescu, M., E. Paulescu, P. Gravila, and V. Badescu. 2012. *Weather Modeling and Forecasting of PV Systems Operation*. UK: Springer Science and Business Media

Riaz, N., B. K. Mohamad, A. Azmi, and M. Shariff. 2014. Iron-doped TiO_2 photocatalysts for environmental applications: fundamentals and progress. *Advanced Materials Research* 925: 689–693

Serpone, N., Y. K. Ah-you, T. P. Tran, R. Harris, E. Pelizzetti, and H. Hidaka. 1997. AMI-stimulated sunlight photo reduction and elimination of Hg(II) and $CH_3Hg(II)$ chloride salts form aqueous suspension of titanium dioxide. *Solar Energy* 39: 491–498

Shannon, Mark A., P. W. Bohn, Menachem Elimelech, John G. Georgiadis, B. J. Mariñas, and Anne M. Mayes. 2008. Science and technology for water purification in the coming decades. *Nature* 452: 301–310

Gonzalez-Martin, A., O. J. Murphy, and C. Salinas. 2000. Photocatalytic oxidation of organics using a porous titanium dioxide and an efficient oxidant. *US patent* 6, pp.186

Spasiano, D., R. Marotta, S. Malato, P. Fernandez-Iba¯nezb, and I. Di Somma. 2015. Solar photocatalysis: Materials, reactors, some commercial, and pre-industrialized applications: A comprehensive approach. *Applied Catalysis: Environmental* 170-171: 90–123

Silva, T. F. C. V., M. E. F. Silva, A. Cristina Cunha-Queda, A. Fonseca, I. Saraiva, R. A. R. Boaventura, and V. J. P. Vilar. 2013. Sanitary landfill leachate treatment using combined solar photo-Fenton and biological oxidation processes at pre-industrial scale. *Chemical Engineering Journal* 228: 850–866

Spieker, W. A. and J. R. Regalbuto. 2001. A fundamental model of platinum impregnation onto alumina. *Chemical Engineering Science* 56(11): 3491–3504

Tanguay, J. F., S. L. Suib, L. Buddiman, and L. L. Chung. 1989. Heterogenous photo degradation using titanium dioxide. *Journal of Catalysis* 117: 335–347

Trailas, M., J. Peral, and X. Doneneck. 2000. Photocatalytic mechanisms photogenerated trialomethyl radicals on TiO_2. *Environmental Science and Technology* 37: 131–136

Truchi, C. S. and D. F. Ollis. 1989. Mixed reactant a photocatalysis in intermediate and mutual rate inhibition. *Journal of Catalysis* 119: 483–496

Yang, J. C., Y. C. Kim, Y. G. Shul, C. H. Shin, and T. K. Lee. 1997. Characterization of photo reduced Pt/TiO$_2$ and decomposition of dichloroacetic acid over photo reduced Pt/TiO$_2$ catalysts. *Applied Surface Science* 121: 525–529

Yawalkar, A. A., D. S. Bhatkahnde, V. G. Pangarkar, and A. A. C. M. Beenackers. 2001. Solar-assisted photocatalytic and photochemical degradation of phenol. *Journal of Chemical Technology and Biotechnology* 7(6): 363–370

Zapata, A., I. Oller, E. Bizani, J. A. Sanchez-Perez, M. I. Maldonado, and S. Malato. 2009. Evaluation of operational parameters involved in solar photo-Fenton degradation of a commercial pesticide mixture. *Catalysis Today* 144: 94–99

Zhou, H. and W. S. Daniel. 2004. Advanced technologies in water and wastewater treatment. *Journal of Environmental Engineering and Science* 1(4): 247–264

21

Advancement in Phase Change Materials for Solar Thermal Energy Storage

R. K. Sharma[a], V. V. Tyagi[b],*, and A. K. Pandey[c]

[a]*Department of Mechanical Engineering, Jaypee University of Engineering and Technology, Guna, Madhya Pradesh 473226*

[b]*School of Energy Management, Shri Mata Vaishno Devi University, Katra, Jammu and Kashmir 182320*

[c]*UMPEDAC, University of Malaya, Kuala Lumpur, Malaysia*

**E-mail: vtyagi16@gmail.com*

21.1 INTRODUCTION

As energy consumption increases worldwide due to industrialization, the economic growth of countries and our standards of living also undergo change. Consumption of fossil fuel for energy production is the main cause for depletion of ozone layer and environmental pollution, which is increasing day-by-day. In such a scenario, energy generation from renenwable sources is a good solution to narrow the gap between the demand and supply of energy for useful applications; in particular, for low-temperature requirements, such as heating and cooling of a building, crop drying, water heating, cooking, and so on. Along with the development of renewable energy sources, it is imperative to also think about storage of energy, due to its intermittent availability in nature. Thermal energy storage (TES) systems are able to store energy available in excess, for example, solar energy, and this stored energy can be reused later whenever needed. This stored energy can be in the form of sensible or latent heat. Although storage of sensible heat is quite popular in industrial uses, the storage of latent heat is much more beneficial due to the fact that latent heat thermal energy storage (LHTES) systems have higher storage density at a small container size. Depending upon the uses of stored energy, they can be categorized as short-term or long-term energy storage, on the basis of their usage which can vary from a few hours or for a few months (Sharma, Tyagi, Chen, and Buddhi 2009; Sharma, Ganesan, Tyagi, *et al.* 2015).

Energy storage in any form reduces wastage and saves expensive fuels, thus leading to a cost-effective system. Proper storage of thermal energy by means of phase change materials (PCMs) provides an effective solution for increasing the efficiency of the system and allows for the use of energy in a number of industrial and domestic applications, such as refrigeration, solar water/air heating, solar drying, among others (Pielichowska and Pielichowski 2014). Studies carried out in the past show that PCMs are capable of storing about 3–14 times more heat per unit volume than what is stored as sensible heat (Mehling and Cabeza 2008). The main requirements for the design of a TES system are higher energy density storage materials, better heat transfer between the heat transfer fluid and the PCM, high thermal reliability and chemical stability of the PCM, low thermal losses during the storage period, and easy control. TES systems come with a large number of benefits when implementing storage in an energy system, in terms of economics, efficiency, less pollution, and better reliability.

21.2   PHASE CHANGE MATERIALS

Thermal energy storage materials, which store energy in the form of latent heat, are known as PCMs. They use their chemical bond for storing latent heat during the phase transition process. They possess high latent heat of fusion and store higher energy than sensible energy storage materials due to their compactness and high energy density. When a PCM melts, it absorbs a large amount of energy and when it freezes, it releases the stored energy. The storage capacity of LHTES devices is derived from (G.A. Lane 1983)

$$Q = \int_{T_i}^{T_m} mC_p dT + ma_m \Delta h_m + \int_{T_m}^{T_f} mC_p dT \qquad ...(21.1)$$

$$Q = m\left[C_{sp}(T_m - T_i) = a_m \Delta h_m + C_{ip}(T_f - T_m)\right] \qquad ...(21.2)$$

Where Q is the energy stored, C_p is the specific heat, T_i is the initial temperature, T_m is the medium, T_f is the freezing temperature, and h is the enthalpy.

21.2.1   PCM Classifications

Phase change materials are primarily categorized as inorganic, organic, and their eutectics, as shown in Figure 21.1. A comprehensive classification

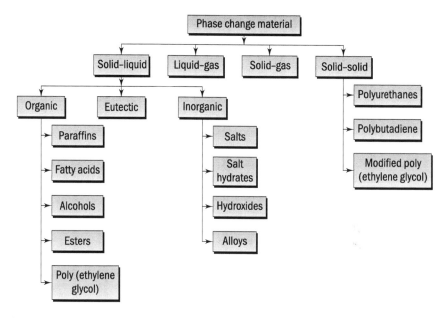

Fig. 21.1 *Classifications of PCMs*
Source *Sharma, Ganesan, Tyagi, et al. (2015)*

can be found in Abhat, Heine, Heinisch, *et al.* (1981); Sharma, Tyagi, Chen, *et al.* (2009); Sharma, Ganesan, Tyagi, *et al.* (2015); Zalba, Marin, Cabeza, *et al.* (2004). A large number of TES materials are available in the commercial market in almost every temperature range. Despite many PCMs' melting/freezing temperature range lying in the required temperature range as stipulated in domestic and industrial applications, their usage is limited because they lack sufficient thermo-physical properties that are required for a TES device to be effective. Therefore, only the available materials with prevailing properties are being used. Emphasis should be given to improving the thermo-physical properties of TES devices. Improvisation of the properties is also necessary due to the fact that no single material can possess all the properties required for TES. For example, thermal conductivity of organic PCMs is improved by adding metal fillers and the subcooling problem of inorganic PCMs is minimized by using a nucleating agent.

Organic phase change materials (O-PCMs) are generally classified as paraffin, fatty acids, and polyethylene glycols. Paraffin consists of n-alkanes chain ($CH_3-(CH_2)-CH_3$). Fatty acids are made up of straight chain hydrocarbons, are relatively expensive, and have a combustible nature. O-PCMs offer congruent melting and negligible or low subcooling. Paraffins are relatively safe, inexpensive, and stable below 500°C. Apart from these many advantages, O-PCMs come with the disadvantage of having low thermal conductivity which limits their usage in many energy storage applications.

Inorganic phase change materials (I-PCMs) generally include hydrated salts and possess higher thermal conductivity than O-PCMs. These materials are able to maintain latent heat of fusion after a large number of melt/freeze cycles, but they melt incongruently. Thermal energy storage properties of inorganic PCMs such as Glauber salt ($Na_2SO_4.H_2O$) were investigated by Telkes (1952). Glauber salt showed a high amount of latent heat (of the order of 254 J/g) and a phase transition temperature of 32.4°C. It was also noticed that Glauber salt is highly prone to subcooling and phase segregation. The use of nucleating agents such as Borax helps to overcome the subcooling problem. However, at the same time, they affect the heat transfer rate by reducing the thermal conductivity. A detailed list of organic and inorganic PCMs tested in past studies is presented in Table 21.1.

Table 21.1   Thermo-physical properties of organic and inorganic PCMs

Compound	T_m (°C)	H_f (kJ/kg)	C_p (kJ/ kg K)	k (W/ (m K))	ρ (kg/m³)	References
Dodecane	−9.6	216	-	-	745 (l)	Himran, Suwono, and Mansoori (1994)
Tridecane	−5.4	196	2.21 (l)	-	753 (l)	Himran, Suwono, and Mansoori (1994)
H_2O	0	333	-	0.612	-	Abhat (1983)
Paraffin C14	4.5	165	-	-	-	Abhat (1983)
Tetradecane	5.5	227	2.07 (s)	0.15	825 (s)	Himran, Suwono, and Mansoori (1994)
Paraffin C15-C16	8	153	2.2 (s)	-	-	Abhat (1983)
$LiClO_3.3H_2O$	8.1	253	-	-	1720	Zalba, Marín, Cabeza, and et al. (2003)
Paraffin C18	28	244	2.16	0.15	814	Abhat (1983)
Nonadecane	32	222	-	-	785	Paris, Falardeau, and Villeneuve (1993)
Eicosane	36.6	247			788	Paris, Falardeau, and Villeneuve (1993)
Heneicozane	40.2	213	-	-	791	Paris, Falardeau, and Villeneuve (1993)
Paraffin C20–C33	48-50	189	2.1	0.21	769 (l) 912 (s)	Abhat (1983)

Contd...

Table 21.1   *Contd...*

Compound	T_m (°C)	H_f (kJ/kg)	C_p (kJ/ kg K)	k (W/ (m K))	ρ (kg/m³)	References
Paraffin C22-45	58-60	189	2.1	0.21	795 (IC) 920 (s)	Abhat (1983)
Napthelene	80	147.7	2.8	0.132 (l) 0.341 (s) 0.310 (s)	976 (l) 1145 (s)	Lane (1980) Durupt, Aoulmi, Bouroukba, et al. (1995)
Caprylic	16 16.5	148.5 149	-	0.149 (l) 0.148 (l)	862 (l) 1033 (s) 981 (s)	Abhat (1983) Dincer and Rosen (2002)
Capric	31.5 32	153 152.7	-	0.149 (l) 0.153 (l)	886 (l) 878 (l)	Abhat (1983)
Lauric acid	42–44 44	178 177.4	1.6	0.147 (l)	870 (l) 862 (l) 1007 (s)	Abhat (1983)
Myristic	54 58 49-51	187 186.6 204.5	1.6 (s) 2.7 (l)	-	844 (l) 990 (s)	Abhat (1983) Sarı and Kaygusuz (2001)
Palmitic	63 61 64	187 203.4 185.4	- - -	0.165 (l) 0.159 (l) 0.162 (l)	874 (l) 847 (l) 850 (l)	Abhat (1983) Sari and Kaygusuz (2002)
Stearic	70 69 60–61 69.4	203 202.5 186.5 199	2.35 (l)	0.172 (l)	941 (l) 848 (l)	Abhat (1983) Sari and Kaygusuz (2001)
Acetamide	81	241	-	-	-	Hale, Hoover, and O'Niell (1971)

Eutectic materials are mixtures of two or more organic or inorganic materials. Eutectics contain O-PCMs that melt/freeze congruently and chances of phase segregation are almost nil (Sharma, Tyagi, Chen, et al. 2009). The main advantage of preparing eutectics is that their thermo-physical properties and melting temperature can be adjusted by playing with the different percentages of the PCMs; for example, a mixture of palmitic and lauric acid, myristic and palmitic acid, lauric and capric acid.

Table 21.2   Desired thermophysical properties of phase change materials

Thermal properties	Physical properties	Chemical properties	Economic properties
• High latent heat of fusion • High specific heat • Suitable phase change temperature • High solid and liquid thermal conductivity	• High material density • Negligible subcooling during freezing • Low vapour pressure • Small phase transition volume change	• Longer chemical stability • Capsule material compatibility • Non-toxic, non-flammable, and non-explosive	• Easy availability • Inexpensive

21.2.2 Thermo-physical Properties

As the thermal and physical properties of PCMs vary from one manufacturer to another, the PCMs must have certain physical, chemical, and economic properties so that they can be used as TES materials (Abhat 1983; Hasnain 1998; Regin, Solanki, and Saini 2008).

A suitable melting temperature and little subcooling offer many domestic and industrial applications for PCMs. High specific heat helps to store more sensible heat. High latent heat increases compactness, enhances heat storage capacity, and minimizes the physical size of PCM containers. Higher thermal conductivity increases the heat transfer rate which eventually increases the effectiveness of TES. If the materials possess higher density, they require smaller containers, and negligible subcooling limits the temperature range and keeps it at a minimum. Low vapour pressure and small volume change in the PCM help to reduce the complexity of geometry of the container. PCMs are subjected to continuous melt/freeze cycles which may hamper their chemical properties. It is highly desirable that PCMs maintain their chemical structure for longer uses of TES materials. Encapsulation is one of the best ways to avoid these chemical reactions that may occur between PCMs and the atmosphere. In addition to these requirements, the inexpensive and abundantly available materials are always preferred for thermal energy storage systems.

21.3 APPLICATION OF PHASE CHANGE MATERIAL

Due to high latent heat storage capacity of PCMs, they are useful in a large number of domestic and industrial applications. Some of these applications have been listed and discussed next to highlight the importance of these materials.

21.3.1 Buildings

For more than three decades, organic and inorganic PCMs have been used for heating and cooling buildings. Electricity consumption in buildings is not fixed because of variable energy demand during different weather conditions. Due to this variable demand for electricity, the prices of electricity consumption also vary. If a suitable TES system is installed which could store the abundantly available solar energy in the daytime and release it during off-peak hours, such as at night and on cloudy days, the consumption of electricity can be significantly reduced. To achieve this goal, walls, ceilings, floors, and windows can be equipped with PCMs. The American Society of Heating, Refrigerating and Air-Conditioning Engineers (ASHRAE) has recommended a human comfort temperature range between 23.5°C and 25.5°C in the summer and between 21.0°C and 23°C in the winter (Zhou, Zhao, and Tian 2012). When this is applied to buildings, a PCM temperature range of 20–30°C is preferred.

21.3.2 Walls and Wallboards

In buildings, PCMs can be effectively used in walls and wallboards since wallboards are easily available and their installation is easy. A wallboard is generally capable of holding 30% of the total PCM and can store energy very effectively in a passive solar system (Stovall and Tomlinson 1995). Athienitis, Liu, Hawes, *et al.* (1997) evaluated the thermal performance of a wallboard equipped with butyl stearate as the PCM, both numerically and experimentally. They reported that by using the PCM wallboard, the maximum temperature of the room could be reduced by up to 4°C. In another study, the thermal performance of paraffin- and fatty-acids-impregnated wallboards was investigated by Neeper (2000). The results of this study revealed that when the phase transition temperature of the PCM is nearly equal to the room temperature, the wallboard can store maximum diurnal energy and this energy decreases if the phase change occurs over a range of temperatures (Figure 21.2).

Shilei, Guohui, Neng, *et al.* (2007) and Shilei, Neng, and Guohui (2006) used a eutectic mixture of capric and lauric acid as TES material for wallboards and investigated its thermal performance. The results of this study revealed that this mixture was able to sustain its thermal and chemical properties even after a large number of melt/freeze cycles and PCM-impregnated wallboards could effectively reduce the cost of energy consumed. A test room which had a wallboard integrated with PCM was built and investigated by Kuznik and Virgone (2009), as shown in Figure 21.3. The investigation was carried out in the summer, winter,

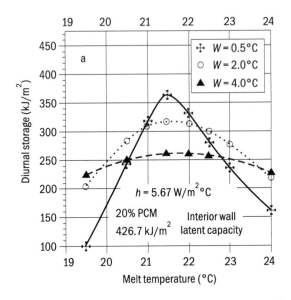

Fig. 21.2 *Diurnal energy storage with melting temperature for an interior wall*
Source *Neeper (2000)*

and on a mid-season day. The results of this study showed that the PCM wallboard increased natural convection and the room temperature reduced by 4.2°C. Liu and Awbi (2009) investigated the effect of a PCM wallboard numerically and found that it significantly contributed in reducing the internal temperature of a room and absorbed/released a high amount of solar energy. Lai and Hokoi (2014) used a microencapsulated PCM

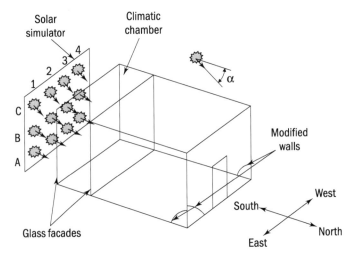

Fig. 21.3 *Test cell*
Source *Kuznik and Virgone (2009)*

and aluminium honeycomb structure to construct a PCM wallboard in their experimental study. They mixed the PCM with two different supporting materials—expanded graphite and iron wire for thermal conductivity enhancement of the base PCM—and investigated the effects of these materials on wallboard performance. The results of the study showed the enhanced thermal performance of the PCM wallboard over a wallboard without PCM. A nano-PCM mixture of paraffin and graphite nano-sheets was prepared and impregnated into a gypsum wallboard by Biswas, Lu, Soroushian, *et al.* (2014). A 1.3-cm-thick nano-PCM wallboard was impregnated with 20% nano PCM by weight which was uniformly dispersed inside the gypsum board. The results of this study revealed that the nano-PCM wallboard was able to reduce the energy consumption for space heating significantly. Zhu, Liu, Hu, *et al.* (2015) investigated a wallboard consisting of a shape-stabilized PCM. The wallboard was designed in a sandwich form. The external and internal layers consisted of PCM and the middle layer was of gypsum. The results of the investigations showed that the recommended thickness of a wallboard is 30–60 mm and with this thickness the annual energy savings could range between 3.4% and 3.9%.

21.3.3 Floors and Ceilings

Floors and ceilings of buildings can also be effectively used for energy storage using PCM impregnation. Bénard, Gobin, and Gutierrez (1981) developed a solar chicken brooder impregnated with paraffin as the TES material. 42 kg of paraffin was filled in a semi-circular disc and

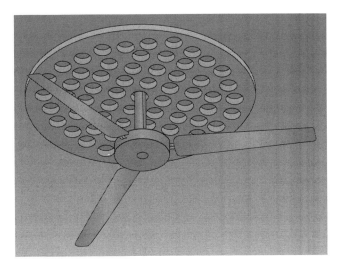

Fig. 21.4 *Developed ceiling fan*
Source *Stalin, Krishnan, Barath, et al. (2013)*

laced below the roof of a building. The results revealed that this design was excellent for maintaining temperatures between 22°C and 30°C. Turnpenny, Etheridge, and Reay (2000) developed a cooling system for reducing the use of air-conditioning in buildings. A numerical model of a ceiling in-built with PCM was developed and a parametric study was carried out. This system could store energy coolness at night-time and release it during the day. Stalin, Krishnan, and Barath (2013) developed a ceiling fan in-built with PCM as shown in Figure 21.4. A circular disc filled with paraffin was placed along the fan's overhead portion, and fitted with

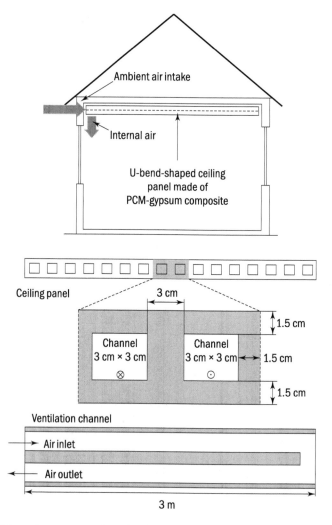

Fig. 21.5 *Schematic representation of ceiling panel in building ventilation*
Source *Jaworski, Lapka, and Furmanski (2014)*

aluminium tubes that received incoming water from an overhead tank with an outlet to the atmosphere. This design was found to be effective for space-cooling purposes. Jaworski, Łapka, and Furmanski (2014) offered a new concept of a ceiling panel, which was made up of gypsum board and in-built with PCM, as shown in Figure 21.5. The thermal performance of such a ceiling panel was investigated experimentally which was supported by numerical validation. Prepared test modules were made using gypsum mortar and microencapsulated PCM, and had a melting point of 22.8°C. The results of this study showed that the total amount of PCM embedded into the panel did not melt or solidify but increased the comfort level inside the room space.

A floor in-built with a composite PCM of paraffin and polyethylene was developed by Lin, Zhang, Xu, et al. (2005). This system was able to charge itself by using electricity at night and released the stored energy during the daytime, and could effectively cool the space. Royon, Karim, and Bontemps (2014) developed a floor impregnated with a paraffin–polymer blend and investigated its thermal performance numerically and optimized the amount of PCM. A PCM-impregnated floor and radiant ceiling system was developed by Belmonte, Eguía, Molina, et al. (2015) and a parametric study investigating the thermal performance of such a system was carried out numerically. The results of this study revealed that if this system is accompanied by an air-to-air heat recovery system, then this can reduce the total energy load by up to 50%. In another study, Cheng, Xie, Zhang, et al. (2015) developed a PCM-inbuilt floor as shown in Figure 21.6. Paraffin was used as the TES material and polyethylene/expanded graphite was used as the supporting material for enhancing the thermal conductivity of the PCM. The results revealed that such a floor could be effectively used for controlling the temperature inside the room space but the effect of shape-stabilized PCMs diminishes when the thermal conductivity exceeds 1.0 W/mK.

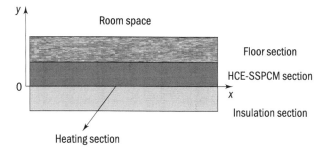

Fig. 21.6 *Sketch of under-floor heating system*
Source *Cheng, Xie, Zhang, et al. (2015)*

21.3.4   Other Building Parts

Phase change materials have proven to be promising candidates for TES in various sections of building envelopes such as trombe walls, windows, tiles, and building blocks. Figure 21.7 shows the typical sketch of a trombe wall, in which a thick south-facing wall has been presented. This wall consists of PCMs and is normally used in winter to keep the building space warm. The exterior side of the wall is glazed using a thick layer of PCM. During the day, this glazed wall absorbs the energy from solar radiation and at night or during off-peak hours, it releases the stored energy and keeps the space warm. A previous study (Eshraghi, Narjabadifam, Mirkhani, *et al.* 2014) has confirmed that the trombe wall is able to provide up to 42% of the total required load.

The effect of the trombe wall in the building space was experimentally investigated by Castellón, Castell, Medrano, *et al.* (2009) in Llieda, Spain. In this experiment, nine cubicles were made with different kinds of bricks and three other types of cubicles, namely concrete, conventional and alveolar, were also prepared. One cubicle of each type was integrated with a PCM. Both concrete walls were equipped with a trombe wall, and a domestic heat pump was attached to all the brick cubicles. The results of this experiments revealed that the fluctuations in the temperature of concrete blocks were reduced by up to 4°C. Trigui, Karkri, Boudaya, *et al.* (2013) and Trigui, Karkri, and Krupa (2014) prepared a composite of paraffin and resin, and impregnated it into a trombe wall. They observed that the prepared composite material possessed a high energy storage capacity.

Window shutters are also exterior devices that are used to shut sunrays from getting inside the room and in turn provide cooling. If they are impregnated with PCMs, it enhances their TES capacity. During the day, they are kept unrolled, allowing them to absorb solar radiation, while

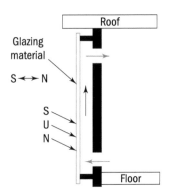

Fig. 21.7 *Schematic representation of PCM-infused trombe wall*

at night they are kept rolled up to minimize heat loss and keep the space warm. Alawadhi (2012) investigated the performance of PCM in-built window shutters and carried out a parametric study numerically. Different PCMs such as *n*-Octadecane, *n*-Eicosane, and paraffins were used in the study. The results showed that P116 could reduce the heat gain by almost 16%. The study also showed that if the paraffin thickness is kept at 3 cm, a maximum heat gain of 23% can be achieved by this system.

Use of PCM in building blocks also enhances the thermal performance of the room and increases the comfort level inside the room space. Hawes and Feldman (1992) examined the heat absorbed by a PCM-infused concrete wall and a parametric study related to the effect of temperature, viscosity of PCM, density of concrete, and hydrogen bonding on PCM penetration, was carried out. The results revealed that these parameters significantly affected the performance of the blocks. The thermal performance of a concrete block impregnated with micronal of melting point 26°C was experimentally investigated by Cabeza, Castellón, Nogués, *et al.* (2007). The authors reported that the PCM-impregnated concrete wall had improved thermal properties compared to the conventional concrete without PCM. Navarro, de Gracia, Castell, *et al.* (2015) designed and developed an active slab system which could work as both passive and active energy storage systems, as shown in Figure 21.8. A solar air collector with a surface area of 1.3 m^2 and a power rating of 1330 Wp (AIRSOL-20), which integrated a fan connected to a photovoltaic panel, was implemented in the south facade as heat supply for the winter mode and it was coupled to the rest of the active system. This system had 14 channels, in which PCM RT-21 was filled. The results of this study revealed that during summer, for almost 70% of the total days, the PCM was 100% melted. The authors claimed

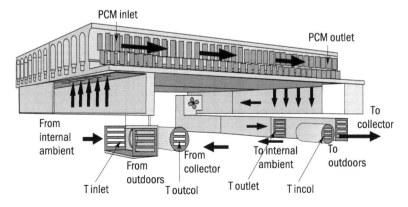

Fig. 21.8 *Schematic representation of active slab system*
Source *Navarro, de Gracia, Castell, et al. (2015)*

that a solar air collector of 30% efficiency is enough to have successfully charged processes for the entire winter season. This designed slab was tested in March and June of 2014 and the results showed that this system had potential for use.

21.3.5   Water Heating

Solar energy is an unlimited source of energy. When stored effectively, it can be utilized during off-peak hours, at night, and on cloudy days. Solar energy storage using PCMs has found a large number of industrial and domestic applications including solar air and water heating, and in solar cookers and dryers. The general principle of solar energy storage using PCMs and technologies has been summarized by Dincer and Rosen (2002); Garg, Mullick, and Bhargava 1985; and Lane (1983, 1986). For solar thermal applications, TES materials such as hydrated salt and organic PCMs in the melting range between 0°C and 150°C are preferred. Kenisarin and Mahkamov (2007) and Sharma, Tyagi, Chen, *et al.* (2009) have presented a comprehensive review of PCMs tested for different kinds of solar TES systems.

Solar water heating is considered to be a relatively inexpensive process. During an early study, Barry (1940) developed a solar water heating system as shown in Figure 21.9. It consists of a copper coil fitted inside a dome-shaped shell. The coil at the bottom of the container is connected to the inlet. The one at the top of the container is connected to the outlet and this discharges the hot water. Integration of PCM in solar water heating systems enhances their effectiveness because with the inbuilt PCM, these sytems can be used in off-peak hours and at night-time as well. Another solar water heating system was designed by Prakash, Garg, and Datta (1985), and this contained a layer of PCM at the bottom. This layer helped water to get heated even during no-sunshine hours. The performance of

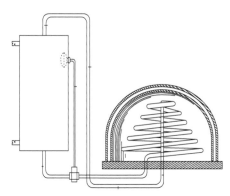

Fig. 21.9 *Schematic representation of solar water heater*
Source *Barry (1940)*

this system was analysed for two different PCM thicknesses and flow rates. The results of this study revealed that this system had significant potential for reducing energy consumption.

Al-Hinti, Al-Ghandoor, Maaly, *et al.* (2010) investigated the effect of capsules filled with paraffin on a solar water heating system. They designed a system with four south-facing flat plate collectors of dimensions 1.94 m × 0.76 m × 0.15 m and it worked on the principle of an open- and

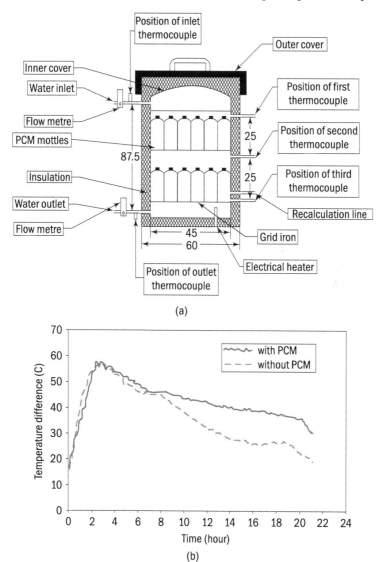

(a)

(b)

Fig. 21.10   *(a) Cross-sectional view of the storage tank and (b) temperature variations with time*

Source *Al-Hinti, Al-Ghandoor, Maaly, et al. (2010)*

closed-loop system (Figure 21.10). 38 containers, each filled with 1 kg paraffin, were attached to the storage tank of length 675 mm, and diameter 450 mm. The results of this study revealed that during the test period of 24 hours, the temperature of water was maintained at up to 30°C more than that of its ambient temperature. Khalifa, Suffer, and Mahmoud (2013) also designed a similar storage tank that consisted of six copper pipes, each of diameter 80 mm, filled with paraffin for TES. The thermal performance of this system was investigated by analysing the various pertinent parameters such as top loss coefficient, water-useful heat gain, and heat transferred between water and paraffin. The results of this study revealed that the temperature of plate increased up to a certain distance of 2.5 m from the entrance which became steady for the remaining 7.6 m length.

Recently, Mahfuz, Anisur, Kibria, *et al.* (2014) experimentally investigated the thermal behaviour of a paraffin-integrated solar water heating system. Their proposed system was made of three major components—a solar collector unit, a shell and tube TES, and an insulated water storage tank. During the sunny hours, valve 1 was open and valve 2 remained closed. The cold water from the water storage tanks passed through the solar collector and gained heat and flowed back to the storage tank. A part of this hot water went through the TES tank for charging of the PCM. The excess water automatically flowed out of this tank and moved towards the main water storage tank. At night when there was no sunlight, valve 2 was opened to allow the water to pass through the PCM tank so that it would extract heat from the PCM, get heated up, and flow back to the main storage tank. The results showed that when the water flow rate is 0.033 kg/min, the energy efficiency of such a system is 63.88% while it is 77.41% when the flow rate is 0.167 kg/min. For the first flow rate, the total life cycle cost was calculated as $654.61, while for the later one the total cost was calculated to be $609.22. A logical interpretation is that as the flow rate increases, the life cycle cost decreases. Chaabane, Mhiri, Bournot, *et al.* (2014) carried out a numerical study on a PCM-integrated solar water heating system. They used one organic PCM, myristic acid, and one organic-inorganic mixture of Rubitherm 42-graphite for this investigation. The results showed that the myristic acid-integrated water heating system performed better than others under the same environmental conditions.

21.3.6 Air Heating

Solar air heating is a technique used to heat or condition the air for use in a building space or other applications. The use of PCMs is an effective way to make the solar air heating system more sustainable (Lane 1980). Morrison, Khalik and Jurinak, in their different studies (Jurinak and

Abdel-Khalik 1978, 1979a, 1979b; Morrison and Abdel-Khalik 1978) numerically investigated the performance of solar energy-based air heating systems integrated with organic and inorganic PCMs. They determined the effect of latent heat and the melting temperature of the PCM on the performance of an air heating system. The effect of semi-congruent melting of the PCM on system performance was also examined. Results showed that the PCM should be selected based on melting point rather than latent heat.

The performance of a PCM-integrated solar air heating system was also investigated by Enibe (2002) as shown in Figure 21.11. Paraffin as TES material was filled in thin rectangular blocks, which was placed in the collector as shown in Figure 21.11(b). The space between the two modules was utilized as an air heater. Cold ambient air flowed into the collector and passed through the spaces between the PCM modules which

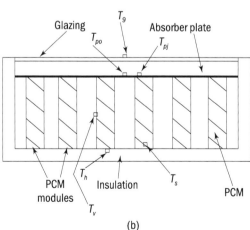

(a)

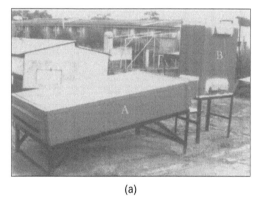

(b)

Fig. 21.11 *(a) Air heating system; A – collector assembly with PCM and air-heating subsystem; B – heated space, (b) Arrangement of PCM modules in the collector*
Source *Enibe (2002)*

Fig. 21.12   *Four test boxes at the site*
Source *Entrop, Brouwers, and Reinders (2011)*

served as air heater. The ambient air got heated up and flowed to the space chamber. This system was found beneficial for drying crops like herbs and medicinal plants. Later on in his work Enibe (2003) carried out a transient numerical investigation for the same setup. The numerically predicted results of the system were compared with the experimental data of the no-load condition between the temperatures of 19°C and 41°C. The radiation was kept between 4.9 and 19.9 MJm^{-2}. Numerically predicted values were well matched with the experimental data.

The effect of microencapsulated paraffin of phase change temperature 23°C on space heating was experimentally evaluated by Entrop, Brouwers, and Reinders (2011). Their experimental setup consisted of four insulated boxes (1130 mm × 725 mm × 690 mm each) having a window facing south to allow solar radiation to enter the space, as shown in Figure 21.12. A data acquisition unit was also installed for monitoring and recording the data. The floors of the two boxes were equipped with microencapsulated PCMs and two others were kept without PCM. The results of this investigation revealed that the boxes with PCM were able to maintain space temperature in a better way than the boxes without PCM, as shown in Table 21.3.

Table 21.3   Minimum and maximum temperatures (±0.5°C) in the boxes per day

Date	Test box 1 Min–Max (°C)	Test box 2 Min–Max (°C)	Test box 3 Min–Max (°C)	Test box 4 Min–Max (°C)
24-6	20.9–24.4	23.0–29.2	19.9–25.2	20.2–28.9
25-6	23.2–25.2	25.1–29.8	22.4–25.7	22.1–28.8
26-6	23.1–25.6	23.8–30.3	21.1–26.2	20.1–29.4
27-6	23.7–27.7	24.8–32.2	22.1–28.6	21.3–31.7
28-6	24.7–30.0	27.0–33.9	23.9–31.0	23.7–33.5
29-6	26.3–30.3	28.3–33.9	24.4–30.6	24.4–32.5
30-6	25.7–30.2	27.3–33.4	24.0–29.2	23.1–30.7
1-7	24.4–28.5	25.4–31.4	23.2–28.5	22.0–30.7

Source Entrop, Brouwers, and Reinders (2011)

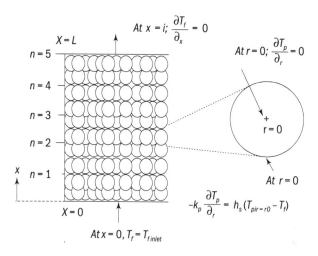

Fig. 21.13   *Arrangement of the packed bed model*
Source *Karthikeyan, Solomon, Kumaresan, et al. (2014)*

In a recent study, Karthikeyan, Solomon, Kumaresan, *et al.* (2014) conducted a numerical investigation on a packed bed storage unit integrated with paraffin as TES, as shown in Figure 21.13. A parametric study of the effect of flow rate on heat transfer fluid (that is, air), temperature at inlet, ball-sized PCM capsule and on effective thermal conductivity on the system, was carried out. A list of chosen parameters in this study is given in Table 21.4. The results of this study showed that at a constant flow rate of air, 0.035 kg/s, inlet air temperature was 70°C and effective thermal conductivity was 0.7 W/m K. The charging/discharging time of the PCM was continuously increased with increasing ball diameter, increasing mass flow rate of air. It was also observed that the effect of thermal conductivity beyond 1 ΩW/m K did not produce significant heat transfer enhancement.

Tyagi, Pandey, Kaushik, *et al.* (2012) performed an experimental investigation of a solar heater embedded with paraffin wax and hytherm oil. A total of 12 evacuated tube collectors (ETCs), four filled with paraffin, four filled with oil, and the remaining four without any storage material,

Table 21.4   Parameters considered and their range of values

Parameters	Values
PCM ball size (mm)	60, 70, 80, 100
HTF inlet fluid temperature (°C)	67, 70, 75, 80
Mass flow rate of HTF (kg s⁻¹)	0.05, 0.035, 0.015
Effective thermal conductivity of PCM (W m⁻¹ K⁻¹)	0.4, 1.0, 2.0

Source Karthikeyan, Solomon, Kumaresan, *et al.* (2014)

were used in the ETC-based solar collector. The glass length exposed to sunlight was 172 cm and was inclined at 45°. The results of this study showed that the outlet temperature of the heater with PCM was more than that of the one without PCM. As the mass flow rate increased, the outlet temperature in all the three considered cases increased. Later Tyagi, Pandey, Giridhar, *et al.* (2012) presented a comparative energy and exergy analysis of a solar air heater integrated with paraffin wax as PCM and hytherm oil. They observed that both energy and exergy efficiencies of the air heating system was higher when PCM was used. Another solar air heater was built, with a mixture of sand and granular carbon used as TES material by Saxena, Srivastava, and Tirth (2015). A total of two halogen lights, each of 300 W were used at inlet and outlet of duct for increasing the exhaust of the heater. A schematic representation of this system is

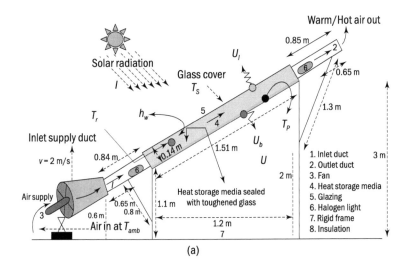

(a)

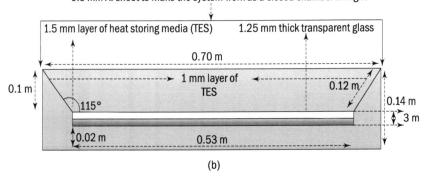

(b)

Fig. 21.14 *Schematic of solar air heater with TES materials (a) heat transfer process (b) TES material spread on absorber plate (Saxena, Srivastava, and Tirth 2015)*

Source *Saxena, Srivastava, and Tirth 2015*

shown in Figure 21.14. This system was placed towards the south at 43°. A simple fan was used for air inflow. The results of this study reflected that the total efficiency of this system increased by almost 41% after using PCM.

Bouadila, Kooli, Lazaar, *et al.* (2013) and Bouadila, Lazaar, Skouri, *et al.* (2014) performed the energy and exergy analysis of a solar air heater which is integrated with a packed-bed latent heat storage material in capsule form. The effect of pertinent parameters such as solar radiation and mass flow rate were investigated. The results revealed that in the presence of PCM, the water outlet temperature was around 20°C throughout the night. Krishnananth and Murugavel (2013) investigated the thermal performance of a solar water heater equipped with paraffin as the TES material, and the results of this study revealed that the developed air heater was able to discharge the outlet water throughout the day at a higher temperature than that of an inlet. A solar air heater inbuilt with an LHTES system was developed by Charvát, Klimeš, Ostrý, *et al.* (2014) in which RT 42 was used as the TES material. The design consisted of 100 compact storage modules filled with paraffin for the TES and the results showed that the PCM-integrated system performed better than the system without PCM.

21.3.7   Solar Cookers

Use of solar energy in cooking is one of the more common ways in which this energy is utilized. As in the evenings, the intensity of solar energy diminishes, it is not possible to use solar cookers at that time. But the use of PCMs in solar cookers makes cooking possible even at night by using heat stored during the day. Buddhi and Sahoo (1997) experimentally investigated the feasibility of solar cookers inbuilt with TES using stearic acid (m.p. 55.1°C) as the PCM, as shown in Figure 21.15, and compared the results with an ordinary solar cooker studied by Thulasi Das, Karmakar, and Rao (1994). It was assumed that the temperature of the PCM/food

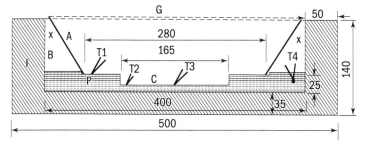

Fig. 21.15   *Schematic representation of box of solar cooker with PCM. A – Abriber tray, B – PCM tray, C – pot container, G – double glass lid, I – glasswool insulation, P – PCM*
Source *Buddhi and Sahoo (1997)*

was 95°C at 3:00 pm and 55°C at 9:00 pm. Tray B was filled with stearic acid and the space between Tray B and the casing was filled with glass wool for the purpose of insulation. The orientation of the solar cooker was adjusted after every 15 minutes so that it received maximum radiation. As the ambient temperature increased, the plate and PCM temperature also increased until the PCM attained its melting temperature. During the melting process, the enhancement in the plate temperature was relatively small because the TES material absorbed maximum radiation while melting. After completing the melting cycle, both the temperature of plate and the PCM increased further. The results showed that even at night, the temperature of the plate was 57.2°C. Using a PCM, experiments for cooking rice were performed in two batches. For the first batch, the rice was cooked from 10:30 am to 1:30 pm and for the second batch, the same was performed from 4:00 pm to 8:00 pm. Finally it was concluded that an ordinary solar cooker can be used two times a day with the use of PCMs.

Hussein, El-Ghetany, and Nada (2008) investigated the performance of a solar cooker integrated with magnesium nitrate hexahydrate (m.p. 89°C) as the TES material. This design consisted of an outdoor flat plate solar collector, an indoor PCM cooking unit, and a heat pipes network. The results of this study revealed that the aforementioned cooker was able to cook and warm the meal at night as well. It was also able to warm food the next morning. A numerical study of the box-type solar cooker integrated with a TES system was carried out by Chen, Sharma, Tyagi, et al. (2008). Different PCMs such as magnesium nitrate hexahydrate, stearic acid, acetamide, acetanilide, and erythritol were used for TES purposes in this cooker. The results indicated that the high thermal conductivity of heat exchanger materials did not contribute significantly to the enhancement of the solar cooker's performance. Out of the tested PCMs, organic materials such as acetamide, and stearic acid affected the performance significantly when compared to other PCMs.

Out of the many types of solar cookers available, the box-type solar cooker has been found to be the most suitable for mass-level use as it can produce temperatures of up to 100°C. The performance of a box-type solar cooker integrated with fins and paraffin as PCM was experimentally investigated by Geddam, Dinesh, and Sivasankar (2015). A parametric study was carried out to investigate the effect of optical efficiency and heat capacity on the cooker. The results revealed that fins helped in reducing cooking time and the PCM helped in cooking food at night [For a comprehensive review of solar cookers integrated with PCM, see Sharma, Chen, Murty, et al. (2009).]

21.3.8 Solar Dryers

A solar dryer is a device which uses solar radiation to dehydrate substances in general, and food in particular. Depending on whether solar radiation heats food directly or indirectly, the solar dryer is classified as direct or indirect. For preservation of agricultural products, especially fruits and vegetables, the use of a dryer with a moderate temperature range (40–75°C) is essential, and the use of TES in this temperature range has gained immense popularity in the last few decades due to its high energy storage capacity. Paraffin is one of the most widely used organic PCMs in solar dryers serving as an energy storage medium. Butler and Troeger (1981) constructed a solar collector-cum-rock-bed storage for peanut drying and experimentally evaluated its performance. They found that in the drying time range of 22–24 hours, the moisture level was reduced from 20% to a safe storage moisture level. VijayaVenkataRaman, Iniyan, and Goic (2012) presented a detailed review of solar drying technologies with and without PCMs. This review lists the technological developments in developing countries, and the various designs of solar dryers, their classification, and performance analysis were also compiled in this paper. Recently, Shalaby, Bek, and El-Sebaii (2014) presented their research on a solar dryer system integrated with PCMs as TES storage media. They also presented the various techniques used for thermal conductivity enhancement of PCMs such as carbon fibres, expanded graphite, and graphite form. A novel indirect solar dryer (Figure 21.16) integrated with paraffin wax as PCM with a melting temperature of 49°C was designed by Shalaby and Bek (2014). This system consisted of two identical solar heaters, one drying compartment, and a blower. The solar radiation on a tilted plate was measured in September 2013 using a Pyranometer and

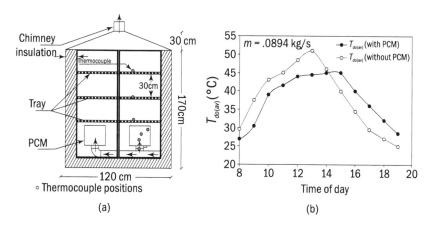

Fig. 21.16 *(a) Drying compartment (b) Variation in the temperature of the drying air*
Source *Shalaby and Bek (2014)*

the temperature of drying air was monitored. A three-phase induction motor of 1.75 HP and 2610 rpm served the purpose of a blower for the dryer compartment. The experiments were conducted with and without PCM and it was observed that after using PCM, the temperature of drying air was 2.5–7.5°C higher than ambient air for a minimum of 5 hours after sunset. Figure 21.16(b) shows the hourly variation of the drying air temperature and it is observed that the average drying air temperature when using PCM is less as compared to when not using it between 8:00 am and 2:00 pm, which indicates the PCM's capability to store a large amount of energy. It was also observed that this design allowed the dryer to operate at temperatures 3.5–6.5°C higher when used with the PCM.

Jain and Tewari (2015) developed a solar crop dryer which was integrated with TES for night use. This dryer consisted of a flat plate solar collector integrated with PCM, and a drying plenum with crop tray. The results of this investigation showed that the PCM-equipped dryer was able to dry food until the midnight hour and the temperature of the drying chamber was 6°C higher than that of the ambient chamber. Economic analysis shows that the return on capital duration was 0.65 years and payback period was 1.5 years. Reyes, Mahn and Vásquez (2014) also developed a hybrid solar dryer with a flat collector of area 10 m^2 for drying mushrooms cut into 8 mm and 12 mm slices. Paraffin wax was used as the TES material. The hybrid solar dryer, as shown in Figure 21.17, consisted of a solar panel (3 m × 1 m) which contained a 5 mm thick glass sheet and a black zinc plate. 14 kg of PCM was distributed in 100 copper pipes of 144 mm inner diameter and these pipes were placed in the solar energy accumulator. Mushrooms were soaked in a diluted 10% benzalkonium chloride solution for 5 min and then rinsed. To maintain the temperature of drying air at 60°C, the energy was taken from the solar panel or the electrical resistance. After 6 pm, when the solar radiation started weakening, the energy was taken from the energy storage device, that is, the paraffin. The use of a low air recycle and thin slices of mushrooms is recommended in such a drying system. The maximum moisture content after rehydration was 1.91 ± 0.24. The thermal efficiency of this system varied from 0.22 to 0.67 and the maximum energy fraction supplied by the accumulator was found to be 0.20.

21.4   CONCLUSION

Research on PCMs and their applications in the last three decades has proven the potential of these materials in a large number of domestic and industrial applications. Although PCMs may seem an attractive option, there is still much to be explored and improved. An insight into a few of the major applications of PCMs have been given in this chapter. The thermo-

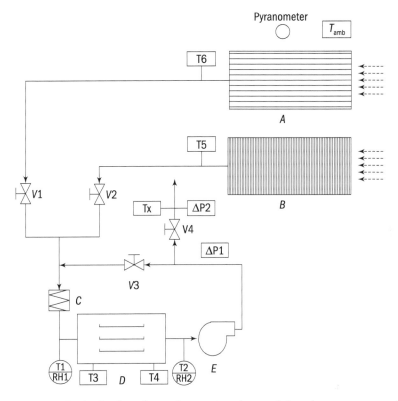

Figure 21.17 *Hybrid-solar dryer for mushroom: A – solar panel, B – solar energy accumulator, C – electrical heater, D – chamber drying, and E – centrifugal fan*
Source *Reyes, Mahn, and Vásquez (2014)*

physical properties of PCMs and a list of PCMs for different temperature range applications have been discussed here.

The building sector consumes 40% of the total energy wordwide. PCMs have potential to store hot/cool energy for building application through different sections. Walls and floors are the two major sections where PCMs are used in the form of wallboards and ceilings to enhance the comfort level inside buildings. Other applications such as water/air heating systems are also an important application for industrial and domestic purposes. PCMs are extremely useful, as solar energy is intermittent in nature and cannot be used round the clock. PCMs are also playing an important role in enhancing the performance as well as time availability of these applications through a storage-based system. Solar cookers are a very cheap and developed technology but not popular today due to the unavailability of sunlight at night hours. If by any means the energy can be stored in the daytime and used in the night-time, the same solar cooker can also be used after sun sets.

REFERENCES

Abhat, A. 1983. Low temperature latent heat thermal energy storage: Heat storage materials. *Solar Energy,* 30(4): 313–332

Abhat, A., D. Heine, M. Heinisch, A. N. Malatidis, and G. Neuer. 1981. Development of a modular heat exchanger with integrated latent heat energy store *Report no. BMFT FBT 81-050.* Institut fuer Kerntechnik und Energiewandlung e.V., Stuttgart, Germany

Al-Hinti, I., A. Al-Ghandoor, A. Maaly, I. Abu Naqeera, Z. Al-Khateeb, and O. Al-Sheikh. 2010. Experimental investigation on the use of water-phase change material storage in conventional solar water heating systems. *Energy Conversion and Management* 51(8): 1735–1740

Alawadhi, Esam M. 2012. Using phase change materials in window shutter to reduce the solar heat gain. *Energy and Buildings* 47(0): 421–429

Athienitis, A. K., C. Liu, D. Hawes, D. Banu, and D. Feldman. 1997. Investigation of the thermal performance of a passive solar test-room with wall latent heat storage. *Building and Environment* 32(5): 405–410

Barry, Edward J. 1940. Solar water heater. *US Patent 2,213,894*

Belmonte, J. F., P. Eguía, A. E. Molina, and J. A. Almendros-Ibáñez. 2015. Thermal simulation and system optimization of a chilled ceiling coupled with a floor containing a phase change material (PCM). *Sustainable Cities and Society* 14:154–170

Bénard, C., D. Gobin, and M. Gutierrez. 1981. Experimental results of a latent-heat solar-roof, used for breeding chickens. *Solar Energy* 26(4): 347–359

Biswas, Kaushik, Jue Lu, ParvizSoroushian, and Som Shrestha. 2014. Combined experimental and numerical evaluation of a prototype nano-PCM enhanced wallboard. *Applied Energy*. Details available at https://ideas.repec.org/a/eee/appene/v131y2014icp517-529.html

Bouadila, Salwa, Sami Kooli, Mariem Lazaar, Safa Skouri, and Abdelhamid Farhat. 2013. Performance of a new solar air heater with packed-bed latent storage energy for nocturnal use. *Applied Energy* 110: 267–275

Bouadila, Salwa, Sami Kooli, Mariem Lazaar, Safa Skouri, and Abdelhamid Farhat. 2014. Energy and exergy analysis of a new solar air heater with latent storage energy. *International Journal of Hydrogen Energy* 39(27): 15266–15274

Buddhi, D. and L. K. Sahoo. 1997. Solar cooker with latent heat storage: Design and experimental testing. *Energy Conversion and Management* 38(5): 493–498

Butler, J. L. and J. M. Troeger. 1981. Drying peanuts using solar energy stored in a rock bed. Vol. I, Solar Energy, Selected Papers and Abstracts, ASAE Publication, St Joseph, MI

Cabeza, Luisa F., Cecilia Castellón, Miquel Nogués, Marc Medrano, Ron Leppers, and Oihana Zubillaga. 2007. Use of microencapsulated PCM in concrete walls for energy savings. *Energy and Buildings* 39(2): 113–119

Castellón, C., A. Castell, M. Medrano, I. Martorell, and L. F. Cabeza. 2009. Experimental Study of PCM Inclusion in Different Building Envelopes. *Journal of Solar Energy Engineering* 131(4): 041006–041006

Chaabane, Monia, Hatem Mhiri, and Philippe Bournot. 2014. Thermal performance of an integrated collector storage solar water heater (ICSSWH) with phase change materials (PCM). *Energy Conversion and Management* 78(0): 897–903

Charvát, Pavel, LubomírKlimeš, and Milan Ostrý. 2014. Numerical and experimental investigation of a PCM-based thermal storage unit for solar air systems. *Energy and Buildings* 68, *Part A*(0): 488–497

Chen, C. R., Atul Sharma, S. K. Tyagi, and D. Buddh. 2008. Numerical heat transfer studies of PCMs used in a box-type solar cooker. *Renewable Energy* 33(5): 1121–1129

Cheng, Wenlong, Biao Xie, Rongming Zhang, Zhiming Xu, and Yuting Xia. 2015. Effect of thermal conductivities of shape-stabilized PCM on under-floor heating system. *Applied Energy* 144: 10–18

D. V. Hale, M. J. Hoover, and M. J. O'Niell. 1971. *Phase Change Materials Handbook*. AL: NASA CR-61363, Marshal Space Flight Center

Dincer, I. and M. A. Rosen. 2002. *Thermal energy storage, Systems and Aplications.* Chichester (England): John Wiley & Sons

Durupt, N., A. Aoulmi, M. Bouroukba, and M. Rogalski. 1995. Heat capacities of liquid polycyclic aromatic hydrocarbons. *Thermochimica Acta* 260(0): 87–94

Enibe, S. O. 2002. Performance of a natural circulation solar air heating system with phase change material energy storage. *Renewable Energy* 27(1): 69–86

Enibe, S. O. 2003. Thermal analysis of a natural circulation solar air heater with phase change material energy storage. *Renewable Energy* 28(14): 2269–2299

Entrop, A. G., H. J. H. Brouwers, and A. H. M. E. Reinders. 2011. Experimental research on the use of micro-encapsulated Phase Change Materials to store solar energy in concrete floors and to save energy in Dutch houses. *Solar Energy* 85(5): 1007–1020

Eshraghi, Javad, Nima Narjabadifam, Nima Mirkhani, Saghi Sadoughi Khosroshahi, and MehdiAshjaee. 2014. A comprehensive feasibility study of applying solar energy to design a zero-energy building for a typical home in Tehran. *Energy and Buildings* 72(0): 329–339

Garg, H. P., S. C. Mullick, and A. K. Bhargava. 1985. *Solar Thermal Energy Storage*. Dordrecht: Reidel Publishing Company

Geddam, Sunil, G. Kumaravel Dinesh, and Thirugnanasambandam Sivasankar. 2015. Determination of thermal performance of a box type solar cooker. *Solar Energy* 113: 324–331

Hasnain, S. M. 1998. Review on sustainable thermal energy storage technologies, Part I: heat storage materials and techniques. *Energy Conversion and Management* 39(11): 1127–1138

Hawes, D. W. and D. Feldman. 1992. Absorption of phase change materials in concrete. *Solar Energy Materials and Solar Cells* 27(2): 91–101

Himran, Syukri, Aryadi Suwono, and G. Ali. Mansoori. 1994. Characterization of alkanes and paraffin waxes for application as phase change energy storage medium. *Energy Sources* 16(1): 117–128

Hussein, H. M. S., H. H. El-Ghetany, and S. A. Nada. 2008. Experimental investigation of novel indirect solar cooker with indoor PCM thermal storage and cooking unit. *Energy Conversion and Management* 49(8): 2237–2246

Jain, Dilip, and Pratibha Tewari. 2015. Performance of indirect through pass natural convective solar crop dryer with phase change thermal energy storage. *Renewable Energy* 80: 244–250

Jaworski, Maciej, Piotr Łapka, and Piotr Furmański. 2014. Numerical modelling and experimental studies of thermal behaviour of building integrated thermal energy storage unit in a form of a ceiling panel. *Applied Energy* 113(0): 548–557

Jurinak, J. J. and S. I. Abdel-Khalik. 1978. Properties optimization for phase-change energy storage in air-based solar heating systems. *Solar Energy* 21(5): 377–383

Jurinak, J .J. and S. I. Abdel-Khalik. 1979a. On the performance of air-based solar heating systems utilizing phase-change energy storage. *Energy* 4(4): 503–522

Jurinak, J. J. and S. I. Abdel-Khalik. 1979b. Sizing phase-change energy storage units for air-based solar heating systems. *Solar Energy* 22(4): 355–359

Karthikeyan, S., G. Ravikumar Solomon, V. Kumaresan, and R. Velraj. 2014. Parametric studies on packed bed storage unit filled with PCM encapsulated spherical containers for low temperature solar air heating applications. *Energy Conversion and Management* 78(0): 74–80

Kenisarin, Murat and Khamid Mahkamov. 2007. Solar energy storage using phase change materials. *Renewable and Sustainable Energy Reviews* 11(9): 1913–1965

Khalifa, Abdul Jabbar N., Kadhim H. Suffer, and Mahmoud Sh. Mahmoud. 2013. A storage domestic solar hot water system with a back layer of phase change material. *Experimental Thermal and Fluid Science* 44(0): 174–181

Krishnananth, S. S. and K. Kalidasa Murugavel. 2013. Experimental study on double pass solar air heater with thermal energy storage. *Journal of King Saud University: Engineering Sciences* 25(2): 135–140

Kuznik, Frédéric and JosephVirgone. 2009. Experimental assessment of a phase change material for wall building use. *Applied Energy* 86(10): 2038–2046

Lai, Chi-ming, and Shuichi Hokoi. 2014. Thermal performance of an aluminum honeycomb wallboard incorporating microencapsulated PCM. *Energy and Buildings* 73: 37–47

Lane, G. A. 1983. *Solar Heat Storage:Latent Heat Materials, Vol. I* (Vol. I). Florida: CRC Press, Inc.

Lane, G. A. 1986. *Solar Heat Storage:Latent Heat Materials, Vol. II, Technology.* Florida: CRC Press

Lane, George A. 1980. Low temperature heat storage with phase change materials. *International Journal of Ambient Energy* 1(3): 155–168

Lin, Kunping, Yinping Zhang, Xu, Xu, Hongfa Di, Rui Yang, and Penghua Qin. 2005. Experimental study of under-floor electric heating system with shape-stabilized PCM plates. *Energy and Buildings* 37(3): 215–220

Liu, Hongim and Hazim B. Awbi. 2009. Performance of phase change material boards under natural convection. *Building and Environment* 44(9): 1788–1793

Mahfuz, M. H., M. R. Anisur, M. A. Kibria, R. Saidur, and I. H. S. C. Metselaar. 2014. Performance investigation of thermal energy storage system with Phase Change Material (PCM) for solar water heating application. *International Communications in Heat and Mass Transfer* 57(0): 132–139

Mehling, Herald and Luisa F. Cabeza. 2008. *Heat and Cold Storage with PCM: An up to date introduction into basics and applications.* Berlin Heidelberg: Springer-Verlag

Morrison, D. J. and S. I. Abdel-Khalik. 1978. Effects of phase-change energy storage on the performance of air-based and liquid-based solar heating systems. *Solar Energy* 20(1): 57–67

Navarro, Lidia, Alvarode Gracia, Albert Castell, Servando Álvarez, and Luisa F. Cabeza. 2015. PCM incorporation in a concrete core slab as a thermal storage and supply system: Proof of concept. *Energy and Buildings* 103: 70–82

Neeper, D. A. 2000. Thermal dynamics of wallboard with latent heat storage. *Solar Energy* 68(5): 393–403

Paris, Jean, Michel Falardeau, and CÉCile Villeneuve. 1993. Thermal storage by latent heat: a viable option for energy conservation in buildings. *Energy Sources* 15(1): 85–93

Pielichowska, Kinga and Krzyszt of Pielichowski. 2014. Phase change materials for thermal energy storage. *Progress in Materials Science* 65(0): 67–123

Prakash, J., H. P. Garg, and G. Datta. 1985. A solar water heater with a built-in latent heat storage. *Energy Conversion and Management* 25(1): 51–56

Regin, A. Felix, S. C. Solanki, and J. S. Saini. 2008. Heat transfer characteristics of thermal energy storage system using PCM capsules: A review. *Renewable and Sustainable Energy Reviews* 12(9): 2438–2458

Reyes, Alejandro, Andrea Mahn, and Francisco Vásquez. 2014. Mushroom dehydration in a hybridsolar dryer, using a phase change material. *Energy Conversion and Management* 83(0): 241–248

Royon, L., L. Karim, and A. Bontemps. 2014. Optimization of PCM embedded in a floor panel developed for thermal management of the lightweight envelope of buildings. *Energy and Buildings* 82: 385–390

Sari, Ahmet and Kamil Kaygusuz. 2002. Thermal performance of palmitic acid as a phase change energy storage material. *Energy Conversion and Management* 43(6): 863–876

Sari, Ahmet and Kamil Kaygusuz. 2001. Thermal energy storage system using stearic acid as a phase change material. *Solar Energy* 71(6): 365–376

Sarı, Ahmet and Kamil Kaygusuz. 2001. Thermal performance of myristic acid as a phase change material for energy storage application. *Renewable Energy* 24(2): 303–317

Saxena, Abhishek, GhanshyamSrivastava, and Vineet Tirth. 2015. Design and thermal performance evaluation of a novel solar air heater. *Renewable Energy* 77: 501–511

Shalaby, S. and M. A. Bek. 2014. Experimental investigation of a novel indirect solar dryer implementing PCM as energy storage medium. *Energy Conversion and Management* 83(0): 1–8

Shalaby, S. M., M. A. Bek, and A. A. El-Sebaii. 2014. Solar dryers with PCM as energy storage medium: A review. *Renewable and Sustainable Energy Reviews* 33(0): 110–116

Sharma, Atul, C. R. Chen, V. V. S. Murty, and Anant Shukla. 2009. Solar cooker with latent heat storage systems: a review. *Renewable and Sustainable Energy Reviews* 13(6): 1599–1605

Sharma, Atul, V. V. Tyagi, C. R. Chen, and D. Buddhi. 2009. Review on thermal energy storage with phase change materials and applications. *Renewable and Sustainable Energy Reviews* 13(2): 318–345

Sharma, R. K., P. Ganesan, V. V. Tyagi, H. S. C. Metselaar, and S. C. Sandaran 2015. Developments in organic solid-liquid phase change materials and their applications in thermal energy storage. *Energy Conversion and Management* 95(0): 193–228

Shilei, Lv, Feng Guohui, Zhu Neng, and Dongyan Li. 2007. Experimental study and evaluation of latent heat storage in phase change materials wallboards. *Energy and Buildings* 39(10): 1088–1091

Shilei, Lv, Zhu Neng, and Feng Guohui. 2006. Eutectic mixtures of capric acid and lauric acid applied in building wallboards for heat energy storage. *Energy and Buildings* 38(6): 708–711

Stalin, M. Joseph, S. Mathana Krishnan, and P. Barath 2013. *Cooling of room with ceiling fan using phase change materials*. Paper presented at the Proceeding of the World Congress on Engineering, Vol III

Stovall, T. K. and J. J. Tomlinson 1995. What are the Potential Benefits of Including Latent Storage in Common Wallboard? *Journal of Solar Energy Engineering* 117(4): 318–325

Telkes, Maria. 1952. Nucleation of Supersaturated Inorganic Salt Solutions. *Industrial & Engineering Chemistry* 44(6): 1308–1310

Thulasi Das, T. C., S. Karmakar, and D. P. Rao. 1994. Solar box-cooker: Part I—Modeling. *Solar Energy* 52(3): 265–272

Trigui, Abdelwaheb, Mustapha Karkri, Chokri Boudaya, Yves Candau, and Laurent Ibos. 2013. Development and characterization of composite phase change material: Thermal conductivity and latent heat thermal energy storage. *Composites Part B: Engineering* 49(0): 22–35

Trigui, Abdelwaheb, Mustapha Karkri, and Igor Krupa. 2014. Thermal conductivity and latent heat thermal energy storage properties of LDPE/wax as a shape-stabilized composite phase change material. *Energy Conversion and Management* 77(0): 586–596

Turnpenny, J. R., D. W. Etheridge, and D. A. Reay. 2000. Novel ventilation cooling system for reducing air-conditioning in buildings.: Part I: Testing and theoretical modelling. *Applied Thermal Engineering*, 20(11): 1019–1037

Tyagi, V. V., A. K. Pandey, G. Giridhar, B. Bandyopadhyay, S. R. Park, and S. K. Tyagi 2012. Comparative study based on exergy analysis of solar air heater collector using thermal energy storage. *International Journal of Energy Research* 36(6): 724–736

Tyagi, V. V., A. K. Pandey, S. C. Kaushik, and S. K. Tyagi. 2012. Thermal performance evaluation of a solar air heater with and without thermal energy storage. *Journal of Thermal Analysis and Calorimetry* 107(3): 1345–1352

Vijaya Venkata Raman, S., S.Iniyan, and Ranko Goic. 2012. A review of solar drying technologies. *Renewable and Sustainable Energy Reviews* 16(5): 2652–2670

Zalba, Belén, José M. Marìn, Luisa F. Cabeza, and Harald Mehling. 2004. Free-cooling of buildings with phase change materials. *International Journal of Refrigeration* 27(8): 839–849

Zalba, Belén, José Ma Marìn, Luisa F. Cabeza, and Harald Mehling. 2003. Review on thermal energy storage with phase change: materials, heat transfer analysis and applications. *Applied Thermal Engineering* 23(3): 251–283

Zhou, D., C. Zhao, and Y. Tian. 2012. Review on thermal energy storage with phase change materials (PCMs) in building applications. *Applied Energy* 92(0): 593–605

Zhu, Na, Pengpeng Liu, Pingfang Hu, Fuli Liu, and Zhangning Jiang. 2015. Modeling and simulation on the performance of a novel double shape-stabilized phase change materials wallboard. *Energy and Buildings* 107: 181–190

22

Earth to Air Heat Exchanger Systems for Small Houses and Industrial Buildings in Changing Indian Climatic Conditions

Anil Kumar Misra

Department of Civil and Environmental Engineering,
THE NorthCap University (Formerly ITM University), Gurgaon
E-mail: anilmishra@ncuindia.edu, anilgeology@gmail.com

22.1 INTRODUCTION

Worldwide, climate change has assumed an alarming proportion. It is both directly or indirectly influenced by an increasing global energy demand. It has become imperative to search and adopt the systems that do not contribute to global warming and are feasible and economically affordable. Earth–air heat exchanger (EAHE) system is the system that fulfils such criteria.

Temperature within the ground surface varies with depth. This variation in the temperature depends on a number of factors (Popiel, Wojtkowiak, and Biernacka 2001), such as physical and structural properties of the subsurface soil/rock, surface cover of ground, and climate interaction. These changes in subsurface temperature can be utilized in cooling of buildings, cold storages, offices, laboratories, and so on and will aid in minimizing the impact of climate change through reducing the energy demand and consumption. Several researchers have worked and are still working in this area throughout the world. Underground buildings were constructed in different parts of Spain and Tunisia to minimize the impact of arid climatic conditions (Petherbridge 1976). In China, excavations in houses were carried out to cope with freezing climatic conditions (Rudofsky 1977). Studies have found shaded ponds (roof pond systems), also known as "sky-therm", to be very effective in decreasing and maintaining the indoor building temperature (Yadav and Rao 1983). The classification of natural cooling systems is based on their source of coolness, nature of the material, and modes of fluid flow and heat transfer (Bahadori 1984). The technique of heat transfer depends on the fact that the soil at subsurface can be used as a heat sink or as a heat source because the impact of seasonal variations on temperature is minimum in the subsurface. A study carried out by Choudhury and Misra (2014) revealed that materials such as bamboo and hydraform (cement and soil) can be used to make underground ducts for the EAHE system with very low cost, and the temperature variation of around 10°C can be achieved between indoor and outdoor.

In India, the utilization of a EAHE system is limited, and it is rarely used in residential buildings as people have reservations on several aspects of the system, such as (i) the construction of the EAHE system requires a large open area, (ii) its performance is low compared to other systems, (iii) it requires regular maintenance and, thus, is very expensive, and (iv) it can cause indoor air pollution and promote bacterial growth. A study carried out on these aspects (Misra, Gupta, Lather, et al. 2015) indicated that the EAHE system can be constructed in small houses at very low cost using low-cost materials and the maintenance cost is also very low. The temperature of earth is stable below 1 m depth (ASHRAE 1982). This finding has motivated engineers to use the earth subsurface as a source for storage and dissipation of heat for passive heating and cooling applications (Tombazis, Argiriou, and Santamouris 1990, Santamouris, Balaras, Dascalaki, et al. 1994). Below 1 m depth, the soil temperature remains in the range of 23–28°C (Khatry, Sodha, and Malik 1978), and it is usually higher in winter and lower in summer as compared to the ambient air temperatures.

As a rapidly developing country, energy requirements of India are likely to increase (25%–30%) by 2020. At present, India is the world's sixth largest consumer of energy, and for sustaining its growth rate, we have to either increase the energy production or decrease the total energy consumption. The increasing pressure of population and increased use of energy in industrial, domestic, and public sectors are areas of concern. Heating and cooling processes in the buildings also increase the energy consumption manifold. Thus, by adopting a suitable design of the EAHE system that can be easily installed in small houses within the foundation, energy consumption in both urban and rural areas can be reduced.

22.2   CHANGING INDIAN CLIMATIC CONDITIONS

In India, there are different types of climatic regions, which range from tropical in the south region to alpine and temperate regions in the extreme north of the Himalayas. The Indian climatic conditions are directly or indirectly influenced by the Thar Desert and Himalayan ranges. The Himalayan mountain ranges act as a very strong barrier for the winds that flow from the central part of Asia and are responsible for keeping a major part of India warm compared to other locations at the same latitudes. The northern plains of India show a continental type of climate that has both hot summer and cold winter conditions.

In the coastal parts of India, warm conditions are unvarying and rain occurs frequently. India is strongly influenced by two seasons of rainfall, with seasonal reversal of winds from January to July. In winters, cold and dry air blows from the north to the north-east direction over the Indian subcontinent. Owing to the very hot summer months, northern plains become very hot and draw winds that are moist in nature over the oceans, which cause the wind reversal over the region known as monsoon of the south-west.

It is an important phenomenon as it controls the climate of India and about 70%–80% of the rainfall that occurs annually in a short span of time, that is, from June to September (four months). This short span of rainfall has strong impact on rivers, water resources, reservoirs, hydropower generation, irrigation and productivity, economic stability, and ecosystems of India. In the Indian subcontinent, the climatic variations are more prominent compared to other parts of the world. Throughout the country, there are large variations in the amounts of total rainfall received in different parts and states of the country; for example, the total rainfall occurring in the western Rajasthan is less than 13 cm (average annual rainfall), while Mawsynram in Meghalaya receives around 1141 cm of rainfall occurs annually. These rainfall patterns directly or indirectly affect the climatic regimes of the country, which vary from humid in the northeast to arid in Rajasthan.

A report published by India Meteorological Department shows that temperature (annual mean) of the entire India increased by 00.56°C (Figure 22.1) from 1901 to 2009 (Attri and Tyagi 2010). It is to be noted that generally temperature (annual mean) has been recorded above normal since 1990. The study has found that such warming is owing to the increase in temperature (maximum) throughout India (Figure 22.2). It also shows that temperature (minimum) is rising steadily from 1990 onwards (Figure 22.3), and the rate of increase is slightly more compared to the increase in maximum temperature (IMD 2009). Intergovernmental Panel on Climate Change (IPCC) in 2007 has reported a warming trend in the order of 0.74°C over the globe. Further, the trends (spatial pattern) of the temperature (annual mean) (Figure 22.4) show a positive (increasing) change in a majority of the region of the country except some parts of

Gujarat, Bihar, and Rajasthan, where negative trends are observed (decreasing) (IMD 2009).

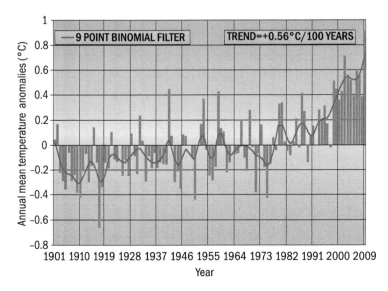

Fig. 22.1 *Entire India temperature (annual mean) anomalies from 1901 to 2009*
Note: *Observations are based on the average between 1961 and 1990, represented as vertical bars. The curve shown with solid grey line represents the timescale variations (sub-decadal) smoothed with the help of binomial filter.*
Source *Attri and Tyagi (2010)*

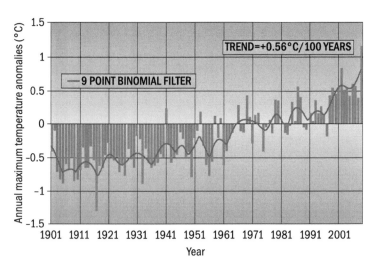

Fig. 22.2 *Entire India temperature (maximum) anomalies from 1901 to 2009*
Note: *Observations are based on the average between 1961 and 1990, represented as vertical bars. The curve shown with solid grey line represents the timescale variations (sub-decadal) smoothed with the help of binomial filter.*
Source *Attri and Tyagi (2010)*

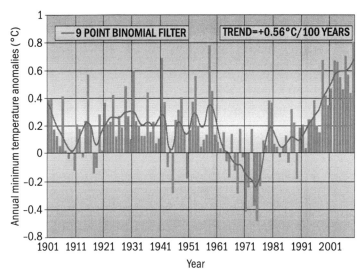

Fig. 22.3 *Entire India temperature (minimum) anomalies from 1901 to 2009*

Note: *Observations are based on the average between 1961 and 1990, represented as vertical bars. The curve shown with the solid grey line represents the timescale variations (sub-decadal) smoothed with the help of binomial filter.*

Source *Attri and Tyagi (2010)*

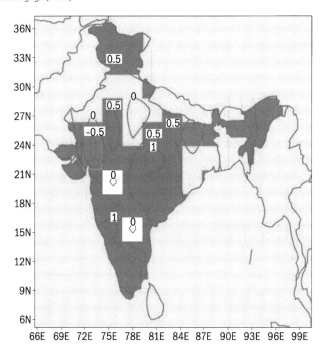

Fig. 22.4 *Mean annual temperature anomalies showing the spatial pattern of trend (°C/100 years) from 1901 to 2009*

Note: *Dark grey (warming) and light grey (cooling) areas are showing significant trends.*

Source *Attri and Tyagi (2010)*

22.3   INDIAN CLIMATE AND EAHE SYSTEMS

Consumption of energy within the houses and buildings for providing cooling and heating comfort, is escalating continuously, and owing to changing climatic conditions, the demand of low-energy techniques is also growing. The EAHE system can be designed with both open loop and closed loop using various materials, but its efficiency and efficacy are largely influenced by the climatic conditions of the area. Therefore, the selection of design for the EAHE system and the materials for making an underground loop system should be made very carefully. The type of design and selection of materials for varying climatic conditions in the context of Indian climate (Figure 22.5) are discussed next.

22.3.1   EAHE System for Highland Climatic Conditions

Highland climatic conditions represent the northern-most part of India. In this area, the climate ranges from nearly tundra above the snowline to tropical in the foothills. In these areas, usually the temperature ranges from +20°C (in summers) to −20°C (in winters). Houses and buildings in these areas need constant heating as there is a drastic fall in temperature during night. The EAHE system with closed-loop design will be most suitable for such areas because this system circulates the same air within the loop again and again, and thus more heating can be achieved. The precipitation rate in these areas is also very high, and thus any galvanized metallic pipe with antimicrobial coating (inside) will be most suitable.

22.3.2   EAHE System for Subtropical Humid Conditions

Subtropical humid climatic conditions persist in majority of north-eastern states and most of north India. These areas usually have both hot summer from March to September and cold winter from November to February. The temperature in summer ranges from 35°C to 48°C, while it ranges from 0°C to 20°C in winter. These areas need the EAHE system for both heating and cooling. Each system with open-loop design can easily serve the purpose as weather conditions are moderate throughout the year. In such areas, metallic pipes, polyvinyl chloride (PVC) pipes, concrete pipes, and bamboo mesh with antimicrobial coating (inside) can be used for constructing the duct system.

22.3.3   EAHE System for Tropical Wet and Dry

These areas experience constantly warm to very warm climatic conditions, and even in winter, the temperature does not fall below 20°C. The rainfall rate is usually high to very high, but despite that the temperature ranges between 20°C and 45°C throughout the year. These areas need cooling most of the time, and the EAHE system with open-loop design can be most suitable in this case, with galvanized metallic pipes or PVC pipes, or concrete pipes along with antimicrobial coating (inside) used for making the loop system.

22.3.4   EAHE System for Arid Climatic Conditions

Arid climatic conditions prevail in majority of the extreme western parts of Rajasthan and Gujarat in India. In these areas, the rate of moisture loss via evaporation and transpiration is very high, and the temperature ranges between 25°C and 50°C throughout the year. Usually summers are very long compared to winters. During winters also, the temperature does not fall below 25°C. These areas need constant cooling throughout the year. The EAHE system with open-loop design will be most suitable for such climatic conditions, as it is easy to cool and re-cool the same air in extreme summer conditions. Galvanized metallic pipes and concrete pipes with antimicrobial coating (inside) will be the most suitable materials for loop systems as these materials show the least dimensional changes compared to other materials in extreme hot conditions.

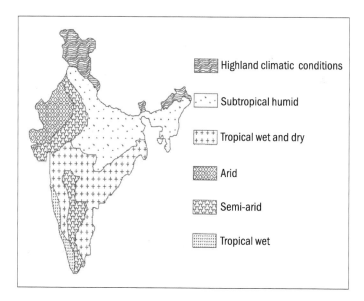

Fig. 22.5 *Indian climatic conditions and design suitability of EAHE systems*
Note: *Highland climatic conditions require constant heating (EAHE system with closed-loop design)*
Subtropical humid require seasonal heading and cooling (EAHE system with open-loop system)
Tropical wet and dry require cooling system (EAHE system with open-loop design)
Arid require constant cooling (EAHE system with closed-loop design)
Semi-arid require cooling (EAHE system with open-loop design)
Tropical wet require cooling (EAHE system with open or closed design)

22.3.5   EAHE System for Semi-arid Climatic Conditions

Semi-arid climatic conditions prevail in most of the western and southern central parts of India. The temperature in these areas varies between 25°C and 45°C. These areas also need cooling most of the time, but the temperature is usually low as compared to arid regions. The EAHE system with open-loop design can efficiently work in these areas, and galvanized metallic pipes and PVC pipes with antimicrobial coating (inside) can be used for loop systems.

22.3.6   EAHE System for Tropical Wet

Such climatic conditions are found in the western part of southern India. Owing to high precipitation, this area does not experience extreme hot and cool climatic conditions. Temperature ranges between 30°C and 40°C throughout the year. During summer, for short duration (March–June), these regions may require cooling. Both open- and closed-loop systems can effectively perform in such conditions. Moreover, galvanized metallic pipes, concrete pipes, and PVC pipes with antimicrobial coating (inside) can be used for duct systems.

22.4   DEMAND FOR ENERGY IN INDIA

A study carried out by British Petroleum (Dudley 2014) related to world energy consumption in 2014 predicted that energy consumption in India would increase by 7.1% in 2014, reaching an all-time high and accounting for 34.7% of the global consumption increment in 2014. Domestic energy consumption in India reached an all-time high in 2014, with the fastest growth for the past 5 years. Several international and national agencies have made projections related to India's energy demands.

Table 22.1 gives the state-wise and source-wise potential of renewable power as per the National Statistical Organization (2015) estimated as on 31 March 2014 in India.

Table 22.1   Source-wise and state-wise renewable power estimation potential of India till 2014 (in MW)

State/union territory	Power generation (wind)	Hydro-power (small)	Power generation (biomass)	Cogeneration bagasse	Energy generated through wastes	Total energy (MW)
Estimated potential				Distribution (%)		
A	B	C	D	E	F	G	H
Andhra Pradesh	14,497	978	578	300	123	16,476	11.16
Arunachal Pradesh	236	1,341	8	0	0	1,585	1.07

Contd...

Table 22.1 *Contd...*

State/union territory	Power generation (wind)	Hydro-power (small)	Power generation (biomass)	Cogeneration bagasse	Energy generated through wastes	Total energy (MW)
	Estimated potential			Distribution (%)		
A	B	C	D	E	F	G	H
Assam	112	239	212	0	8	571	0.39
Bihar	144	223	619	300	73	1,359	0.92
Chhattisgarh	314	1,107	236	0	24	1,681	1.14
Goa	0	7	26	0	0	33	0.02
Gujarat	35,071	202	1,221	350	112	36,956	25.04
Haryana	93	110	1,333	350	24	1,910	1.29
Himachal Pradesh	64	2,398	142	0	2	2,606	1.77
Jammu and Kashmir	5,685	1,431	43	0	0	7,159	4.85
Jharkhand	91	209	90	0	10	400	0.27
Karnataka	13,593	4,141	1,131	450	0	19,315	13.08
Kerala	837	704	1,044	0	36	2,621	1.78
Madhya Pradesh	2,931	820	1,364	0	78	5,193	3.52
Maharashtra	5,961	794	1,887	1,250	287	10,179	6.90
Manipur	56	109	13	0	2	180	0.12
Meghalaya	82	230	11	0	2	325	0.22
Mizoram	0	169	1	0	2	172	0.12
Nagaland	16	197	10	0	0	223	0.15
Odisha	1,384	295	246	0	22	1,947	1.32
Punjab	0	441	3,172	300	45	3,958	2.68
Rajasthan	5,050	57	1,039	0	62	6,208	4.21
Sikkim	98	267	2	0	0	367	0.25
Tamil Nadu	14,152	660	1,070	450	151	16,483	11.17
Tripura	0	47	3	0	2	52	0.04
Uttar Pradesh	1,260	461	1,617	1,250	176	4,764	3.23
Uttarakhand	534	1,708	24	0	5	2,271	1.54
West Bengal	22	396	396	0	148	962	0.65
Andaman and Nicobar	365	8	0	0	0	373	0.25
Chandigarh	0	0	0	0	6	6	0
Dadra and Nagar Haveli	0	0	0	0	0	0	0

Contd...

Table 22.1 *Contd...*

State/union territory	Power generation (wind)	Hydro-power (small)	Power generation (biomass)	Cogene-ration bagasse	Energy generated through wastes	Total energy (MW)
	Estimated potential			Distribution (%)		
A	B	C	D	E	F	G	H
Daman and Diu	4	0	0	0	0	4	0
Delhi	0	0	0	0	131	131	0.09
Lakshadweep	0	0	0	0	0	0	0
Puducherry	120	0	0	0	3	123	0.08
Others	0	0	0	0	1022	1022	0.69
All-India total	**102,772**	**19,749**	**17,538**	**5,000**	**2,556**	**147,615**	**100**
Distribution (%)	**69.620**	**13.380**	**11.880**		**3.390**	**1.730**	**100**

[a]Industrial waste

Source National Statistical Organization (2015); Ministry of New and Renewable Energy

In India, the total electricity consumption has increased from 41,1887 GWh (2005–06) to 882,592 GWh (2013–14), indicating a compound annual growth rate (CAGR) of 8.84%. The total increase in the consumption of electricity from 2012 to 2013 (824,301 GWh) is around 7.07% compared to that of 2013–14 (882,592 GWh). Of the total consumption during 2013–14, the industrial sector accounted for the maximum consumption of electricity (43.83%), followed by domestic sector (22.46%), agriculture sector (18.03%), and commercial sector (8.72%) (Figure 22.6). Consumption of electricity by different sectors from 2013 to 2014 is shown in Figure 22.7.

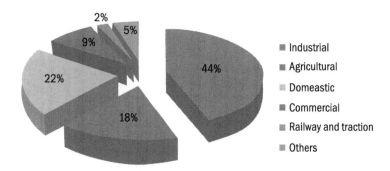

2%
5%
9%
22%
44%
18%

- Industrial
- Agricultural
- Domeastic
- Commercial
- Railway and traction
- Others

Total consumption: 882,592 GWh

Fig. 22.6 *Consumption of electricity in India during 2013–14 by different sectors*

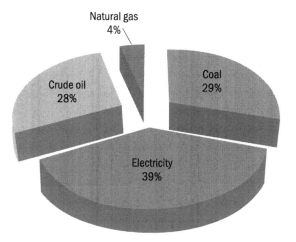

Total consumption: 47,264 petajoules

Fig. 22.7 *Consumption of electricity by different sectors from 2013 to 2014*

The estimation indicates that the consumption of electricity in the commercial sector (8.82%) and industry sector (10.97%) has escalated at a faster rate compared to other sectors from 2005–06 to 2013–14. Moreover, the energy intensity and per capita energy consumption (PEC) are important policy indicators at both international and national levels. In India, as such no data are available for the non-conventional energy consumption from different sources, especially for villages (rural sectors). Both the indicators are computed on the basis of conventional energy consumption.

The energy consumption was calculated in petajoule for coal and lignite that contributed approximately 41.29% of the overall consumption between 2013 and 2014. Petroleum (crude oil) was the second largest source of energy consumption (38.70%) during the period, which was followed by electricity (14.47%). The overall energy consumption from conventional sources escalated from 23,903 petajoules during 2012–13 to 24,071 petajoule from 2013 to 2014, with an increase of 0.70% (Table 22.2).

Table 22.2 Conventional Energy Consumption Trends in India

Period (year)	Coal and lignite	Petroleum (crude oil)[a]	Gas (natural)	Electricity[b]	Total
1	2	3	4	5	6 = 2 to 5
2005–06	7,009	5,448	1,207	1,483	15,146
2006–07	7,459	6,136	1,186	1,641	16,421

Contd...

Table 22.2 *Contd...*

Period (year)	Coal and lignite	Petroleum (crude oil)[a]	Gas (natural)	Electricity[b]	Total
2007–08	7,926	6,536	1,213	1,839	17,514
2008–09	8,476	6,732	1,223	2,026	18,457
2009–10	9,137	8,071	1,792	2,233	21,233
2010–11	9,207	8,248	1,974	2,464	21,892
2011–12	9,325	8,547	1,790	2,721	22,383
2012–13	9,909	9,178	1,532	3,283	23,903
2013–14 (p)	9,939	9,316	1,334	3,482	24,071
Growth rate from 2013–14 to 2012–13 (%)	0.310	1.500	–12.930	6.040	0.700
CAGR from 2005–06 to 2013–14 (%)	3.96	6.14	1.12	9.95	5.28

[a]Crude oil in terms of refinery's crude throughput.
[b]Includes thermal, hydro, and nuclear electricity from utilities.
Note: Energy consumption is calculated in terms of petajoules.
Sources: National Statistical Organization (2015); Office of Coal Controller, Ministry of Coal; Ministry of Petroleum and Natural Gas; Central Electricity Authority

Increase in electricity consumption by domestic sector in India is highest (22.46% in 2013–14) after industrial sector electricity consumption (43.83% in 2013–14). The demand increased in 2014–15 at the same pace, and thus focus should be given to minimize the energy consumption of the domestic sector by adopting and promoting low energy-efficient systems used for the cooling and heating of houses. If the EAHE system is used on a large scale, it can drastically reduce the domestic sector consumption and the energy thus saved can be utilized by other sectors.

22.5   ADVANTAGES OF EAHE SYSTEM

There are several advantages of using the EAHE system over other available techniques. Some of the most important advantages of the EAHE system are as follows:

- As such, there is no standard configuration and equipment required for designing the EAHE system. All design types require almost the same kind of equipment.
- This system can help to reduce the energy crisis in India by decreasing the domestic electricity demand, especially during summers.

- Indirectly, the EAHE system can contribute in reducing the greenhouse gas emissions, such as carbon dioxide, methane, chlorofluorocarbons (CFCs), and the load on thermal power plants.
- The EAHE system is completely pollution free and eco-friendly.
- This system efficiently performs in the arid and semi-arid zones, and hence it can be installed in western India (Rajasthan and Gujarat).
- This system can also be beneficial to rural people owing to the less installation and maintenance cost, and it causes no environmental degradation. Hence, it can be easily adopted by villagers.
- The EAHE system is capable of effectively maintaining the cooling in cold storages, and thus it can be largely used in all types of buildings.
- The construction cost of the EAHE system is less compared to other cooling systems and it also requires low maintenance.
- This system can be used for both cooling and heating, and, thus, it can be used in any part of India.
- Materials used for making the loop system have the least impact on the overall performance of the EAHE system, and, thus, low-cost materials can also be used to make underground loops.

22.6   DESIGN OF EAHE SYSTEMS

The EAHE system can be formed by using long plastic, concrete, or metallic pipes that are laid underground at a certain depth (depending upon the type of lithology). One end of the EAHE system is used for inlet and the other end that is inside the building is used as the outlet. It provides some pre-conditioning of air—either preheating in the winter or pre-cooling in the summer. During summers, the temperature of soil surrounding the pipes, absorb the heat from air passing through the pipes to soil resulting in cooled air. In winters or at nights, the reverse process takes place. Thus, the EAHE system can be used for cooling in summer and heating in winter.

The EAHE system usually consists of vertically or horizontally buried loop(s) of pipe in the ground. Vertical loops are constructed in a building having less space and these are buried deeper, while horizontal loops are buried at shallow depths (1–4 m), depending usually on the type of lithology and the depth of water-bearing strata. Earth temperature at the depth between 1 m and 4 m or beyond is usually stable, with no diurnal fluctuation and only a small seasonal or annual variation. This stability is the result of a natural physical phenomenon.

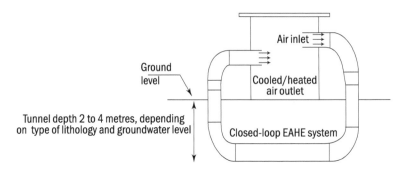

Fig. 22.8 *Closed-loop design of the EAHE system*

The EAHE system can be constructed using the following two different types of loop configurations:

1. **Closed-loop EAHE system:** In this system, a U-shaped loop of pipe (30–150 m) is installed in the ground and both inlet opening and outlet openings are within the system (Figure 22.8). The air from inside the building is blown (air inlet point) through the U-shaped loop and at a depth (below the ground) it cools down to near earth's temperature before returning to the building. This system is considered more efficient (during the high temperature) compared to the open system because it cools and re-cools the same air. This system can be easily installed in small houses and is suitable for Indian conditions.

2. **Open-loop EAHE system:** In this system, configuration of loops is different and is considered more suitable for buildings with a large space area. In this system, air from outside the building (inlet system) is blown into the duct system and it passes through an air filtration unit, which is usually installed within the duct system, and filtered and cooled air reaches the building through an outlet system (Figure 22.9).

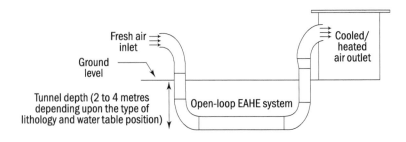

Fig. 22.9 *Open-loop design of the EAHE system*

22.7 COMPONENTS OF EAHE SYSTEM

Some of the most important components of the EAHE system are as follows:

- **Fan:** It blows the fresh/ambient air into the underground air tunnel.
- **Air inlet point:** This is the point through which the fresh air from outside is blown into the underground duct system, where it is passed through a filtration unit for the removal of pollen and dust particles.
- **Air outlet point:** This is the point from which filtered and cooled/heated air is passed into the building.
- **Underground pipes:** These are the pipes that form the underground duct system. These could be of plastic, metal, or concrete usually coated with inner antimicrobial layers. The diameter of these pipes varies between 120 mm and 700 mm and the pipes are generally buried at 1–4 m depth below ground, where the ambient earth temperature is between 10°C and 23°C throughout the year.
- **Temperature sensors:** These sensors are used to measure the temperature of circulating air, soil, and pipe surfaces at different locations within the duct system. Copper–constantan thermocouple PT-100 is generally used.
- **Humidity sensors:** Measurement of humidity is done by using humidity sensors.

22.8 PERFORMANCE OF EAHE SYSTEM

The EAHE system is considered suitable for residential, commercial, and industrial buildings having some open area for the construction of the earth–pipe–air heat exchanger tunnel system. However, a majority of the houses in India, especially in urban areas, are without any open space and are fully covered from three sides and open from only one side; therefore, the design of the loop system should be small that can be easily installed within the foundation of the buildings. A study carried out by Misra, Gupta, Lather, *et al.* (2015) on a prototype model constructed in a metal tray (length 99 cm, width 99 cm, and height 10.5 cm) using PVC pipes for the duct system (diameter 25 mm) with the total length of 25 m demonstrated the efficiency of the prototype model. The entire loop system of the model was filled with hard, compact, and clay-rich soil taken from a depth of 1 m, and inlet and outlet points of the loop system were designed within the model. Performance of the model was evaluated by measuring

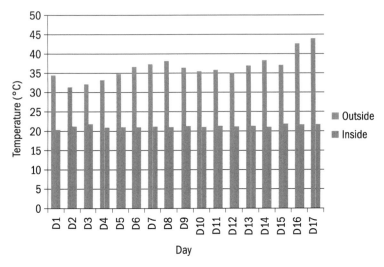

Fig. 22.10   *Performance of EAHE system using the PVC pipe*
Source *Misra, Gupta, Lather, et al. (2015)*

the air temperature variation at inlet and outlet points of the loops using two PT-100 sensors at the range of 0–199°C, with the least count of 0.1°C. Testing on the model was performed for three weeks and it showed the average temperature variation of 15°C; that is, the indoor air temperature was recorded 15°C less compared to the outdoor air temperature. The minimum energy efficiency ratio (EER) (W/W) calculated for the prototype model was around 3.0, which is equivalent to an Energy Star 5 rating, the most efficient system. Figure 22.10 shows the overall performance of the model. The findings of this model have proved that the EAHE systems can be constructed in small houses and are suitable to Indian conditions.

A study of the EAHE system was performed in North Eastern Regional Institute of Science and Technology (NERIST), Nirjuli, Arunachal Pradesh

Fig. 22.11   *Design of the EAHE tunnel system constructed using bamboo mesh*

(Kumar, Choudhury, Kumar, *et al.* 2007). In this study, the duct system of the EAHE system was constructed using bamboo mesh and cement and soil plaster (Figure 22.11) and the findings revealed the indoor and outdoor average temperature variation of 10°C. It demonstrates that the materials used in making the duct system (loops) have least impact on the overall performance of the EAHE system.

The performance of the EAHE system directly or indirectly depends on the velocity of the air flow within the tunnel and the diameter of the duct (loop) system (Krarti and Kreider 1996). Experiments conducted using different pipe diameters and wind velocities supported the findings reported by Krarti and Kreider (1996). The curve plotted in Figure 22.12 shows the relationship between temperature of air and diameter of pipe. It illustrates that the temperature drop increases with increase in the diameter of pipes. It is because with the increase in the pipe diameter, the amount of air exposed for heat transfer also increases, and, hence, more heat transfer occurs, resulting into more temperature drop.

A curve between air temperature and air velocity is shown in Figure 22.13. At higher velocities, the final temperature is more (hotter) compared to that at lower velocities. This is because at a slow speed, the air would be in contact with the soil for a longer period of time, and hence more heat exchange will take place between air and soil. In the case of higher speed, the amount of air reaching the system would be more, and hence there will be a lower rate of cooling.

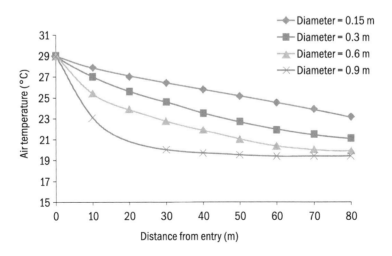

Fig. 22.12 *Tube diameters and their effect on air temperature at various positions along the tube*

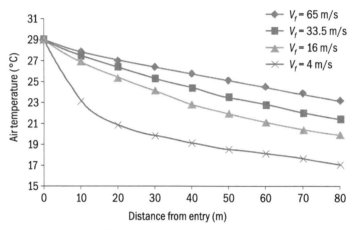

Fig. 22.13 *Air velocities and their effect on air temperature at various positions along the tube*

Moreover, the curve also indicates that the temperature decreases as the moving air distance from the inlet point increases. It is because as the length of pipes or tunnels increases, the time for the heat transfer between the air inside the pipe and the earth also increases. So there is a decrease in temperature with the length of pipes.

22.9   CONCLUSION

A worldwide energy crisis has become a major challenge to meet the continuously escalating demand of energy. Energy conservation by opting energy efficient systems is one of the solutions. The EAHE system can drastically reduce the energy demand in urban and rural areas and can be easily constructed in small houses with and without open space. The design and operation of the EAHE system is easy and the system requires low maintenance as revealed by several experimental studies of this system. The overall findings on the EAHE system demonstrate the following:

- The EAHE systems can reduce the energy demand of buildings and can be easily installed in small houses with and without any open area.
- These systems can easily provide an indoor and outdoor temperature variation of 10°C if they are designed and operated carefully.
- By controlling wind velocity and duct diameter, the efficiency of the EAHE system can be enhanced manifold.
- The overall impact of the duct (loop) material on the performance of the EAHE system is minimal.

Depending on the Indian climatic condition, the EAHE systems can be used both in summers for cooling and in winter for heating all types of buildings. The EAHE system is capable of reducing the energy consumption between 50% and 60% which is consumed by cooling (air conditioning) and warming systems. Directly or indirectly, the EAHE system aids in environmental management and impact minimization of climate change by reducing the consumption of coal, hydrocarbons, and other biofuels used in generating power in thermal power plants and in houses for cooking and heating purposes.

REFERENCES

Attri, S. D. and A. Tyagi. 2010. Climate profile of India, Met Monograph No. Environment Meteorology, 01/2010, India Meteorological Department, Government of India, New Delhi

ASHRAE. 1982. *Handbook of Application.* Atlanta, GA: American Society of Heating, Refrigerating, and Air Conditioning Engineers

Bahadori, M. N. (1984) Natural cooling systems and an economic feasibility study of long-term storage of coolness, *Energy* 9(7): 587–604

Choudhury, T. and A. K. Misra. 2014. Minimizing changing climate impact on buildings using easily and economically feasible earth to air heat exchanger technique. Mitigation and Adaptation Strategies for Global Change 19(7): 947–954

Dudley, B. (2014) B P Statistical review of World energy, 2014; World Petroleum Conference, Moscow, 16 June 2014

IPCC. 2007. *Climate Change 2007: The Physical Science Basis.* Cambridge, UK: Cambridge University Press

IMD (India Meteorological Department). 2009. *Annual Climate Summary.* Pune: National Climate Centre

Kumar, R., T. Choudhury, P. Kumar, S. Mazumdar, and J. Das. 2007. Experimental Investigation of Earth to Air Heat Exchanger at NERIST, B.Tech Project Report, Department of Civil Engineering, North Eastern Regional Institute of Science and Technology, Nirjuli, Arunachal Pradesh

Khatry, A. K., M. S. Sodha, and M. A. S. Malik. 1978. Periodic variation of ground temperature with depth and time. *Solar Energy* 20: 425–427

Krarti, M. and J. F. Kreider. 1996. Analytical model for heat transfer in an underground air tunnel. *Energy Conversion and Management* 37: 1561–1574

Misra, A. K., M. Gupta, M. Lather, and H. Garg. 2015. Design and performance evaluation of low cost earth to air heat exchanger model suitable for small buildings in aid and semi-arid regions. *KSCE Journal of Civil Engineering* 19(4): 853–856

National Statistical Organization. 2015. Energy Statistics. New Delhi: National Statistical Organization

Popiel, C.O., J. Wojtkowiak, and B. Biernacka. 2001. Measurement of temperature distribution in ground. *Experimental Thermal and Fluid Science* 25: 301–309

Petherbridge, Guy T. 1976. Vernacular Architecture in the Maghreb: some historical and geographical factors. *Maghreb Review* 1(3): 12–17

Rudofsky, B. 1977. The Prodigious Builders. New York: Harcourt Brace Jovanovich

Santamouris, M., C.A. Balaras, E. Dascalaki, and M. Vallindras. 1994. Passive solar agricultural greenhouse: a worldwide classification evaluation of technologies and systems used for heating purposes. *Solar Energy* 53(5): 411–426

Tombazis, A., A. Argiriou, and M. Santamouris. 1990. Performance evaluation of passive and hybrid cooling components for a hotel complex. *International Journal of Solar Energy* 9(9): 1–12

Yadav, R. and D. P. Rao. 1983. Digital simulation of indoor temperature of buildings with roof ponds. Solar Energy 1(2): 205–215

23

Development in Metal Oxide Nanomaterial-based Solar Cells

Bal Chandra Yadav[a,b,*], Praveen Kumar[a], Satyendra Singh[c], and Richa Kothari[b,d]

[a]*Department of Applied Physics, Babasaheb Bhimrao Ambedkar University, Lucknow, Uttar Pradesh 226025*
[b]*DST Centre for Policy Research, Babasaheb Bhimrao Ambedkar University, Lucknow, Uttar Pradesh 226025*
[c]*Department of Physics, University of Allahabad, Allahabad, Uttar Pradesh 211002*
[d]*Department of Environmental Science, Babasaheb Bhimrao Ambedkar University, Lucknow, Uttar Pradesh 226025*
E-mail: balchandra_yadav@rediffmail.com

23.1 INTRODUCTION

There is a persistent demand for energy from sustainable resources. At present it is becoming increasingly difficult to meet the gap between demand and supply of clean energy and at the same time avoid exploiting these resources. Scientists are continuously searching for green materials, that is materials which are widely used without degrading the environment. This is the time of science and technology revolution, where reduction in energy consumption techniques and maximum use of renewable energy sources are needed. Use of solar-powered products is one of the best possible ways to save exploitation of fossil fuels. This is not only an eco-friendly technique but has many other advantages as well.

A solar cell or photovoltaic cell converts sunlight directly into electricity by photovoltaic effect. A solar photovoltaic cell consists of a p–n junction in a semiconductor, through which voltage is developed. Photovoltaic cells work on the theory of charge separation at an interface of two materials that have dissimilar conduction mechanism. Solid-state junction devices, which are usually made up of silicon dominate the market and are made by the semiconductor industry. Despite playing a significant role in the fulfilment of energy requirements, historically photovoltaic cells have been considered very expensive and not very technologically developed. An ideal solar cell material may comprise the following characteristics:

- Band gap of 1.1 eV to 1.7 eV
- Direct band structure
- Easy availability
- Non-toxic
- Easy reproducible deposition technique, suitable for large area production
- Excellent photovoltaic conversion efficiency
- Long-term stability

According to solid-state physics, silicon is not the ideal material for photovoltaic conversion. For example, 1 mm of GaAs (a direct semiconductor) has a capacity of 90% light absorption, while for the same, 100 mm of crystalline silicon material is required. Researchers are continuously searching for new suitable materials. The third generation of photovoltaic cells, such as nanocrystalline and conducting polymer films, are emerging technologies and they are challenging the silica-based solid-state junction devices. Low-cost fabrication and attractive features ease their market reach. Therefore, an electrolyte (solid, liquid, gel—forming a photo-electrochemical cell), holds the possibility of replacing classical devices. Progress in the fabrication and characterization of nanocrystalline materials creates new opportunities for these systems. Devices based on interpenetrating networks of microscopic semiconductors have high conversion efficiency, in comparison to conventional devices. Dye-sensitized solar cells (DSSCs) are prototype devices of semiconductor materials, which have a nanocrystalline structure that provides a photo-absorbing activity, by which optical absorption and charge separation processes take place.

An important consideration when developing a photovoltaic solar cell is reduction in manufacturing cost and improving the conversion efficiency by the use of nanotechnology.

23.2 HISTORY OF SOLAR CELL

The term "photovoltaic" has two words: the Greek words *'phos'*, meaning light, and "*Volta*", Italian physicist after whom volt/voltage is named. Hence, it literally translates to light and electricity. In 1839, Alexandre-Edmond Becquerel, the French physicist, reported the first photovoltaic effect. He observed that when silver chloride was placed in an acidic solution, it illuminated and when connected to platinum electrodes there was generation of voltage and current. Charles Fritt was the first to successfully convert light into electricity in 1883. He coated the semiconductor material selenium with an extremely thin layer of gold to make a junction. The resulting cells had a very low light-to-electricity conversion rate of about 1%.

Russell Ohl in 1941 invented the first silicon solar cell. It was made publicly known only five years later. The advancement in solar cells and the modern age of solar power technology began in 1954. Bell Laboratories, working on semiconductors, found that when silicon is doped with some definite impurities, it can respond to light. This precious knowledge helped us achieve very useful results, especially in the production of solar cells, with a sunlight energy conversion rate of about 6%.

After achieving this milestone of reliable power supply, geostationary communications satellites also started using it. Sputnik 3 ("Satellite-3"), launched on 15 May 1957 by USSR, was the first satellite to use solar photovoltaic system. After reaching this level of advancement in solar photovoltaic system, many governments started funding research to improve solar cell quality.

23.3   METAL OXIDE NANOMATERIALS

Nanomaterials can be metals, ceramics, polymeric materials, or composite materials. They are particles between the range of 1–100 nm in size. 1 nm spans 3 to 5 atoms lined up in a row. In contrast, the diameter of a human hair is about 5 orders of magnitude larger than a nanoscale particle. Metal oxides form an important group of functional materials with diverse structural, electronic, magnetic, and optical properties. At room temperature, these materials behave like metallic, semiconductor, or insulating materials as per their band gaps. Applications of metal oxides include magnetic memory, sensors, field emission, corrosion-resistant coatings, catalysis, piezoelectric devices, microelectronic circuits, fuel cells and photo-electrochemical cells. The physico-chemical properties of metal oxides are largely influenced by their particle size, especially by the 'nano' domain (10^{-9} m). As the particle size decreases from bulk to tens of nanometres, the ratio of surface to bulk atoms also increases resulting in larger surface area with respect to the volume of the material. A higher number of the chemically active sites and modified density of electronic states at the surface of the particles make metal oxides more important for solar cell.

Figure 23.1 explains the relation between conductivity observed and an electronic energy band; it also explains a model of electronic energy band in a bulk material. Direct band gap is preferred for crystalline C–Si

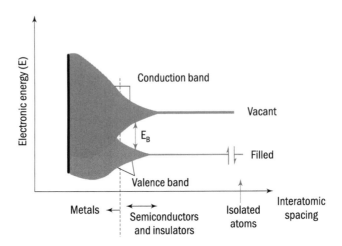

Fig. 23.1 *Generation of band structure in solids from atomic orbitals of isolated atoms*

solar cells such as Si: 1.1 eV, a-Si: 1.7 eV (Soga Tetswa 2006), CdTe: 1.5 eV (Katagiri, Sasaguchi, Hondo, *et al.* 1997), CIGS: 1.1–1.5 eV (Hartman, Johnson, Bertoni, *et al.* 2011; Reddy, Reddy, and Miles 2006), while an indirect band gap semiconductor requires a photon which can complete the transition. In the case of a direct band gap semiconductor, a photon is not required. In Figure 23.2, band positions of several metal oxides and sulphides in contact with an aqueous electrolyte at pH 1 are shown. The edge of the conduction band (red colour) and upper edge of the valence band (green colour) are presented along with the band gap in electron volts. The energy scale is indicated in electron volts using either normal hydrogen electrode or the vacuum level as the reference. Ordinate presents internal energy and not free energy. The free energy of an electron-hole pair is smaller than the band gap energy due to transnational entropy of the electrons and holes in the conduction and valence band, respectively. On the right-hand side, the standard potential of several redox couples are presented against the standard hydrogen electrode potential.

23.4   THREE GENERATIONS OF DEVELOPMENT

The first generation of photovoltaic cells are made up of silicon wafer and, therefore, known as silicon wafer-based solar cells. They occupy a large area and have a single layer of p–n junction diode. These cells have the capacity to generate electrical energy from artificial light that have same wavelengths of solar light. Solar cells are commercially produced using this technology. About 86% of these cells make the total solar cell market.

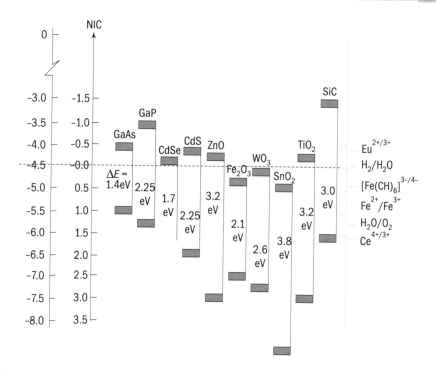

Fig. 23.2 *Band position of several metal oxides and sulphides*

The second generation of photovoltaic cells uses thin-film deposition of semiconducting materials. These are made from multiple-junction photovoltaic cells and are designed to produce higher efficiency. The second generation photovoltaic cells are based on amorphous; poly-crystalline; and micro-crystalline silicon, telluride, copper indium selenite/sulphide, cadmium (Cd). The benefit of using thin-film material is that it leads to a reduction in the material consumption for cell production. This helps in reducing the cost of manufacturing thin-film solar cells. Other benefits from this reduction is easy handling, and the fact that it requires less caution when being placed on a rooftop, in walls, open barren lands, poles, etc. These panels are also set on light and flexible surfaces including textiles. At present there are several semiconductor materials/technologies undergoing investigation and production.

Organic semiconductor solar cells and DSSCs are examples of third generation of photovoltaic solar cells (Precker and da Silva 2002; Hoppe and Sariciftci 2004; Wohrle and Meissner 1991). The third-generation photovoltaic solar cells are completely different from the first and second generations of solar cells as they do not work on the theory of p-n junctions.

These cells have separate photo-generated charge carriers. Polymer solar cells and photo-electrochemical cells are examples of this type of cells.

23.5   NANOMATERIAL-BASED SOLAR CELL RESEARCH

Metal oxide semiconductors are environment-friendly materials that are used in photovoltaics either as photoelectrodes in dye solar cells (DSCs) or to build metal oxide p–n junctions. Several recent reports have indicated that inorganic one-dimensional (1D) nanomaterials, such as TiO_2 (Green, Emery, Hishikawa, et al. 2012), WO_3 (Han, Shin, Seo, et al. 2009), SiO_2 (Wang, Buyanova, Chen, et al. 2006), V_2O_5 (Keis, Bauer, Boschloo, et al. 2002), ZnO (Kumara, Okuya, Kaneko, et al. 2003; Keis, Bauer, Boschloo, et al. 2002; Gonzalez-Volls and Lira-Cantu 2009; Keis, Linndgren, Schubert, et al. 2000; Guillen, Casanueva, Anta, et al. 2008; Dentani, Nagasaka, Funabiki, et al. 2008; Otsuka, Funabiki, Sugiyama, et al. 2006; Birkel, Lee, Koll, et al. 2012; Sayama, Sugihara, and Arakawa 1998; Ou, Rani, Field, et al. 2012), SnO_2 (LeViet, Jose, Reddy, et al. 2010), and Nb_2O_5 (Zhang, Wang, Yang, et al. 2012; Rao and Dutta 2007; Precker and da Silva 2002) have been extensively investigated.

Earlier, working electrodes of photochemical systems employed group IV or III–V semiconductors. Under light irradiation conditions, narrow band gap semiconductors of group IV or III–V caused corrosion problems in the electrode. This led to the development of oxide semiconductors with wide band gaps. Examples of oxide semiconductors include zinc oxide (ZnO), titanium dioxide (TiO_2), and tin oxide (SnO_2). These electrode materials started being used from 1960 onwards. However, these wide band gap oxides showed poor response towards light in the visible region. When the coating of the dye on the wide band gap oxides increased, it increased the optical absorption into the visible region. The use of photoelectrode films with rough surface, such as sintered ZnO particles and textured TiO_2 DSSCs have increased the efficiency by up to 3% approximately. But when nanomaterial-sized TiO_2 particles sensitized with a trimeric ruthenium complex were used in DSSCs, the overall conversion efficiency increased by up to 7.1%–7.9% (Oregan and Gratzel 1991). This was the result of using newly developed dyes such as RuL (μ-(CN)Ru(CN)L$'$)$_2$, L = 2, 2$'$ bipyridine-4,4-dicarboxlic acid, and L$'$ = 2, 2$'$-bipyridne) that had high extinction coefficients and broad absorption spectrum with the onset at 750 nm. However, the use of nanocrystalline nanoparticles to form photoelectrode films is also a critical reason because they yield an externally large internal surface for adsorption (Nazeeruddin, Kay, Rodico, et al. 1993). After this porous (50%–65%) nanoparticle film was used in DSSCs that allowed a 2000-fold increase in the surface area. The problem of interfacial charge recombination also arises with increase in internal surface area due to the use of nanoparticle. This is because of the reaction between photo-generated electrons and redox species in the electrolyte. This affects open circuit voltage by decreasing the concentration of electrons in the conduction band of the semiconductor (Nazeeruddin, Kay, Rodico, *et al.* 1993; Cehen, Hodes, Gratzel, *et al.* 2000; Diamant, Chen, Melamed, *et al.* 2003). Since the small size of individual nanoparticles only allows limited band bending at the semiconductor surface, no electric field can assist the separation of electrons in the semiconductor (Palomars, Clifford, Haqe, *et al.* 2004; Chen, Chappel, Diamant, *et al.* 2001). To remove the interfacial charge recombination problem, core shell structures were developed and applied to DSSCs. These nanostructures consist of a core which is made up of nanomaterial (nanowire/nanotubes) and the shell is covered with a coating layer covering on the surface of the core nanomaterials. But there were several uncertainties and complications in terms of the practical utility of the core shell structure and affects the solar cell performance. When non-porous TiO_2 film coated with oxides such as Nb_2O_5, ZnO, $SrTiO_3$, ZrO_2, Al_2O_3, and SnO_2 was used (Diamant, Chen, Melamed, *et al.* 2003; Palomars, Clifford, Haqe, *et al.* 2004; Diamant, Chappel, Chen, *et al.* 2004) instead of bare TiO_2 nanoparticle electrodes, there was increased open-circuit voltage and short-circuit current, thus increasing the efficiency of the solar cell by up to 37%. Nanostructure films do not have identical composition, the non-ideal electron transport results in lack of microscopic electrostatic potential gradient in the film. The film is permeated with a concentrated electrolyte (Diamant, Chappel, Chen, *et al.* 2004). Therefore, the electron transport in the nanoparticle film is affected by the process of diffusion instead of drift. It gives rise to nanowired structures of different metal oxides in the DSSC. In a DSSC, ZnO nanowires demonstrated that a one-dimensional nanostructure might provide direct pathways for electron transport (Hagfeldt and Gratzal 2000). TiO_2 nanowires have also attracted a lot of interest with regard to their application in DSSCs despite TiO_2 and ZnO having an extremely long electron diffusion length of up to 100 μm (Jennings, Ghicov, Peter, *et al.* 2008). Currently, most of the dye-sensitized solar cells made of one-dimensional nanostructures, show efficiency much lower than those obtained by TiO_2 nanoparticles. This is due to the limit in the surface area of the one-dimensional nanostructure. There are some difficulties in their fabrication, in terms of avoiding bottom attachment of nanowire/nanotube array when the length is over 10 μm. The process becomes even more complex in the case of a transparent conductive substrate.

23.6   THIN FILM SOLAR CELLS

Thin-film solar cell technology is very important for the mankind. However, silicon solar cells are costly to fabricate and have limited efficiency (about 14% in natural conditions and up to 25% in controlled conditions). Cost comparison per unit power production between fossil fuel combustion and silicon solar cells was found to yield a large difference as well. The main component of these cells is a thin material coating. Fabrication of these solar cells requires a very low quantity of material. Compared to silicon wafers, it is only 1%. This is the main reason why this technology is cheaper to some extent. The majority of this type of cells is made from amorphous silicon that are devoid of a crystalline structure. These are much cheaper fabricators, but have a very low efficiency of about 8%. Incorporation of nanomaterials into thin-film solar cells enhances conversion efficiency at a lesser cost. Nowadays, many nanomaterials are being investigated for photovoltaic applications. The following are the three benefits of using nano-structured layers in thin-film solar cells:

(a) The effective optical absorption path becomes larger than actual film thickness because of multiple reflections.

(b) There is a reduction in recombination losses due to the shorter path followed by light-generated electrons and holes. The absorber layer thickness in traditional thin-film solar cells is in micrometres, while in nano-structured solar cells it can be as thin as 150 nm.

(c) Energy band gap of different layers, made for desired design value, uses nanoparticles of various sizes. This allows for more design flexibility in the absorber and window layers of the solar cells. 'Al' and 'W' are used as impurities for enhancing photovoltaic properties, Al-doped titanium dioxide electrodes increase the open-circuit voltage (V_{oc}), but reduce the short-circuit current (I_{sc}).

23.6.1   Dye-sensitized Solar Cells

The phenomenon of illuminated organic dyes having the capability of generating electricity at oxide electrodes in electrochemical cells was discovered in 1960. In University of California at Berkeley, efforts were made to understand the processes of photosynthesis by extracting chlorophyll from spinach. A new principle was born when these experiments opened the way for electric power generation with the help of DSSCs. In 1972, it was recognized that DSSCs are very unstable. Work is currently in progress to enhance the efficiency of solar cells for optimization of the porosity of the electrodes. For this, fine oxide powders are used, but they are also quite unstable. Modern technology uses a porous layer of TiO_2 nanoparticles that are covered by molecular dye and absorbs light. In the presence of a platinum-based catalyst (Pt), TiO_2 is immersed in an electrolyte solution; for instance, alkaline battery (TiO_2) and cathode (Pt) are placed in a liquid conductor (the electrolyte). Transparent electrodes placed allow light coming from the sun to pass towards the layer of dye. When the light reaches the dye layer, it excites the electrons. These excited electrons then start flowing towards the transparent electrode of TiO_2 and collected for powering a load. After flowing through the external circuit, they are re-introduced into the cell on a metal electrode on the back, after which they flow into the electrolyte. The electrolyte then transports the electrons back to the dye molecules.

In a traditional solar cell, silicon is used as the source of photoelectrons and for providing the electric field which separates the charges and current flows. In a DSSC, a dye-sentized n-type semiconducting oxide film deposited on a transparent conducive glass substrate acts as a working electrode. The counter electrode is a platinum-coated glass substrate that is kept parallel to the working electrode with a face-to-face configuration. A DSSC carries charge only through semiconductors; photoelectrons come from photosensitive dye and separation of charge occurs at the surface between the electrolyte, semiconductor, and dye. The nano-sized dye molecules are much smaller in size. To capture incidence light the layer of dye is made thicker than the molecules. To tackle this situation, the nanomaterial provides a temporary platform to hold dye molecules in a 3-D matrix and increases the number of molecules for any given surface area of the cell.

23.6.2 Construction

The uppermost surface of a solar cell is made up of transparent material (SnO_2: F top contact) (Figure 23.3). Sunlight enters through it and reaches

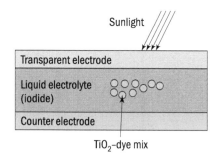

Fig. 23.3 *Structure of dye-sensitized solar cell*

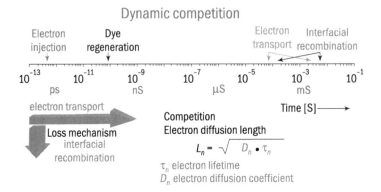

Fig. 23.4 *Dynamics of redox processes involved in the conversion of light to electric power by dye-sensitized solar cells*

the TiO_2 layer. Here sunlight strikes the dye. Photons with adequate energy are absorbed and the dye enters into an excited state from which an electron can be "injected" directly into the conduction band of TiO_2. There is an electron concentration gradient so it moves by diffusion to the clear anode on top. The dye molecule temporarily loses an electron and if another electron does not get attached to the dye molecule, it becomes either inactive or decays. In liquid electrolyte of iodide, dye strips (TiO_2-dye mix) are present and oxidize iodide to tri-iodide. This happens very quickly, in about the time taken for the injected electron to recombine. This is shown in Figure 23.4.

23.6.3   Efficiency

Ecole Polytechnique Federale de Lausanne (EPFL) scientists developed a solid state DSSC which was fabricated by only a two-step process. These were capable of raising efficiency by up to 15% without losing stability. The current efficiency of different solar cells is shown in Figure 23.5.

23.7   APPLICATIONS AND IMPLEMENTATIONS

Polycrystalline photovoltaic solar cells are connected and encapsulated in a module that is laminated by backing materials. The photovoltaic module has semiconductor wafers that are covered by a glass sheet with resin barriers behind it. This glass sheet is transparent. It allows sunlight to enter and protects the cell from the environment. Gerald Pearson, Calvin Fuller and Daryl Chapin, a group of American researchers, used silicon as a semiconductor and designed a solar cell which had a conversion rate of 6% (Chaplin, Fuller, and Pearson 1954). The first solar panel was developed

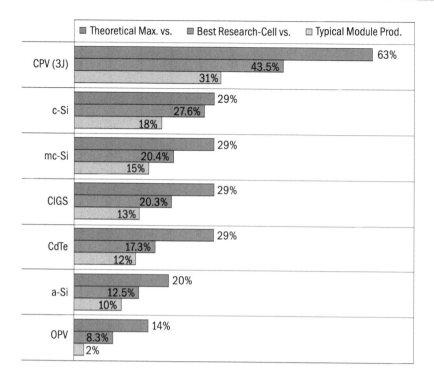

Fig. 23.5 *Current status of solar cell efficiency*

by these three inventors. Commercial solar cells have been available in the market since 1956.

Solar cells during that time were very expensive, about $300 per watt, and were used in toys and radios only. Bell Telephone Laboratories launched the Telstar in 1962 which was the first active telecommunications satellite. It had solar cells of 14 watts. In 1970, prices started to drop and reached up to $20 per watt due to energy crisis. The invention of flexible solar shingles, a type of roofing material which has the capacity to convert sunlight into electricity, made by Subhendu Guha and his team in 1998, changed the whole scenario and till 1999 the world solar electricity generation reached 1000 MW. In the present day, in a controlled situation conversion, efficiency reaches up to 24.7% for single crystal silicon (Green 2001). The High-Efficiency Metamorphic Multijunction Concentrator Solar Cell (HEMM solar cell) was created in 2007 by the combined attempt of Boeing Spectrolab and National Renewable Energy Laboratory. It had a conversion efficiency of 40% (double that of the silicon solar cell). At present cells with an efficiency in the range of 17.2%–17.5% can be routinely obtained using screen print technology across the industry.

23.8   POLYMER SOLAR CELLS

Organic and polymer solar cells are made up of organic semiconductors with thin films of about 100 nm thickness. These are mainly composed of polymers and molecular compounds such as carbon fullerenes, copper phthalocyanine (a blue-green organic pigment), and polyphenylene vinylene. These cells may give very good results where disposability and mechanical flexibility are concerned. Lower conversion efficiency and stability problems are the main barriers for large-scale power production with polymer solar cells.

23.9   PHOTOVOLTAIC INDUSTRY GOAL: REDUCE COST OF ENERGY

Reducing the cost of energy is possible if the performance, reliability, and durability of the solar cell can be reformed. Performance may be controlled by "$/kW" which is dictated by materials and the process cost per unit area and module efficiency per unit area. Reliability and durability may be controlled by "h" which is dictated by failure rate over time (obsolete reliability) and degradation rate over time (underperformance, durability). Different institution reports are depicted in Table 23.1.

23.10   CONCLUSION

Dye-sensitized solar cells are devices that give us hope for the future; they give us a chance to reduce active material for fabrication. Large-scale production of solar cells needs several semiconductor materials. However, appropriate materials are not being found in adequate amount and are very expensive too. With the increase in knowledge structure, chemical and optoelectronic characters of thin-film materials and utilization of properties and functions such as generation of surface electric fields, passivation, activation, graded band gaps, photon scattering/recycling, the scenario can change. Further study is needed in the field of audited account of photons, excited carriers, carrier mobility in the device and absorption coefficient, and extinction coefficient of the material used, along with understanding of the electronic role of interfaces in the layered structure. This may help bridge the gap between theoretical and achieved efficiency.

Table 23.1 Different institution reports

Classification	Efficiency (%)	Area (cm²)	Voc (mA/cm²)	Jsc (mA/cm²)	FF (%)	Test centre	References
Si (crystalline)	25 ± 0.5	4	0.706	42.7	82.8	Sandia	Natsuhara and Matsumoto (2006)
Si (thin-film transfer)	19.1 ± 0.4	3.983	0.65	37.8	77.6	FhG-ISE	Zhao, Wang, Green, et al. (1998)
Si (thin-film submodule)	10.5 ± 0.3	94	0.492	29.7	72.1	FhG-ISE	Schultz, Glunz, and Willeke (2004)
GaAs (multicrystalline)	18.4 ± 0.5	4.011	0.994	79.7	79.7	NREL	Keevers, Young, Schubert, et al. (2007)
InP (crystalline)	22.1 ± 0.7	4.02	0.878	85.4	85.4	NREL	Enkatasubramanian, Quinn, Hills, et al. (1997)
Thin-film chalcogenide	19.6 ± 0.6	0.9927	0.713	34.81	79.2	NREL	Keavney, Haven, and Vernon (1990)
CIGS (cell)CIGS	17.4 ± 0.5	4.011	0.6815	33.84	75.5	FhG-ISE	Repins, Contreras, Egaas, et al. (2008)
Amorphous/ nanocrystalline, Si (amorphous)	10.1 ± 0.3	1.036	0.886	16.75	67	NREL	Benagli, Borrello, Vallat, et al. (2009)
Si nanocrystalline)	10.1 ± 0.2	1.199	0.539	24.4	76.6	JQA	Benagli, Borrello, Vallat, et al. (2009)
Photochemical	11 ± 0.2	1.007	0.714	21.93	70.3	AIST	Yamamoto, Toshimi, Suzuki, et al. (1998)
Organic thin film	10 ± 0.3	1.021	0.899	16.75	66.1	AIST	Morooka, Ogura, Orihashi, et al. (2009)
InGaP/GaAs/InGaAs	37.5 ± 1.3	1.046	3.105	14.56	85.5	AIST	Banerjee, Su, Beglau, et al. (2011)
a-Si/nc-Si/nc-Si (thin-film cell)	12.4 ± 0.7	1.05	1.936	8.96	71.5	NREL	Yoshimi, Sasaki, Sawada, et al. (2003)
a-Si/nc-Si (thin-film submodule)	11.7 ± 0.4	14.23	5.462	2.99	71.3	AIST	Jorgensen, Norrman, Gevorgyan, et al. (2012)

REFERENCES

Banerjee, A., T. Su, D. Beglau, G. Pietka, F. Liu, G. DeMaggio, S. Almutawalli, B. Yan, G. Yue, J. Yang, and S. Guha. 2011. High efficiency, multi-junction nc-Si:h based solar cells at high deposition rate. *37th IEEE PVSC*, Seattle.

Benagli, S., D. Borrello, E. S. Vallat, J. Meier, U. Kroll, J. Hotzel, J. Spitznagel, J. Steinhauser, and Y. Djeridane. 2009. High-efficiency Amorphous Silicon Devices on LPCVD-ZNO TCO Prepared in Industrial KAI-M R and D Reactor. *24th European Photovoltaic Solar Energy Conference*, Hamburg

Birkel, A., Y. G. Lee, D. Koll, X. Van Meerbeek, S. Frank, M. J. Choi, Y. S. Kang, K. Char, and W. Termel. 2012. Highly efficient and stable dye-sensitized solar cells based on SnO_2 nanocryatals prepared by microwave-assisted synthsis. *Energy & Environmental Science* 5: 5392–5400

Cehen, D., G. Hodes, M. Gratzel, J. F. Gullemoles, and I. Riess. 2000. Nature of photovoltaic action in dye-sensitized solar cells. *J Phys. Chem.* B 104: 2053–2059

Chapin, D. M., C. S. Fuller, and G. L. Pearson. 1954. A new silicon p - n junction photocell for converting solar radiation into electrical power. *J. Appl. Phys.* 25: 676

Chen, S. G., S. Chappel, Y. Diamant, and A. Zaban. 2001. Preparation of Nb_2O_5 coated TiO_2 nanoporous electrodes and their application in dye-sensitized solar cells, *Chemistry of Materials* 13: 4629–4634

Dentani, T., K. Nagasaka, K. Funabiki, J. J. Jin, T. Yoshida, H. Minoura, and M. Matsui. 2008. Flexible zinc oxide solar cells sensitized by styryl dyes. *Dyes and Pigments* 77: 59–69

Diamant, Y., S. Chappel, S. G. Chen, O. Melamed, and A. Zabon. 2004. Core–shell nanoporous electrode for dye sensitized solar cells: the effect of shell characteristics on the electronic properties of the electrode. *Coorination Chemistry Reviews* 248(13): 1271–1276

Diamant, Y., S. G. Chen, O. Melamed, and A. Zaban. 2003. Core-shell nanoporous electrode for dye sensitized solar cells: the effect of the SrTiO3 shell on the electronic properties of the TiO_2 core. *J. Phys. Chem.*B 107: 1977–1981

Dou, Letian, Jingbi You, Jun Yang, Chun-Chao Chen, Youjun He, Seiichiro Murase, Tom Moriarty, Keith Emery, *et al.* 2012. Tandem polymer solar cells featuring a spectrally matched low band gap polymer. *Nature Photonics* 6(3): 180

Enkatasubramanian, R., B. C. Quinn, J. S. Hills, P. R. Sharps, M. L. Timmons, J. A. Hutchby, H. Field, A. Ahrenkiel, and B. Keyes. 1997. 18.2% (AM1.5) efficient GaAs solar cell on optical-grade polycrystalline Ge substrate. *Conference Record*, 25th IEEE Photovoltaic Specialists Conference, 31–36; Washington, USA.

Guillen, E., F. Casanueva, J. A. Anta, A. V. Post, G. Oskam, R. Alcantara, C. F. Lorenzo, and J. M. Calleja. 2008. Photovoltaic performance of nanostructured zinc oxide sensitised with xanthene dyes. *Journal of Photochemistry and Photobiology A: Chemistry* 200: 364–370

Gonzalez-Volls, I. and M. Lira-Cantu. 2009. Vertically-aligned nanostructures of ZnO for excitonic solar cells: a review. *Energy & Environmental Science* 2: 19–34

Green, M. A. 2001. Multiple band and impurity photovoltaic solar cells: general theory and comparison to tandem cells. *Prog. Photovoltaics Res. Appl.* 9: 137–144

Green, M. A., K. Emery, Y. Hishikawa, W. Warta, and E. D. Dunlop. 2012. *Progress in Photostatic Research and Applications* 20: 606–614

Hagfeldt, A. and M. Gratzal. 2000. Molecular photovoltaics. *Accounts of Chemical Research.* 33: 269–277

Han, S., W. S. Shin, M. Seo, D. Gupta, S. J. Moon, and S. Yoo. 2009. Improving performance of organic solar cells using amorphous tungsten oxides as an interfacial buffer layer on transparent anodes. *Organic Electronics* 10: 791–797

Hartman, K., J. L. Johnson, M. I. Bertoni, D. Recht, M. J. Aziz, M. A. Scarpulla, and T. Buonassisi. 2011. SnS thin-films by RF sputtering at room temperature. *Thin Solid Films* 519: 7421–7424

Hoppe, H. and N. Sariciftci. 2004. Organic solar cells: An overview. *Journal of Material Research* 19: 1924–1945

Jennings, J. R., A. Ghicov, L. M. Peter, P. Schmuki, and A. B. Walker. 2008. Dye-sensitized solar cells based on oriented TiO2 nanotube arrays: transport, trapping, and transfer of electrons. *Journal of American Chemical Society* 130: 13364–13372

Jorgensen, M., K. Norrman, S. A. Gevorgyan, T. Tromholt, B. Andreasen, and F. C. Krebs. 2012. Stability of polymer solar cells. *Advanced Materials* 24: 580–612

Katagiri, H., N. Sasaguchi, S. Hondo, S. Hoshino, J. Ohasi, and T. Yokota. 1997. Preparation and evaluation of Cu_2ZnSnS_4 thin films by sulfurization of E B evaporated precursors. *Solar Energy Material and Solar Cells* 49: 407–414

Keavney, C. J., V. E. Haven, and S. M. Vernon. 1990. Emitter structures in MOCVD in P solar cells. *Conference Record*, 21st IEEE Photovoltaic Specialists Conference, 141–144, Kissimimee

Keevers, M. J., T. L. Young, U. Schubert, and M. A. Green. 2007. 10% Efficient CSG Minimodules. *22nd European Photovoltaic Solar Energy Conference*, Milan

Keis, K., J. Linndgren, S. E. Lindquist, and A. Hagfeldt. 2000. Studies of the adsorption process of Ru complexes in nanoporous ZnO electrodes. *Langmuir* 16: 4688–4694

Keis, K., E. Magnusson, H. Lindstrom, S. E. Liindquist, and Hagfeldt. 2002. A 5% efficient photoeletrochemical solar cell based on nanostructured ZnO electrodes. *Solar Energy Materials and Solar cells* 73: 51–58

Keis, K., C. Bauer, G. Boschloo, A. Hagfeldt, K. Westermark, H. Rensmo, and H. Siegbahn. 2002. Nanostructured ZnO electrodes for dye-sensitized solar cell applications. *Journal of Photochemistry and Photobiology A: Chemistry* 148: 57–64

Kumara, G. R. R. A., K. Okuya, S. Kaneko, I. R. M. Kottegoda, P. K. M. Bandaranayake, A. Konno, M. Okuya, S. Kaneko, and K. Murakami. 2003. Efficient dye-sensitized photoelectrochemical cells made from nanocrystalline tin (iv) oxide-zinc oxide composite films. *Semiconductor Science and Technology* 18: 312–318

Le Viet, A., M .V. Reddy, R. Jose, B. V. R. Chowdari and S. Ramakrishna. 2010 Electrospun niobium pentoxide nanofiber polymorphs for rechargeable lithium ion batteries. *Journal of Physical Chemistry* C 114: 664-671

Lee, Sang-Hyeun Misook Kang, Sung M. Cho, Gui Young Han, Byung-Woo Kim, Ki June Joon, and Chan-Hwa Chung 2001. Synthesis of TiO$_2$ photocatalyst thin film by solvothermal method with a small amount of water and its photocatalytic performance. *Journal of Photochemistry and Photobiology* A: Chemistry 146 (1–2): 121–128

Morooka, M., Ogura, R., Orihashi, M., and Takenaka, M. 2009. Development of dye-sensitized solar cells for practical applications. *Electrochemistry* 77: 960–965

Natsuhara, H. and K. Matsumoto. 2006. TiO$_2$ thin films as protective material for transparent conducting oxides used in Si thin film solar cells. *Solar Energy Materials and Solar Cells* 90: 2867–2880

Nazeeruddin, M. K., Kay, A., Rodico, I., Humphry-paker, R., Muller, E., Liska, P., Vlachopouls, N., and Gratzel, M. 1993. *JACS* 115: 6382–6390

Oregan, B. and M. Gratzel. 1991. A low-cost, high efficiency solar cell based on dye-sensitized colloidal TO2 films. *Nature* 353: 737–740

Otsuka, A., K. Funabiki, N. Sugiyama, T. Yoshida, and H. Mastui Minoura. 2006. *Dye sensitization of ZnO by unsymmetrical squaraine dyes suppressing aggregation.* Chemistry Letters 35: 666–667

Ou, J. Z., R. A. Rani, M. H. Field, Y. Zheng, P. Reece, S. Zhuiykov, S. Sriram, M. Bhaskaran, R. B. Kanee, and K. Kalanter-Zadeh. 2012. *Elevated temperature anodized Nb2O5: a photoanode material with exceptionally large photoconversion efficiencies.* ACS Nano 6: 4045–4053

Palomars, E., J. N. Clifford, S. A. Haqe, T. Lutz, and J. R. Durrant. 2004. *Control of charge recombination dynamics in dye sensitized solar cells by the use of conformally deposited metal oxide blocking layers.* Journal of the American Chemical Society 125: 475–482

Precker, J. W. and M. A. da Silva. 2002. Experimental estimation of the band gap in silicon and germanium from the temperature-voltage curve of diode thermometers. *American Journal of Physics* 70: 1150–1153

Rao, A. R. and V. Dutta. 2007. Low-temperature synthesis of TiO_2 nanoparticles and preparation of TiO_2 thin films by spray deposition. *Solar Energy Materials and Solar Cells* 91: 1075–1080

Reddy, K. T. R., N. K. Reddy, and R. W. Miles. 2006. Photovoltaic properties of SnS based solar cells. *Solar Energy Materials and Solar Cells* 90: 3041–3046

Repins, I., M. A. Contreras, B. Egaas, C. Dehart, J. Scharf, and C. L. Perkins. 2008. 19.9%-efficient ZnO/CdS/CuInGaSe2 solar cell with 81.2% fill factor. *Progress in Photovoltaics: Research and Applications* 16: 235–239

Sayama, K., H. Sugihara, and H. Arakawa. 1998. Photoelectrochemical properties of a porous Nb2O5 electrode sensitized by a ruthenium dye. *Chemistry of Materials* 10: 3825–3832

Schultz, O., S. W. Glunz, and G. P. Willeke. 2004. Multicrystalline silicon solar cells exceeding 20% efficiency. *Progress in Photovoltaics: Research and Applications* 12: 553–558

Service, R. 2011. Outlook brightens for plastic solar cells *Science* 332 (6027): 293

Soga Tetswa. 2006. Nanostructured Materials for Solar Energy conversion, 181 *Science and technology*, UK

Wang, X. J., I. A. Buyanova, W. M. Chen, M. Izadifard, D. P. Norton, S. J. Pearton, A. Osinky, and A. Dabiran. 2006. Band gap properties of Zn1-x CdxO alloys grown by molecular-beam epitaxy. *Applied Physics Letters* 89: 151909

Wohrle, D. and D. Meissner. 1991. Organic solar cells *Adanced Material* 3: 129–138

Yamamoto, K., M. Toshimi, T. Suzuki, Y. Tawada, T. Okamoto, and A. Nakajima. 1998. Thin -ilm poly-Si solar cell on glass substrate fabricated at low temperature. MRS Spring Meeting, San Francisco.

Yoshimi, M., T. Sasaki, T. Sawada, T. Suezaki, T. Meguro, T. Matsuda, K. Santo, K. Wadano, M. Ichikawa, A. Nakajima, and K. Yamamoto. 2003. High-efficiency thin-film silicon hybrid solar cell module on 1 m2 class large area substrate. *Conference Record*, 3rd World Conference on Photovoltaic Energy Conversion, 1566-1569 Osaka.

Yu-peng ZHANGa, Jun-jie XUa, Zhi-hua SUNa, Chen-zhe LIa, Chun-xu PAN. 2011. Preparation of graphene and TiO2 layer by layer composite with highly photocatalytic efficiency. *Progress in Natural Science Materials International* 21: 46–47

Zhang, H. M., Y. Wang, D. J. Yang, Y. B. Li, H. W. Liu, B. J. Wood, and H. J. Zhao. 2012. Directly hydrothermal growth of single crystal Nb3O7(OH) nanorod film for high performance dye-sensitized solar cells. *Advanced Materials* 24: 1598–1603

Zhao, J., A. Wang, M. A. Green, and F. N. Ferrazza. 1998. 19.8% efficient "honeycomb" textured multicrystalline and 24.4% monocrystalline silicon solar cells. *Applied Physics Letters* 73: 1991–1993

Index

About the Editors

Prof. D. P. Singh, currently the Head and Dean, School for Environmental Sciences, Babasaheb Bhimrao Ambedkar University, Lucknow, is an eminent scholar in the field of Environmental Science. He obtained his PhD from Department of Botany, Banaras Hindu University, Varanasi. He has worked extensively in the areas of wastewater treatment, microbiology, stress physiology, bioremediation, and alternative energy options. He has received several honours and awards to his credit, including more than 100 research publications in high impact factor journals of national and international repute. He has supervised more than 20 PhD students and several MSc and MTech students for their research work. He has delivered invited lectures in different seminars and symposia and served as a principal investigator for several government funded projects. Dr Singh has published three books, that is, *Environmental Microbiology and Biotechnology*, *Stress Physiology*, and *Microbes in Sustainable Management of Soil, Water and Agriculture*.

Dr Richa Kothari is Assistant Professor at School for Environmental Sciences, Babasaheb Bhimrao Ambedkar University, Lucknow. She has done her PhD in Energy and Environment from School of Energy and Environmental Studies, Devi Ahilya University, Indore, Madhya Pradesh. Dr Kothari has been actively involved in research in the areas of algal-based biofuel production, bio-hydrogen production and utilization of renewable energy for wastewater treatment. She is also recipient of prestigious research fellowships from Ministry of New and Renewable Energy (MNRE) and Council of Scientific and Industrial Research (CSIR). She has published more than 40 research papers/book chapters in reputed international peer-reviewed journals with high impact factor and supervised several MSc, MTech and PhD students for dissertation. She has written a textbook titled *Environmental Science for Undergraduates*.

Dr V. V. Tyagi is Assistant Professor at School of Energy Management, Faculty of Engineering, Shri Mata Vaishno Devi University, Katra, Jammu. He has been actively involved in research in the areas of solar thermal energy storage systems and energy policy and photovoltaic/thermal hybrid systems. He did his PhD from School of Energy and Environment Studies, Devi Ahilya University, Indore and further worked as a Postdoctorate Research Fellow at UM Power Energy Dedicated Advanced Centre (UMPEDAC), University of Malaya, Kuala Lumpur, Malaysia; Centre for Energy Studies, Indian Institute of Technology, Delhi; and National Institute of Solar Energy. He has published more than 45 research papers in peer reviewed SCI international journals and conferences. Dr Tyagi has more than 3000 citations of his published research papers and supervised several students at master's and graduate levels. He is also serving as an editorial board member for three reputed international journals.

Printed and bound by CPI Group (UK) Ltd, Croydon, CR0 4YY

17/10/2024

01775681-0016